Fluids in the Crust

Fluids in the Crust

Equilibrium and transport properties

Edited by

K.I. Shmulovich

Institute of Experimental Mineralogy
Russian Academy of Sciences
Russia

B.W.D. Yardley

Department of Earth Sciences
University of Leeds
UK

and

G.G. Gonchar

Institute of Experimental Mineralogy
Russian Academy of Sciences
Russia

Translated by G.G. Gonchar

CHAPMAN & HALL

London · Glasgow · Weinheim · New York · Tokyo · Melbourne · Madras

Published by Chapman & Hall, 2–6 Boundary Row, London SE1 8HN, UK

Chapman & Hall, 2–6 Boundary Row, London SE1 8HN, UK

Blackie Academic & Professional, Wester Cleddens Road, Bishopbriggs, Glasgow G64 2NZ, UK

Chapman & Hall GmbH, Pappelallee 3, 69469 Weinheim, Germany

Chapman & Hall USA, One Penn Plaza, 41st Floor, New York NY 10119, USA

Chapman & Hall Japan, ITP-Japan, Kyowa Building, 3F, 2-2-1 Hirakawa-cho, Chiyoda-ku, Tokyo 102, Japan

Chapman & Hall Australia, Thomas Nelson Australia, 102 Dodds Street, South Melbourne, Victoria 3205, Australia

Chapman & Hall India, R. Seshadri, 32 Second Main Road, CIT East, Madras 600 035, India

---

First edition 1995

Typeset in 10/12 pt Palatino by Pure Tech Corporation, Pondicherry, India.

Printed in Great Britain by St Edmundsbury Press, Bury St Edmunds, Suffolk.

ISBN 0 412 56320 7

A catalogue record for this book is available from the British Library

Library of Congress Catalog Card Number: 94–071815

∞ Printed on acid-free text paper, manufactured in accordance with ANSI/NISO Z39.48-1992 (Permanence of Paper).

Contents

List of contributors ix
Preface xi
List of symbols for rock-forming minerals xiii

CHAPTER 1 An introduction to crustal fluids 1
Bruce W.D. Yardley and Kirill I.Shmulovich

1.1 Introduction 1
1.2 Fluid systems 2
1.3 Mineral solubility and the composition of crustal fluids 2
1.4 Fluid migration through crustal rocks 8
1.5 Conclusions 9
References 10

CHAPTER 2 Fluids in geological processes 13
Vilen A. Zharikov

2.1 Introduction 13
2.2 Equilibrium systems with fluid components 14
2.3 Dynamic phenomena in systems with fluid components 30
Acknowledgements 37
References 37
Glossary of symbols 41

CHAPTER 3 Hydrothermal experimental techniques used at the Institute of Experimental Mineralogy, Russian Academy of Sciences 43
Kirill. I. Shmulovich, Vladimir I. Sorokin and Georgiy P. Zaraisky

3.1 Introduction 43
3.2 Parts of the apparatus for studying hydrothermal processes 45
3.3 Sampling techniques 48
3.4 Apparatus and techniques for experiments on diffusion and infiltration metasomatism 51

3.5 Summary and conclusions 55
References 56

CHAPTER 4 Solubility and complex formation in the systems $Hg–H_2O$, $S–H_2O$, $SiO_2–H_2O$ and $SnO_2–H_2O$ 57
Vladimir I. Sorokin and Tat'yana P. Dadze

4.1 Introduction 57
4.2 Mercury solubility in water 58
4.3 Sulphur solubility in water and the thermodynamics of the aqueous species S^o_{aq}, H_2S_{aq}, SO_{2aq}, $H_2S_2O_3$, HSO_4^- to 440°C 63
4.4 SiO_2 solubility in water and in acid solutions (HCl, HNO_3) at 100–400°C and 1013 bar 75
4.5 Solubility of SnO_2 in water and aqueous electrolyte solutions at 200–400°C and 16–1500 bar 79
4.6 Summary 87
Acknowledgements 89
References 90
Glossary of symbols 93

CHAPTER 5 Experimental studies of the solubility and complexing of selected ore elements (Au, Ag, Cu, Mo, As, Sb, Hg) in aqueous solutions 95
Alexander V. Zotov, Aleksey V. Kudrin, Konstantin A. Levin, Nadezhda D. Shikina and Lidia N. Var'yash

5.1 Introduction 95
5.2 Gold 96
5.3 Silver 102
5.4 Copper 108
5.5 Molybdenum 114
5.6 Arsenic 120
5.7 Antimony 124
5.8 Mercury 127
5.9 Discussion and geological implications 129
Acknowledgements 132
References 132
Glossary of symbols 136

CHAPTER 6 The influence of acidic fluoride and chloride solutions on the geochemical behaviour of Al, Si and W 139
Georgiy P. Zaraisky

6.1 Introduction 139

6.2 Gain and loss of rock-forming components during greisenization: an example from the Akchatau W–Mo deposit 140
6.3 Results of mineral solubility experiments 144
6.4 Silicification of rocks associated with acid metasomatism 152
6.5 Conclusions 156
Acknowledgements 158
References 159

CHAPTER 7 The behaviour of components in complex fluid mixtures under high T–P conditions 163
Michael A. Korzhinskiy

7.1 Introduction 163
7.2 Experimental procedures and techniques 165
7.3 Behaviour of the acid component HCl 167
7.4 Behaviour of the salt components $MgCl_2$, $CaCl_2$ and $FeCl_2$ 178
7.5 Concluding discussion 187
References 190
Glossary of symbols 192

CHAPTER 8 Phase equilibria in fluid systems at high pressures and temperatures 193
Kirill I. Shmulovich, Sergey I. Tkachenko and Natalia V. Plyasunova

8.1 Introduction 193
8.2 Binary H_2O–salt systems 194
8.3 Ternary systems H_2O–CO_2–salt systems 199
8.4 Fluid immiscibility and magmatic crystallization 203
8.5 Fluid in metamorphism and anatectic melting 203
8.6 Phenomena accompanying the two-phase fluid state 207
8.7 Conclusions 211
Acknowledgements 212
References 212

CHAPTER 9 Diffusion of electrolytes in hydrothermal systems: free solution and porous media 215
Victor N. Balashov

9.1 Introduction 215
9.2 The diffusion of the electrolyte $A_{v_1}B_{v_2}$ in dilute solutions 216

9.3 P–T–m dependencies of diffusion coefficients in electrolyte solutions 221
9.4 Theoretical calculations of transport properties of rocks, based on the data on functions of microcrack size distributions 231
9.5 Diffusion of electrolytes in pore solutions: correction for the electric double layer (EDL) effect 235
9.6 Diffusion of NaCl in pore solutions of granite to 600°C 242
Acknowledgments 245
References 246
Glossary of symbols 250

CHAPTER 10 Thermal decompaction of rocks 253
Georgiy P. Zaraisky and Victor N. Balashov

10.1 Introduction 253
10.2 Experimental procedure 254
10.3 Results 259
10.4 Discussion 271
10.5 Conclusions: possible geological implications of the effect of thermal decompaction of rocks 278
References 281
Glossary of symbols 283

CHAPTER 11 Permeability of rocks at elevated temperatures and pressures 285
Vyacheslav M. Shmonov, Valentina M. Vitovtova and Irina V. Zarubina

11.1 Introduction 285
11.2 Key concepts, methods of permeability measurement and experimental procedure 287
11.3 Permeability of magmatic rocks 292
11.4 Permeability of metamorphic rocks 296
11.5 Permeability of sedimentary rocks 305
11.6 Discussion and summary of conclusions 308
Acknowledgements 310
References 310
Glossary of symbols 312

Index of aqueous species 315

Index of systems 317

Subject index 319

Contributors

Victor N. Balashov Institute of Experimental Mineralogy, Russian Academy of Sciences, 142432 Chernogolovka, Moscow district, Russia

Tat'yana P. Dadze Institute of Experimental Mineralogy, Russian Academy of Sciences, 142432 Chernogolovka, Moscow district, Russia

Michael A. Korzhinskiy Institute of Experimental Mineralogy, Russian Academy of Sciences, 142432 Chernogolovka, Moscow district, Russia

Aleksey V. Kudrin Institute of Ore Deposits, Petrography, Mineralogy and Geochemistry, Russian Academy of Sciences, Staromonetny per. 35, 109017 Moscow, Russia

Konstantin A. Levin Institute of Ore Deposits, Petrography, Mineralogy and Geochemistry, Russian Academy of Sciences, Staromonetny per. 35, 109017 Moscow, Russia

Natalia V. Plyasunova Institute of Experimental Mineralogy, Russian Academy of Sciences, 142432 Chernogolovka, Moscow district, Russia

Nadezhda D. Shikina Institute of Ore Deposits, Petrography, Mineralogy and Geochemistry, Russian Academy of Sciences, Staromonetny per. 35, 109017 Moscow, Russia

Vyacheslav M. Shmonov Institute of Experimental Mineralogy, Russian Academy of Sciences, 142432 Chernogolovka, Moscow district, Russia

Kirill I. Shmulovich Institute of Experimental Mineralogy, Russian Academy of Sciences, 142432 Chernogolovka, Moscow district, Russia

Vladimir I. Sorokin Institute of Experimental Mineralogy, Russian Academy of Sciences, 142432 Chernogolovka, Moscow district, Russia

Sergey I. Tkachenko Institute of Experimental Mineralogy, Russian Academy of Sciences, 142432 Chernogolovka, Moscow district, Russia

Lidia N. Var'yash Institute of Ore Deposits, Petrography, Mineralogy and Geochemistry, Russian Academy of Sciences, Staromonetny per. 35, 109017 Moscow, Russia

Valentina M. Vitovtova Institute of Experimental Mineralogy, Russian Academy of Sciences, 142432 Chernogolovka, Moscow district, Russia

Bruce W.D. Yardley Department of Earth Sciences, University of Leeds, Leeds LS2 9JT, UK

Georgiy P. Zaraisky Institute of Experimental Mineralogy, Russian Academy of Sciences, 142432 Chernogolovka, Moscow district, Russia

Irina V. Zarubina Institute of Experimental Mineralogy, Russian Academy of Sciences, 142432 Chernogolovka, Moscow district, Russia

Vilen A. Zharikov Institute of Experimental Mineralogy, Russian Academy of Sciences, 142432 Chernogolovka, Moscow district, Russia

Alexander V. Zotov Institute of Ore Deposits, Petrography, Mineralogy and Geochemistry, Russian Academy of Sciences, Staromonetny per. 35, 109017 Moscow, Russia

Preface

For much of the 20th century, scientific contacts between the Soviet Union and western countries were few and far between, and often superficial. In earth sciences, ideas and data were slow to cross the Iron Curtain, and there was considerable mutual mistrust of diverging scientific philosophies. In geochemistry, most western scientists were slow to appreciate the advances being made in the Soviet Union by D.S. Korzhinskii, who put the study of ore genesis on a rigorous thermodynamic basis as early as the 1930s. Korzhinskii appreciated that the most fundamental requirement for the application of quantitative models is data on mineral and fluid behaviour at the elevated pressures and temperatures that occur in the Earth's crust. He began the work at the Institute of Experimental Mineralogy (IEM) in 1965, and it became a separate establishment of the Academy of Sciences in Chernogolovka in 1969. The aim was to initiate a major programme of high P–T experimental studies to apply physical chemistry and thermodynamics to resolving geological problems.

For many years, Chernogolovka was a closed city, and western scientists were unable to visit the laboratories, but with the advent of perestroika in 1989, the first groups of visitors were eagerly welcomed to the IEM. What they found was an experimental facility on a massive scale, with 300 staff, including 80 researchers and most of the rest providing technical support. Much of the equipment is made in-house, and there are large numbers of pressure vessels capable of taking large samples to extreme conditions. Inevitably, however, much of the sophisticated electronic control and the newest generation of analytical facilities that have become routine in the West in the past 15 years, are lacking.

While some of the experimental work of the Institute has been published in English in the past, most of it is unknown. Furthermore, the Soviet approach to publication was often a barrier to understanding, in that details which western authors expect, are not always included, and the language is not always aimed at clear communication, especially after translation. We decided to produce this volume over a glass of vodka during a visit to Chernogolovka by Bruce Yardley, as a way of bringing together some of the results from Chernogolovka and from colleagues in

Moscow. We also hoped to provide an insight into the ways of approaching problems of crustal fluid processes that have been developed in Russia. We emphasize that the primary focus of this book is the subject matter, not the laboratory; there are many other areas of research at Chernogolovka that we do not touch on. The individual chapters have been written in Russian and edited initially by Kirill Shmulovich, before translation by Galina Gonchar. Bruce Yardley has provided further reviews and editing.

This book would not have been possible without the support of Academician V.A. Zharikov, Director of the Institute of Experimental Mineralogy, who provided active encouragement and allowed Mrs Gonchar to devote herself to the project at the expense of her other duties. We are also indebted to David McGarvie, formerly of Blackies, and Ruth Cripwell, of Chapman & Hall, for their faith and support for the project. Special thanks are also due to Dr Dima Varlamov, who drew the figures for many of the chapters.

At the present time, the future of the Institute of Experimental Mineralogy and other institutes in Russia is clouded by economic uncertainties. We hope that this volume will demonstrate the high calibre of past work on crustal fluids and their interactions with rocks and minerals, and will herald a new and productive phase in such research.

K.I. Shmulovich
B.W.D. Yardley
G.G. Gonchar

Chernogolovka, March 1994

List of symbols for rock-forming minerals

Ak	åkermanite
Ab	albite
And	andalusite
An	anorthite
Bt	biotite
Cal	calcite
Chl	chlorite
Cpx	clinopyroxene
Crn	corundum
Crs	cristoballite
Dsp	diaspore
Di	diopside
Dol	dolomite
Ep	epidote
Fl	fluorite
Grs	grossularite
Grt	garnet
Hbl	hornblende
Hd	hedenbergite
Hem	hematite
Kln	kaolinite
Kfs	K-feldspar
Mag	magnetite
Mc	microcline
Mer	merwinite
Mul	mullite
Ms	muscovite
Ol	olivine
Or	orthoclase
Per	periclase
Pl	plagioclase
Prl	pyrophyllite

Py	pyrite
Qtz	quartz
Ran	rankinite
Sd	siderite
Sil	sillimanite
Spur	spurrite
Toz	topaz
Wo	wollastonite

CHAPTER ONE

An introduction to crustal fluids

Bruce W.D. Yardley and Kirill I. Shmulovich

1.1 INTRODUCTION

Fluids are universally recognized as playing a major role in many geological and geochemical processes in the crust. They act as solvents for the dissolution of metals, and as a medium to transport and concentrate them as ores. At the same time they may fundamentally change the transport properties of their host rocks, by dissolving or infilling pores, or generating hydrofractures. In regions of the crust that are stressed, the presence or absence of fluids exerts a fundamental control on rheology and thus dictates the nature and extent of crustal deformation.

The interest in crustal fluids, recognized in the West by the appearance of the seminal text by Fyfe *et al.* (1978), has served to define problems as much as solve them in the intervening years, and despite the general acceptance of the importance of fluids, all points of detail remain controversial. The cycle of fluids, and in particular whether or not they remain abundant in stable crust after metamorphism, remains the subject of lively disagreement between geophysicists and petrologists. The amounts of fluid required to transport different elements is generally poorly known. Permeabilities of crustal rocks are also largely unknown, and so the duration and extent of fluid transport are difficult to constrain.

Research on hydrothermal fluids has a long tradition in the former Soviet Union, where the work of the late D.S. Korzhinskii has been particularly influential. Korzhinskii, whose ideas are best known in the West from two translated volumes (Korzhinskii, 1959, 1970), made a distinction between mobile and inert components which has provided a conceptual framework for much of the work described in this volume.

Fluids in the Crust: Equilibrium and transport properties.
Edited by K.I. Shmulovich, B.W.D. Yardley and G.G. Gonchar.
Published in 1994 by Chapman & Hall, London. ISBN 0 412 56320 7

Chapter 2 provides an insight into this conceptual basis, which was long in advance of the development of comparable ideas in the western literature, although it is now no more sophisticated than the treatments of the intermediate grades of internal and external buffering that have been developed in the western literature in recent years (e.g. Rice and Ferry, 1982).

In the rest of this book, the aim is to present some of the basic data that will help these types of problem to be resolved. The first part of the book concerns the composition of the fluid phase or phases that produce metasomatic effects and hydrothermal ores, while the second part is concerned with mass transfer through rocks

1.2 FLUID SYSTEMS

Natural fluids in the crust are composed of some combination of water, salts and non-polar gas (CO_2, CH_4 and N_2 primarily).

Even in the absence of the non-polar gases, the activity–composition relationships of electrolyte solutions under supercritical conditions are still not very well known, despite considerable advances in recent years through the work of H. Helgeson, K. Pitzer and their respective co-workers (see Fletcher, 1993 for an introduction to this problem). Chapter 8, by Shmulovich *et al.*, includes new experimental results on salt–H_2O fluids.

The problems of modelling the solution chemistry of fluids that contain mixed non-polar solvents, salts and water are correspondingly more complex. F.U. Franck and co-workers at Karlsruhe have pioneered experimental studies of these fluids, but there are few data. The compositions actually attained depend on a combination of original inputs, exchange equilibria with coexisting minerals, and fluid phase equilibria within this complex system, especially the existence of a miscibility gap between water and non-polar gas which can be extended to include all crustal P–T conditions by the presence of salts. Chapter 8 presents new results on fluid immiscibility in H_2O–CO_2–salt fluids, and demonstrates that the miscibility gap widens with increasing pressure above c. 3 kbar. The results of M.A. Korzhinskiy in Chapter 7, and of Shmulovich *et al.* in Chapter 8 provide a significant addition to the experimental data base on such fluids, and confirm that the presence of CO_2 serves to promote the formation of associated complexes, in much the same way as a decrease in pressure.

1.3 MINERAL SOLUBILITY AND THE COMPOSITION OF CRUSTAL FLUIDS

Mass transfer by fluids depends on their dissolved load. There has been steady progress in accumulating data on the solubility of ore metals

(Helgeson, 1969; Drummond and Ohmoto, 1985; Shock and Helgeson, 1989), but much less progress in quantifying the solution chemistry of metamorphic fluids. A major problem for attempts to quantify crustal fluid chemistries is the uncertainty in estimating the concentrations of ligands available to complex with cations. Species such as chloride behave conservatively, i.e. they remain in the fluid phase and are only incorporated into minerals to a very minor degree. This means that it is very difficult to estimate their past concentration in a fluid phase from analyses of the solid phases that remain, but the chloride content of a fluid controls the concentrations of many metals. For example, the concentration of Na in a saturated solution of albite in sea water is much higher than the concentration of Na in a saturated solution of albite in meteoric water. Some progress has in fact been made in deducing chloride contents of palaeofluids, based on Cl=OH substitution in hydrous minerals (Munoz and Swenson, 1981; Korzhinskiy, 1980; Zhu and Sverjensky, 1991), but the availability of other potentially important but conservative anions, e.g. borate, remains very uncertain. Some of the only constraints are provided by fluid inclusion measurements, which can give a good estimate of fluid salinity but may be difficult to interpret (Crawford and Hollister, 1986) because of the existence of multiple inclusion generations in many types of sample. Only recently has fluid inclusion analysis been developed to the point where data on the concentrations of a range of ligands can be provided (Table 1.1).

A valuable starting point for the investigation of crustal fluid chemistry is the assumption that, in general, fluids are saturated solutions of the rocks in which they reside. This is likely to be a reasonable assumption in metamorphic settings for most of the time, because we know from studies of active geothermal systems that secondary minerals are typically in equilibrium with the fluid from which they are growing, and very local fluid–mineral equilibration has also been reported in very low grade metamorphism on the basis of fluid inclusion studies (Banks *et al.*, 1991). Furthermore, the rates of metamorphic fluid release are governed by rates of heat supply, and are therefore slow (Yardley, 1986). Thus, although local and transient exceptions can occur, fluid is in general released at a rate which will facilitate its equilibration with the host rocks. Nevertheless, the very existence of veins and metasomatic rocks demonstrates that fluids can move into rocks with which they are not in equilibrium, and the consequent reactions produce the hydrothermal rocks that we observe. In the case of magma-derived fluids, or where fluids unmix into liquid and vapour phases, fluids may develop compositions which are quite unlike those expected from rock-buffering (e.g. Shmulovich *et al.*, Chapter 8), and these may be particularly important in ore genesis.

Table 1.1. Chemical composition of crustal fluids: examples of modern well fluids and of ancient fluids preserved as fluid inclusions in quartz

	Modern fluids *Analyses in ppm (1) or* *mg.l*$^{-1}$ (2–4)				*Fluid inclusions* *Analyses in ppm*					
Element	1	2	3	4	5	6	7	8	9	10
Na	50400	63000	61100	18900	17314	57484	29135	39520	182438	78000
K	17500	6150	854	430	863	5219	1218	13960	61409	37000
Ca	28000	44600	28800	63800	793	16687	39004	9620	30723	3200
Mg	54	2770	1830	78	135	1305	1133	120	383	–
Fe	2290	320	338	2.07	18	2770	2372	11950	16310	60000
Mn	1400	60	65	4.57	2.2	456	224	3240	21564	18000
Al	4.2	–	–	1.6	272	–	–	2075	950	–
Li	215	52	–	0.81	27	633	130	215	128	–
Rb	135	–	–	–	–	24	18	427	292	1700
Cs	14	–	–	–	–	–	–	–	146	2300
Sr	400	1770	1820	1580	53	1132	1024	175	784	30
Ba	235	89	56	–	24	638	231	137	345	–
Cu	8	–	–	0.33	3.8	–	–	–	< 10	900
Zn	540	18	217	0.52	1.7	253	77	1569	2207	5200
Pb	102	3	44	3.03	2.6	14	167	–	803	3300
As	12	–	–	–	96	–	–	59	–	480
SiO_2[1]	400	37	59	18	*900*	*1500*	*1500*	*6000*	*7500*	*5400*
B	390	–	–	6.2	432	144	140	3975	438	–
SO_4	5	128	80	223	996	6540	3023	6075	24210	–
H_2S	16	–	–	–	340	–	–	–	–	–
CO_2[2]	> 108	< 22	< 22	42	94500	–	–	–	–	–
F	15	–	–	7	401	< 50	203	817	1770	–
Cl	155000	200400	150700	162700	21389	130639	124875	106360	398940	266000
Br	120	2340	1070	1250	104	721	800	150	146	420
I	18	–	–	7	1.59	18	11	4.8	0.91	–

1. Geothermal brine, Salton Sea, U.S.A., T = 300°C, pH (25°C) = 5.2 (Muffler and White, 1969)
2. Oilfield brine, C. Mississippi, U.S.A., T = 143°C (Carpenter *et al.*, 1974, sample 3)
3. Oilfield brine, C. Mississippi, U.S.A., T = 129°C (Carpenter *et al.*, 1974, sample 39)
4. Shield brine, Sudbury, Canada, T = 22°C, pH = 6.55 (Frape and Fritz, 1987, sample N3646A)
5. H_2O–CO_2 Ore fluid, Au–quartz vein, Brusson, N. Italy, T = c.270°C (Yardley *et al.*, 1993, sample LD659)
6,7. Low grade fluids from alpine thrust faults, Pyrenees, Spain, T = *c*.300°C (Banks *et.al.*, 1991, samples 50176, 50177)
8. Granite-derived tourmaline–topaz–quartz rock, Cornwall, U.K., T = c. 600°C (Bottrell and Yardley, 1988)
9. Granite-derived hypersaline brine, quartz–fluorite vein, Capitan pluton, New Mexico, U.S.A., T = c.650°C (Banks *et al.*, 1994, sample CPU–2, with additional analyses by D.A. Banks from A. Campbell *et al.*, in prep.)
10. Quartz–cassiterite vein fluid, Mole granite, Australia, T = c.550°C (Heinrich *et al.*, 1992)

1 Silica analyses in italics are estimates based on quartz saturation at the approximate conditions of formation.
2 Analyses of modern fluids presented as HCO_3^- have been recalculated.

If a fluid is a saturated solution of a known rock under known conditions, then it should be possible to predict many aspects of its composition from experimental and theoretical data on the solubilities of the minerals present. The difficulties with this approach are twofold. In the first place, coexisting silicate minerals contain components in common and dissolve incongruently in a way which depends on ligand concentrations in the fluid, and these are probably unknown. Secondly, there are the major difficulties in modelling activities in supercritical fluids, especially brines, alluded to above. Nevertheless a number of important sub-systems have been studied and can be used in studies of natural fluid behaviour.

1.3.1 Silica

The solubility of silica is known over a wider range of metamorphic conditions than that of most other minerals, and experimental results for quartz have been integrated into a thermodynamic model by Walther and Helgeson (1977). New data on the solubility of silica are provided here by Sorokin and Dadze in Chapter 4. The key point about quartz solubility is that the silica released into solution combines only with water molecules, to form a notional $H_4SiO_4^0$ aqueous complex. The actual hydration sphere is larger than this under supercritical conditions, so that the complex corresponds to $Si(OH)_4 \cdot 2H_2O$ according to Walther and Orville (1983). As a result, the solubility of quartz is independent of the concentration of ligands such as chloride, which is probably very variable in metamorphic fluids, and also of pH. Only under alkaline conditions, where additional silica may dissolve in the form of alkali silica complexes such as $KH_3SiO_4^0$, does silica solubility become pH-dependent (Anderson and Burnham, 1983). Suitably alkaline fluids are however likely to be very rare in metamorphism, because the presence of even small amounts of chloride is sufficient to prevent high pH values being attained. One additional factor that will, however, have a large effect on silica solubility, is the lowering of water activity by the presence of dissolved non-polar gases (e.g. CO_2) in the fluid. Silica solubility is greatly reduced under these conditions (Walther and Orville, 1983). It is not clear how many other elements similarly dissolve in a way which is independent of most of the common variables of crustal fluid chemistry, but results presented here indicate that cassiterite may behave in this way under oxidizing conditions, and mercury may also exhibit this type of behaviour in the absence of high sulphide concentrations (Sorokin and Dadze, Chapter 4; Zotov *et al.*, Chapter 5).

1.3.2 Cation concentrations in metamorphic fluids

Mineral assemblages are able to buffer the activity ratios of cations in the fluid phase, and also the activity ratios of cations to H^+ through equilibria such as:

$$NaAlSi_3O_8 + K^+ = KAlSi_3O_8 + Na^+, \quad (1.1)$$

$$3KAlSi_3O_8 + 2H^+ = KAl_3Si_3O_{10}(OH)_2 + 2K^+ + 6SiO_2. \quad (1.2)$$

The equilibrium constants for these reactions are:

$$K_1 = aNa^+/aK^+ \qquad K_2 = (aK^+/aH^+)^2,$$

if we take a standard state of the pure solids at the pressure and temperature of interest, and assume that the solids are indeed pure.

These equilibria define the ionic activity ratios between Na^+, K^+ and H^+ for a fluid in equilibrium with albite, K-feldspar, muscovite and quartz at any given P–T for which the values of K_1 and K_2 are known. However the assemblage does not define the absolute concentrations of the cations in the fluid, which must depend on the concentration of available ligands. In consequence, mineral assemblages do not normally buffer pH, unless the salinity of the fluid is also defined. Reaction 1.2 constrains the ratio $aK^+ : aH^+$, but the value of aK^+ increases with the salinity of the fluid. Thus a saline fluid in equilibrium with this model granite system will have a lower pH than a very dilute fluid in equilibrium with the same rock at the same P–T conditions.

Eugster and Gunter (1981) have summarized the relationships between mineral chemistry and fluid chemistry for a number of cation exchange equilibria. They show, for example, that fluids will almost invariably have Na > K and Fe > Mg, even when saturated with K-rich and Mg-rich minerals. A key point emphasized in their study is that, while the ratio between cations of the same valency, buffered by a particular mineral assemblage, is always the same for a given P and T, this is not true of the ratios between cations of different valency. Consider as an example the exchange of cations between plagioclase and chloride fluid, studied experimentally by Orville (1972). The Ca:Na activity ratio in the fluid is related to coexisting plagioclase of fixed composition according to the reaction:

$$CaAl_2Si_2O_8 + 2Na^+ + 4SiO_2 = 2NaAlSi_3O_8 + Ca^{2+}. \quad (1.3)$$

The equilibrium constant $K_3 = aCa^{2+} \cdot (aNa^+)^{-2}$ and therefore does not buffer the simple cation activity ratio. Instead, with increasing salinity of the fluid, and hence increasing Na, the activity of Ca^{2+} increases with the square of the Na activity. A brine in equilibrium with a given plagioclase will have a much higher Ca:Na ratio than a dilute fluid in equilibrium with the same feldspar.

1.3.3 The influence of redox state

Redox state influences metal solubilities in two main ways. The concentrations of transition metal cations in the fluid depends on their valency. For example, Fe is much more soluble in divalent form than when trivalent, and is present in solution in divalent form even under the relatively oxidizing conditions of the hematite–magnetite buffer (Boctor *et al.*, 1980). This means that the precipitation of hematite from a hydrothermal fluid may be written as a reaction dependent on both pH and fO_2:

$$2Fe^{2+} + 2H_2O + 0.5O_2 = Fe_2O_3 + 4H^+. \qquad (1.4)$$

Equation 1.4 demonstrates that the amount of Fe in a solution saturated with hematite will decrease with increasing fO_2, for a constant pH. Another example of a metal whose solubility depends on redox state is Au, which is shown by Zotov *et al.* in Chapter 5 to become more soluble (as a simple cation) with increasing oxidation. This probably accounts for the mobility of Au in gossans.

Redox state also influences metal solubility through its effect on sulphur speciation. Some metals, e.g. Cu, form insoluble sulphides but can remain in solution in the presence of S provided it is in the form of sulphate. Au behaves in the opposite sense; under reducing conditions its solubility is enhanced in the reduced sulphur field by the formation of soluble thiosulphide complexes (Seward, 1973, 1989).

1.3.4 Metal complexing by minor ligands

Not all metals form soluble chlorides, and the concentration of some metals in solution may depend on the presence of additional ligands with which they form soluble complexes. Au, whose solubility under most conditions of ore formation is probably dominated by the formation of thiosulphide complexes, has been noted above, but there are almost certainly others. Aluminium, one of the major rock forming elements, has a very low solubility in aqueous solutions, even at elevated P and T (Ragnarsdottir and Walther, 1985; Woodland and Walther, 1987; Baumgartner and Eugster, 1988), but this is inconsistent with geological evidence for Al-mobility, e.g. from vein assemablages. Experimental results on Al solubility in fluoride-rich fluids, under conditions appropriate to greisen formation, are presented by Zaraisky in Chapter 6, and demonstrate a marked increase in Al-solubility that can be ascribed to fluoride complexing. Evidence for a similar effect at low temperatures was found by Yardley *et al.* (1993), who reported a reasonable correspondence between Al concentrations in low grade ore fluids and the predicted values assuming fluoride complexing (see also Table 1.1). Other metals for which complexing may be dominated by ligands other than chloride include Ti, W and the REE.

1.4 FLUID MIGRATION THROUGH CRUSTAL ROCKS

It is a feature of most current models for fluid behaviour in the crust that fluids released by metamorphism or magmatic crystallization must escape from their source and make their way to areas of lower hydraulic head (see however Chapter 2 for a purist thermodynamic approach based on component mobility which does not require this physical assumption). There has been a considerable difference in emphasis between the way in which most western scientists have begun to tackle the analysis of such flow, and the approaches developed in Russia, and the combination of both approaches is likely to prove particularly fruitful in the future.

The work of Watson and Brennan (1987) has been particularly influential in recent western literature. They pointed out that different types of fluids may have different wetting characteristics for common rock-forming minerals, something which has been borne out by further experimental work subsequently (e.g. Holness and Graham, 1991). So-called wetting fluids have low dihedral angles with mineral grains, and thus form an interconnected pore network which retains a finite permeability down to very low porosities. Wetting fluids can, at least in principle, move through a rock pervasively, and the porosity–permeability relationship can be modelled (von Bargen and Waff, 1986). On the other hand, non-wetting fluids (including those with significant CO_2) occur in isolated pores so that the rock is effectively impermeable to them if the porosity is too low for the pores to interconect (c. 4%). They are likely to move only in fractures, unless anomalously high porosities are present. Interestingly, there is in fact evidence that large porosities can be generated transiently in carbonate rocks, both from stable isotope studies (Tracy *et al.*, 1983) and from textural imaging (Yardley and Lloyd, 1989).

Focussed flow along fractures is the other main mechanism of crustal fluid flow that has received attention in the western literature (Brennan, 1991), and here there is a large body of data and literature because of the implications for major engineering projects, especially for geothermal energy. Metamorphic and hydrothermal veins are generally taken to be sites of former fractures along which fluid flow occurred, although the extent to which such fractures were sites of long-range flow as opposed to local segregation has been controversial (cf. Ferry and Dipple, 1991; Yardley and Bottrell, 1992; Walther, 1990).

Researchers in the former Soviet Union have given much more emphasis to direct experimental studies of fluid flow through rocks at elevated pressures and temperatures. Chapter 11, by Shmonov *et al.*, presents the results of direct measurements of rock permeabilities at 2 kbar and up to 600°C. The rocks studied are natural ones, which have therefore experienced unloading and possible alteration on return to the surface, while permeability has been measured using Ar, thereby precluding changes to rock texture that could be brought about by fluid–rock interactions, but the results represent a first step towards understanding permeability in the deep crust, and also indicate that many crystalline rocks appear to be too permeable to sustain fluid pressures equal to lithostatic as is so commonly assumed.

A second effect recognized by Russian researchers is thermal decompaction, the anomalous expansion of rocks on heating caused by the creation of voids between mineral grains due to their anisotropic expansion and variable orientations. Both experimental and theoretical results are presented by Zaraisky and Balashov in Chapter 10, and they argue that, even when the effects of creep in closing the pores are taken into account, the effect will be an important one in enhancing permeability in rapidly heated thermal aureoles.

A recurrent uncertainty in studies of fluid flow through rocks with fine cracks or small pores, is the extent to which interactions at the mineral surfaces with specific fluid components may change the composition bulk of the fluid itself. In other words, can the crust act as a chromatographic column, separating out fluid components by differentially retarding them as they move through it? Not only may gas species and water become separated, membrane filtration effects are possible in sediments with small pore sizes, which serve to change the salinity of the fluid such that salt tends to remain behind as water leaves the rock. The possibility also exists that different dissolved salts could be differentially retarded. This topic is introduced in Chapter 2, and has been the subject of a detailed theoretical analysis by Balashov, in Chapter 9. Balashov is able to produce a general model for mass transfer through granite, taking into account surface interactions between rock and fluid and their consequences for the rate of transport of different components.

1.5 CONCLUSIONS

The concentrations of some elements in aqueous crustal fluids are a simple function of pressure, temperature and water activity. Significant impurities in the form of dissolved gases or high salt concentrations are necessary before their solubility is changed significantly. However the concentrations of most metals are a function of ligand availability, and this is not always buffered by the rock itself. In particular, the most important anion present in deep fluids, chloride, behaves very conservatively and may occur at very different concentrations in mineralogically similar rocks. Salinity has an indirect effect on the concentrations of other possible complexing ligands, because it controls the concentrations of metals which may serve to precipitate them out of the fluid. The other major uncertainty in attempting to calculate metamorphic fluid chemistries is that there is no adequate model applicable to interpreting activity–composition relationships of strong electrolytes under supercritical conditions, and understanding of solute speciation in real crustal fluids is still in its infancy (Brady and Walther, 1990).

Our understanding of fluid compositions and fluid processes in metamorphism and in ore formation has reached the point where most of the necessary theoretical basis is in place. However it is often the case that the available experimental data is insufficient for tackling real problems in natural systems, and indeed we do not always know enough about natural systems to provide adequate constraints to develop a model. The remaining chapters of this book are intended to fill in some of the gaps in our knowledge, and at the same time to introduce the conceptual framework for considering crustal fluid behaviour that has developed in the former Soviet Union.

REFERENCES

Banks, D.A., Davies, G.R., Yardley, B.W.D., *et al.* (1991) The chemistry of brines from an alpine thrust system in the central Pyrenees: an application of fluid inclusion analysis to the study of fluid behaviour in orogenesis. *Geochimica et Cosmochimica Acta*, v. **55**, p. 1021–30.

Banks, D.A., Yardley, B.W.D., Campbell, A.R. and Jarvis, K.E. (1994) REE composition of an aqueous magmatic fluid: a fluid inclusion study from the Capitan pluton, New Mexico. *Chemical Geology*, v. **112** (in press).

Baumgartner, L.P. and Eugster, H.P. (1988) Experimental determination of corundum solubility and Al-speciation in supercritical H_2O–HCl solutions. *Geological Society of America Abstracts with Programs*, v. **22**, p. A191.

Boctor, N.Z., Popp, R.K. and Frantz, J.D. (1980) Mineral–solution equilibria – IV. Solubilities and the thermodynamic properties of $FeCl_2^0$ in the system Fe_2O_3–H_2–H_2O–HCl. *Geochimica et Cosmochimica Acta*, v. **44**, p. 1509–18.

Bottrell, S.H. and Yardley, B.W.D. (1988) The composition of a primary granite-derived ore fluid from S.W. England, determined by fluid inclusion analysis. *Geochimica et Cosmochimica Acta*, v. **52**, p. 585–8.

Brady, P.V. and Walther, J.V. (1990) Algorithms for predicting ion association in supercritical H_2O fluids. *Geochimica et Cosmochimica Acta*, v. **54**, p. 1555–61.

Brennan, J.M. (1991) Development and maintenance of metamorphic permeability: implications for fluid transport. *Reviews in Mineralogy*, v. **26**, p. 291–320.

Carpenter, A.B., Trout, M.L. and Pickett, E.E. (1974) Preliminary report on the origin and chemical evolution of lead- and zinc-rich oil field brines in central Mississippi. *Economic Geology*, v. **69**, p. 1191–1206.

Crawford, M.L. and Hollister, L.S. (1986) Metamorphic fluids: the evidence from fluid inclusions, in *Fluid–rock Interactions During Metamorphism* (eds J.V. Walther and B.J. Wood), *Advances in Physical Geochemistry*, v. **5**, p. 1–35. Springer-Verlag, New York.

Drummond, S.E. and Ohmoto, H. (1985) Chemical evolution and mineral deposition in boiling hydrothermal systems. *Economic Geology*, v. **80**, p. 126–47.

Eugster, H.P. and Gunter, W.D. (1981) The compositions of supercritical metamorphic fluids. *Bulletin Societe Francaise Mineralogie et Cristallographie*, v. **104**, p. 817–26.

Ferry, J.M. and Dipple, G.M. (1991) Fluid flow, mineral reactions and metasomatism. *Geology*, v. **19**, p. 211–14.
Fletcher, P. (1993) *Chemical Thermodynamics for Earth Scientists*. Longman Scientific & Technical, Harlow, U.K.
Frape, S.K. and Fritz, P. (1987) Geochemical trends for groundwaters from the Canadian Shield, in *Saline Water and Gases in Crystalline Rocks* (eds P. Fritz and S.K. Frape). Geological Association of Canada Special Paper 33, p. 19–38.
Fyfe, W.S., Price, N. and Thompson, A.B. (1978) *Fluids in the Earth's Crust*. Elsevier, Amsterdam.
Heinrich, C.A., Ryan, C.G., Mernagh, T.P. and Eadington, P.J. (1992) Segregation of ore metals between magmatic brine and vapor: a fluid inclusion study using PIXE microanalysis. *Economic Geology*, v. **87**, p. 1566–83.
Helgeson, H.C. (1969) Thermodynamics of hydrothermal systems at elevated temperatures and pressures. *American Journal of Science*, v. **267**, p. 729–804.
Holness, M.B. and Graham, C.M. (1991) Equilibrium dihedral angles in the system H_2O–CO_2–NaCl–calcite, and implications for fluid flow during metamorphism. *Contributions to Mineralogy and Petrology*, v. **108**, p. 368–83.
Korzhinskiy, M.A. (1980) Apatite solid solutions as indicators of the fugacity of HF and HCl in hydrothermal fluids. *Geochemistry International*, v. **3**, p. 45–60.
Korzhinskii, D.S. (1959) *Physicochemical basis of the analysis of the paragenesis of minerals*. Consultants Bureau, New York.
Korzhinskii, D.S. (1970) *Theory of metasomatic zoning*. Oxford University Press, Oxford.
Muffler, L.J.P. and White, D.E. (1969) Active metamorphism of Upper Cenozoic sediments in the Salton Sea geothermal field and the Salton Trough, southeastern California. *Geological Society of America Bulletin*, v. **80**, p. 157–82.
Munoz, J.L. and Swenson, A. (1981) Chloride-hydroxyl exchange in biotite and estimation of relative HCl:HF fugacities in hydrothermal fluids. *Economic Geology*, v. **76**, p. 2212–21.
Orville, P.M. (1972) Plagioclase cation exchange equilibria with aqueous chloride solution: results at 700°C and 2000 bars in the presence of quartz. *American Journal of Science*, v. **272**, p. 234–72.
Ragnarsdottir, V. and Walther, J.V. (1985) Experimental determination of corundum solubilities in pure water between 400–700°C and 1–3 kbar. *Geochimica et Cosmochimica Acta*, v. **49**, p. 2109–15.
Rice, J.M. and Ferry, J.M. (1982) Buffering, infiltration and the control of intensive variables during metamorphism. *Reviews in Mineralogy*, v. **10**, p. 263–354.
Seward, T.M. (1973) Thio complexes of gold and the transport of gold in hydrothermal ore solutions. *Geochimica et Cosmochimica Acta*, v. **37**, p. 379–99.
Seward, T.M. (1989) The hydrothermal chemistry of gold and its implications for ore formation: boiling and conductive cooling as examples, in *The Geology of Gold Deposits: The Perspective in 1988* (eds R. Keays, R. Ramsay and D. Groves). Economic Geology Monograph, v. **6**, p. 398–404.
Shock, E.L. and Helgeson, H.C. (1989) Calculation of the thermodynamic and transport properties of aqueous species at high pressures and temperatures: correlation algorithms for ionic species and equations of state predictions to 5 kb and 1000°C. *Geochimica et Cosmochimica Acta*, v. **52**, p. 2009–36.
Tracy, R.J., Rye, D.M., Hewitt, D.A. and Schiffries, C.M. (1983) Petrologic and stable isotope studies of fluid–rock interactions, south central Connecticut. I. The role of infiltration in producing reaction assemblages in impure marbles. *American Journal of Science*, v. **283-A**, p. 589–616.
von Bargen, N. and Waff, H.S. (1986) Permeabilities, interfacial areas and curvatures of partially molten systems: results of numerical computations of equilibrium microstructures. *Journal of Geophysical Research*, v. **91**, p. 9261–76.

Walther, J.V. (1990) Fluid dynamics during progressive regional metamorphism, in *The Role of Fluids in Crustal Processes*. National Academy Press, Washington D.C., p. 64–71.

Walther, J.V. and Helgeson, H.C. (1977) Calculations of the thermodynamic properties of aqueous silica and the solubility of quartz and its polymorphs at high pressures and temperatures. *American Journal of Science*, v. **277**, p. 1315–51.

Walther, J.V. and Orville, P.M. (1983) The extraction-quench technique for determination of the thermodynamic properties of solute complexes: application to the solubility of quartz in fluid mixtures. *American Mineralogist*, v. **68**, p. 731–41.

Watson, E.B. and Brennan, J.M. (1987) Fluids in the lithosphere, 1. Experimentally determined wetting characteristics of CO_2–H_2O fluids and their implications for fluid transport, host rock physical properties, and fluid inclusion formation. *Earth and Planetary Science Letters*, v. **85**, p. 497–515.

Woodland, A.B. and Walther, J.V. (1987) Experimental determination of the solubility of the assemblage paragonite, albite and quartz in supercritical H_2O. *Geochimica et Cosmochimica Acta*, v. **51**, p. 365–72.

Yardley, B.W.D. (1986) Fluid migration and veining in the Connemara Schists, Ireland, in *Fluid–rock Interactions During Metamorphism* (eds J.V. Walther and B.J. Wood), *Advances in Physical Geochemistry*, v. **5**, p. 109–31. Springer-Verlag, New York.

Yardley, B.W.D. and Bottrell, S.H. (1992) Silica mobility and fluid movement during metamorphism of the Connemara Schists, Ireland. *Journal of Metamorphic Geology*, v. **10**, p. 453–64.

Yardley, B.W.D. and Lloyd, G.E. (1989) An application of cathodoluminescence microscopy to the study of textures and reactions in high grade marbles from Connemara, Ireland. *Geological Magazine*, v. **126**, p. 333–7.

Yardley, B.W.D., Banks, D.A., Bottrell, S.H. and Diamond, L.W. (1993) Post-metamorphic gold-quartz veins from N.W. Italy: the composition and origin of the ore fluid. *Mineralogical Magazine*, v. **57**, p. 407–22.

Zhu, C. and Sverjensky, D.A. (1991) Partitioning of F–Cl–OH between minerals and hydrothermal fluids. *Geochimica et Cosmochimica Acta*, v. **55**, p. 1837–58.

CHAPTER TWO

Fluids in geological processes

Vilen A. Zharikov

2.1 INTRODUCTION

The role of fluids in the Earth's crust and mantle has been a subject of debate by petrologists, geochemists and geophysicists for much of the present century, and an extensive literature exists in both English and Russian.

Observations indicate that a volatile liquid or gas phase of low density is a key participant in many geological processes. This phase, be it liquid, gas or supercritical, is commonly defined as **fluid**, irrespective of its composition. The present work uses the term 'fluid' in this broad sense. When needed, however, the more rigorous physico-chemical definition of fluid, as the supercritical state of a volatile system, is used.

Among the lines of evidence for fluid involvement in geological processes is the regularity of metasomatic processes. Zoned metasomatic rocks develop primarily in tectonically weakend and fractured zones, and may require enormous mass exchange amounting to hundreds of kilograms per cubic metre of rock. Furthermore, abundant gas–liquid inclusions commonly occur in metasomatic minerals.

That the presence of a fluid medium is essential to crustal processes is apparent from physico-chemical data for reaction rates, diffusion rates and melt viscosities at magma temperatures. As far back as 1958, Fyfe, *et al.* pointed out that the scale of mass transport in natural processes was only compatible with diffusion-rate data if temperatures were at least 1200–1500 °C or if the processes had operated in the presence of fluid. The former alternative is clearly incompatible with all the available independent evidence about temperatures

Fluids in the Crust: Equilibrium and transport properties.
Edited by K.I. Shmulovich, B.W.D. Yardley and G.G. Gonchar.
Published in 1994 by Chapman & Hall, London. ISBN 0 412 563207

in the crust and upper mantle, and so there is a strong case for fluid involvement.

The subject of fluid interactions in geochemical processes embraces several important aspects.

1. Equilibrium thermodynamic properties of systems involving volatiles, for example: (a) P–V–T–X properties of fluid systems; (b) mineral solubility in fluids and the properties of volatile–salt systems; (c) specific properties of fluids occupying small volume pores; (d) mineral and phase equilibria in systems with perfectly mobile (in a thermodynamic sense) components; (e) the influence of non-equilibrium states on phase relations, in particular conditions in which fluid pressure differs from total pressure.
2. The effect of fluid on the kinetics of reactions and mass transfer in the system.
3. The dynamics of systems containing fluid components, which deals primarily with: (a) dynamic parameters (viscosity, porosity, etc.) of fluid systems; (b) diffusion of components in fluids; (c) infiltration of homogeneous and heterogeneous fluids, especially in low-porosity rocks; (d) filtration effects resulting in differentiation of solution components due to the varying rate of movement of different components in the flowing fluid; (e) the influence of fluids on the dynamic properties of magmatic systems and on the mass and heat transfer processes in them.
4. The origin of fluids and the identification and quantification of primary (juvenile) and secondary (meteoric) fluid inputs by geochemical methods.

Although this is not an exhaustive list of problems related to 'fluids in geological processes', it includes the most important and timely ones. This chapter provides an outline and discussion of them, while the remaining chapters in this volume provide detailed accounts of specific aspects.

2.2 EQUILIBRIUM SYSTEMS WITH FLUID COMPONENTS

2.2.1 Hydrothermal fluid model

At present there are two approaches available to modelling natural heterogeneous systems. One (specific) allows a model for a particular object or particular process to be calculated from the experimentally determined thermodynamic properties of coexisting phases. A striking pioneering example of this is the classic work of Krauskopf (1957) on the heavy metal content of magmatic vapour. The advantage of this approach is that it yields rapid, concrete results, but the great diversity of objects requires immense experimental work.

The alternative (general) approach is to construct a general thermodynamic model for an idealized hydrothermal fluid and then search for solutions to this model which apply to natural examples. One of the earliest attempts to do this resulted in the construction of qualitative topological schemes for volatile–silicate systems (Zavaritskii, 1926; Niggli, 1937) rather than of a model.

Constructing a general model for a hydrothermal fluid is no easy task. Even when greatly simplified, the fluid composition should be represented by the components: H_2O–CO_2–H_2S–H_2 (or O_2)–chloride salt–silicate–sulphide. In many instances, one has to incorporate other volatile species (e.g. CH_4, N_2) and oxide components. The general model is constructed by first studying simple unary–binary systems followed by studies of more complex multicomponent ones.

The extent to which binary–ternary systems have been studied experimentally is shown in Figure 2.1 with T–P coordinates. Numbers indicate studies carried out in the past 10–12 years (for earlier reviews, refer to: Zharikov *et al.*, 1973b; Zharikov, 1976b; Fyfe *et al.*, 1978; Shmulovich *et al.*, 1982; Carmichael and Eugster, 1987). It can be seen that the systems H_2O–CO_2, H_2O–chloride, H_2O–CO_2–NaCl have received extensive study over wide temperature–pressure ranges. Data for the other binary and ternary systems are fewer. As regards more complex multicomponent systems such as H_2O–CO_2–NaCl–albite, H_2O–CO_2–NaCl–granite or H_2O–CO_2–HF–granite, only theoretical diagrams are available supported by a small number of experiments.

Nevertheless, some key properties of fluid systems definitely emerge. The results from the systems water–CO_2–salt are detailed in the chapters that follow; here we point out some of the key properties of the systems involving volatiles, based in large part on the results of studies specified in Figure 2.1.

1. There is a considerable difference between the pressure dependencies of fugacity coefficients for H_2O and CO_2. The fugacity coefficient of H_2O shows a weak positive correlation with pressure, while that of CO_2 rises sharply with pressure, by a factor of 10 to 100 at pressures greater than 5 kbar, depending on temperature (Shmulovich, 1988; Figure 47).
2. The system H_2O–CO_2 is non-ideal over the entire T–P range of the Earth's crust; activity coefficients increase as the temperature is lowered and the immiscible region approached.
3. In water–salt systems, the critical temperature rises with the concentration of salt along the critical curve. An increase in pressure lowers the activity coefficient of H_2O as a function of solution concentration and type of salt (Sourirajan and Kennedy, 1962).
4. Finally, multicomponent systems such as H_2O–CO_2–salt, H_2O–CO_2–salt–silicate have wide two-phase fields.

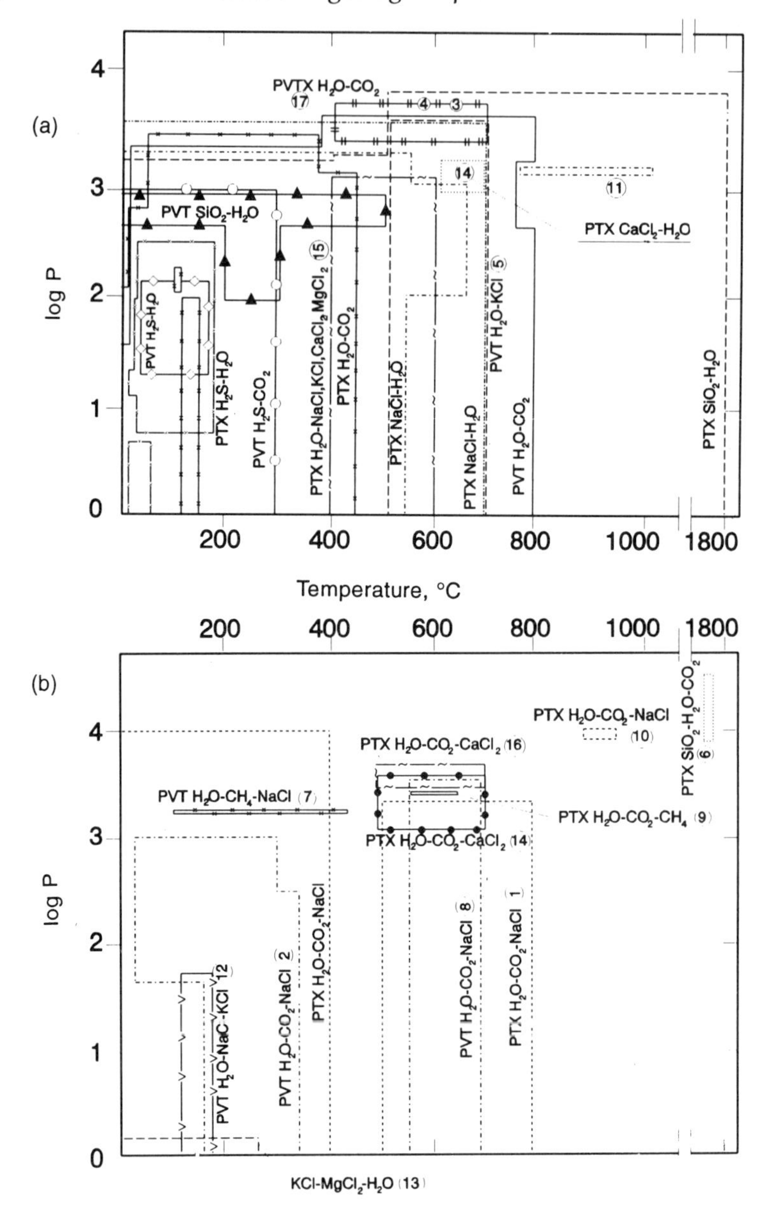

Figure 2.1a, b. Illustration of the P–T ranges over which binary (a) and ternary (b) systems involving volatile components have been studied during the last years. Numbers within circles indicate studies by the following authors: 1 Kotel'nikov and Kotel'nikova (1990); 2 Pampura *et al.* (1981); 3 Plyasunov and Zakirov (1991); 4 Shmulovich *et al.* (1980); 5 Bodnar and Sterner (1985); 6 Boettcher (1984); 7 Franck (1985); 8 Gehrig (1980); 9 Jacobs and Kerrick (1981); 10 Johnson (1991); 11 Ming-Chou (1986); 12 Susarla *et al.* (1987); 13 Voigt and Fanghaenel (1985); 14 Zhang and Frantz (1989); 15 Tkachenko and Shmulovich (1992); 16 Plyasunova (1993); 17 Sterner and Bodnar (1991).

The different characteristics of different fluid systems have important petrological consequences. As long ago as 1940, Korzhinskii identified mineralogical depth facies based on the carbonation reactions $A + CO_2 = B + (Ca, Mg)CO_3$. The reason for the pressure-dependence of such reactions lies in the sharp increase of f_{CO_2} with increasing pressure (see Shmulovich, 1975; Zharikov *et al.*, 1977). A schematic diagram and some major carbonation reactions constituting the basis for the depth facies are shown in Figure 2.2a–c.

Unlike carbonation reactions, hydration reactions are commonly rather independent of pressure, above c.0.3 kbar (Figure 2.2a,b). This is due to the fact that a slight increase in the water fugacity with pressure compensates for the water activity reduction owing to the hydration of dissolved components.

A new development, stemming from the results of experimental studies of the fluid–salt and fluid–silicate systems is the suggestion that heterogeneous fluids (i.e. more than one fluid phase coexisting immiscibly) may be more widely involved in natural processes than has generally been supposed (Ryabchikov, 1975; Bowers and Helgeson, 1983; Crawford *et al.*, 1979; Trommsdorff and Skippen, 1986; Yardley and Bottrell, 1988; see also Chapter 8, this volume). An example is the H_2O–CO_2–NaCl–albite system. Schematic sections for this system, based on data from a number of independent studies, are depicted in Figure 2.3. In the liquidus region, two melts may coexist in this system, provided the salt content is sufficiently large (c. 12–13 mol%). The crystallization of albite results in the development of a two-phase fluid – a CO_2-rich 'gas' phase and an aqueous brine. Provided CO_2 is present, a single fluid phase is possible between the retrograde boiling curve and the immiscible region in the pure H_2O–CO_2 system only when the salinity of the system is low (c. 10 mol%). Systematic studies of T–P–X boundaries outlining the two-phase fluid region present an urgent problem to be addressed at present and in the future. It is essential to consider the possibility of a heterogeneous fluid, since fluid unmixing produces a radical redistribution of fluid components and also induces a complex pattern of heterogeneous infiltration, which is the direct consequence of the separation of the gas constituent from the liquid during flow (see below).

2.2.2 Phase equilibria in open systems

(a) Systems with perfectly mobile components. The thermodynamic treatment of open systems as being comprised of perfectly mobile and inert components was developed by Korzhinskii (1942, 1949, 1959). See also discussions between Korzhinskii (1966, 1967) and Weill and Fyfe (1964, 1967) in *Geochimica et Cosmochimica Acta*, **28**, **30**, **31**. Recall that such systems differ from the classical ones of metamorphic petrology in

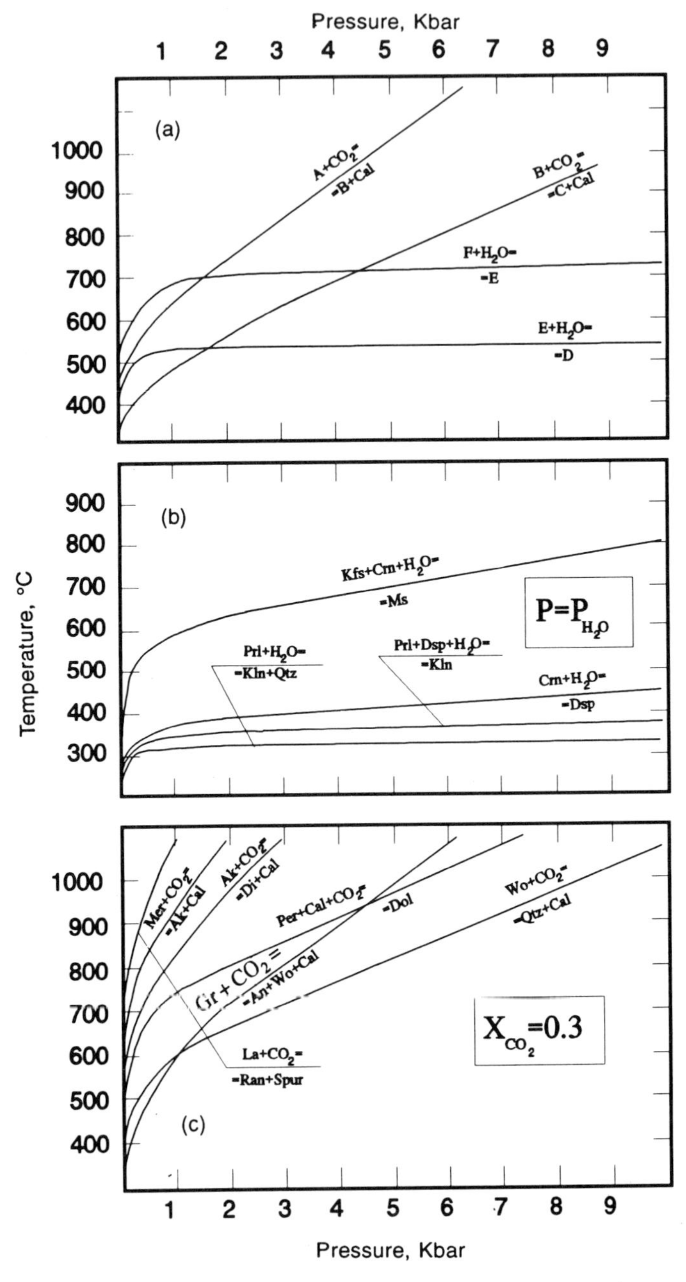

Figure 2.2. Dehydration–decarbonation reactions. (a) Schematic diagram, where A, B, C, D, E, F symbolize abstract minerals. (b) Some dehydration reactions studied experimentally (Zharikov *et al.*, 1972) under conditions of $P = P_{H_2O}$. (c) Some decarbonation reactions studied experimentally (Zharikov *et al.*, 1977) and recalculated for conditions of $X_{CO_2} = 0.3$. At $X_{CO_2} = 1.0$, univariant lines are compressed against the temperature axis.

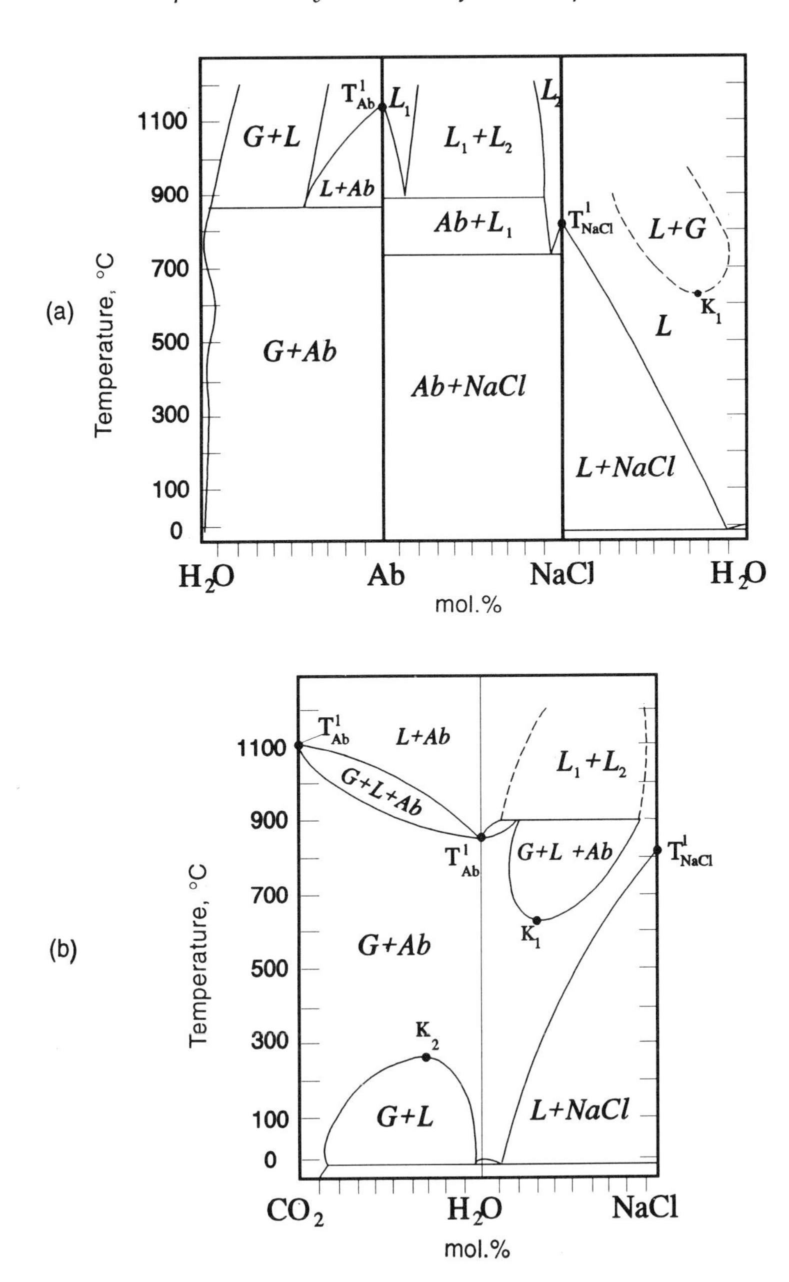

Figure 2.3. Schematic diagrams for portions of the system albite–NaCl–H_2O–CO_2 at P = 1 kbar. (a) The boundary join $NaAlSi_3O_8$–NaCl–H_2O. (b) Projection of the system on to the H_2O–CO_2–NaCl plane. The diagrams are constructed from individual experimental data points (e.g. Chukhrov, F.V. 1974; Ryabchikov, 1975; Valyashko *et al.*, 1984; Johannes, 1985). Equilibria in the low-temperature portion (below 100 °C) are simplified.

that some components may be 'perfectly mobile', i.e. their chemical potential or some other related intensive parameter is independently controlled. Examples include the mole fraction of a component in the fluid, controlled by a buffering equilibrium in some source region upflow, or the control of fluid fugacity or pressure by lithostatic pressure and rock strength. The thermodynamic potential (G_m) of an isothermal–isobaric system involving perfectly mobile components is expressed as:

$$G_m = G - \sum_{j}^{k} \mu_j m_j = \sum_{a}^{i} \mu_a m_a, \qquad (2.1)$$

or, in differential form, as:

$$dG_m = -SdT + VdP + \sum_{a}^{i} \mu_a dm_a - \sum_{j}^{k} m_j d\mu_j - \delta Q_i = 0, \qquad (2.2)$$

where a . . ., i are inert components whose independent parameter is their mass $m_a \ldots, m_i$, and j . . ., k are perfectly mobile components whose independent parameter is their chemical potential $\mu_j \ldots, \mu_k$, while δQ_i stands for uncompensated heat, and $\delta Q_i = 0$ at the equilibrium state.

Variation in the thermodynamic potential as a function of intensive parameters of the systems can be written as:

$$d\Delta G_m = -\Delta SdT + \Delta VdP - \sum_{j}^{k} \Delta m_j d\mu_j = 0. \qquad (2.3)$$

In systems with perfectly mobile components, the number of phases possible is defined by the number of inert components (and other extensive parameters of the system), irrespective of the number of perfectly mobile components.

These relationships have received the name **Korzhinskii's phase rule**:

$$n_{TP\mu} = k_i - r; \quad n_{T,V\mu} = k_i + 1 - r; \quad n_{S,P\mu} = k_i + 1 - r,$$

where n is the degrees of freedom, k_i is the number of inert components and r the number of phases. Subscripts to n denote constant temperature, pressure, volume or entropy and thus define the type of system.

The limiting case in which all the components are perfectly mobile may be achieved in isochoric, isoentropic systems, in which the components form a homogeneous solution (Helmholtz free energy $F_{(V,T\mu)} = -PV$) or a melt ($H_{(S,P\mu)} = TS$), or in a two-phase 'isolated' system ($U_{(S,V\mu)} = TS - PV$). That Korzhinskii's phase rule may be applied successfully to natural systems provides further evidence for the validity of the concept of perfectly mobile components.

H_2O and H_2O–CO_2 are the principle fluid types in natural high temperature fluid systems. Comprehensive petrologic studies of metamorphic and magmatic processes in Archaean rocks from eastern Siberia led Korzhinskii (1940, 1945) to conclude that water and CO_2 act as perfectly mobile components in natural processes. In other words, reactions involving water and CO_2 are not limited by the content of these components in the system but are governed by their independently defined chemical potentials. In this case, the equilibrium change in the thermodynamic potential of the system ($d\Delta G_m$) is expressed as:

$$d\Delta G_m = -\Delta S dT + \Delta V_s\, dP - \sum \Delta m_{H_2O} \cdot RT d\ln\alpha_{H_2O} - \Delta m_{CO_2} \times RT d\ln\alpha_{CO_2} = -\Delta S dT + \Delta V_S\, dP - RT\left[\Delta m_{H_2O}\, d\ln\gamma_{H_2O} \times \varphi_{H_2O} X_{H_2O} P + \Delta m_{CO_2}\, d\ln\gamma_{CO_2}\, \varphi_{CO_2} X_{CO_2} \times P\right] \quad (2.4)$$

for dehydration–decarbonation reactions, where γ, φ and X are, respectively, activity coefficients, fugacity coefficients and mole fractions of H_2O and CO_2 in fluid, while ΔV_s denotes the change in the solid phase volumes.

The melting temperature T^l of solids in systems where H_2O and CO_2 are perfectly mobile is obtained from:

$$dT^l = \frac{\Delta V}{\Delta S} dP - \frac{m^l_{H_2O} - m^s_{H_2O}}{\Delta S} \cdot RT d\ln\alpha_{H_2O} - \frac{m^l_{CO_2} - m^s_{CO_2}}{\Delta S} \times RT d\ln\alpha_{CO_2} = \frac{\Delta V}{\Delta S} dP - RT\left[\frac{m^l_{H_2O} - m^s_{H_2O}}{\Delta S} \cdot d\ln\gamma_{H_2O}\, \varphi_{H_2O} X_{H_2O} P + \frac{m^l_{CO_2} - m^s_{CO_2}}{\Delta S} \cdot d\ln\gamma_{CO_2} \cdot \varphi_{CO_2} X_{H_2O} P\right], \quad (2.5)$$

where ΔS and ΔV are the entropy and volume changes of melting; the other symbols are defined as above.

Examples of the perfectly mobile behaviour of H_2O and CO_2 in magmatic, metamorphic and metasomatic processes have been provided by numerous investigations, but other studies have considered that water released by breakdown of hydrous minerals may be an active agent in metamorphism and granite formation. For example, some authors discuss the process of partial melting in the absence of a low-density fluid due to dehydration–melting reactions of muscovite, biotite or amphibole (Thompson, 1982), (termed fluid-free melting reactions by Grant, 1985). Such systems, involving a fixed amount of water, imply that it behaves as an inert component.

In order to evaluate the likely behaviour of water in metamorphism, it makes sense to discuss separately the different patterns of behaviour to be expected during dehydration and melting reactions for the cases where:

H_2O is a perfectly mobile component in an open system;
H_2O is an inert component in an open or closed system.

(b) Water as a perfectly mobile component. It follows from Equations 2.3 and 2.4 that for the dehydration reaction

$$B = A + H_2O,\ {}^1\Delta G^{R}_{T,P} = (\Delta G^A + \Delta G_{H_2O} - G^B)_{T^0} - \int_{T^0}^{T} (S^A + S_{H_2O} - S^B)dT$$

$$+ \int_{P^0}^{P} (V^A - V^B)\, dP + \int_{P^0}^{P} V_{H_2O}\, dP = 0, \qquad (2.6)$$

where the last term

$$\int_{P^0}^{P} V_{H_2O}\, dP = \Delta G^{T,P}_{H_2O} = \int_{P^0T^0}^{PT} RT d\ln \gamma_{H_2O}\, \varphi_{H_2O}\, X_{H_2O} P,$$

and P^0, T^0 are, respectively, the pressure and temperature of standard or initial conditions.

Clearly, the T–P coordinates of a dehydration reaction are dependent on the mole fraction (X_{H_2O}) and activity of water (a_{H_2O}) in the fluid, defined by the environmental conditions.

Figure 2.4 combines the experimental muscovite dehydration curve for the reaction muscovite = K–feldspar + corundum + H_2O with $P = P_{H_2O}$ (Zharikov *et al.*, 1972) with muscovite dehydration curves calculated for conditions of $a_{H_2O} = 0.3$ and 0.6 at $P_{H_2O} = 1$ bar = constant using the experimental curve and thermodynamic data from Zharikov (1976a) and Robie *et al.* (1978). The univariant lines $P = P_{H_2O}$ and $P_{H_2O} = 1$ bar bound all possible T–P coordinates for muscovite dehydration if water is perfectly mobile.

The melting temperature of the solid phases from Equation 2.3 is defined as:

$$dT^l = \frac{\Delta V}{\Delta S} \cdot dP - \frac{\sum_j^k \Delta m_j \,.\, RT d\ln \alpha_j}{\Delta S}. \qquad (2.7)$$

Rearranging terms and treating water as perfectly mobile, we obtain:

$$d\ln T^l = \frac{\Delta V^l}{\Delta H^l} \cdot dP - \frac{(m^l_{H_2O} - m^s_{H_2O})}{\Delta H_l} \cdot RT d\ln \gamma_{H_2O} \varphi_{H_2O} X_{H_2O} \cdot P, \qquad (2.8)$$

where ΔH^l and ΔV^l are the enthalpy and volume changes of melting, and $m^s_{H_2O}$ and $m^l_{H_2O}$ are the water contents in the solid phase and melt of the same composition.

Equation 2.8 defines the familiar, experimentally determined melting temperature relationships for perfectly mobile behaviour of water in the range from $X^f_{H_2O} = 0$ to $a^f_{H_2O} = X^f_{H_2O} = 1$ and $P_{H_2O} = P$. When $d\ln T^l$ is less than zero, a decrease in the melting temperature occurs relative to the

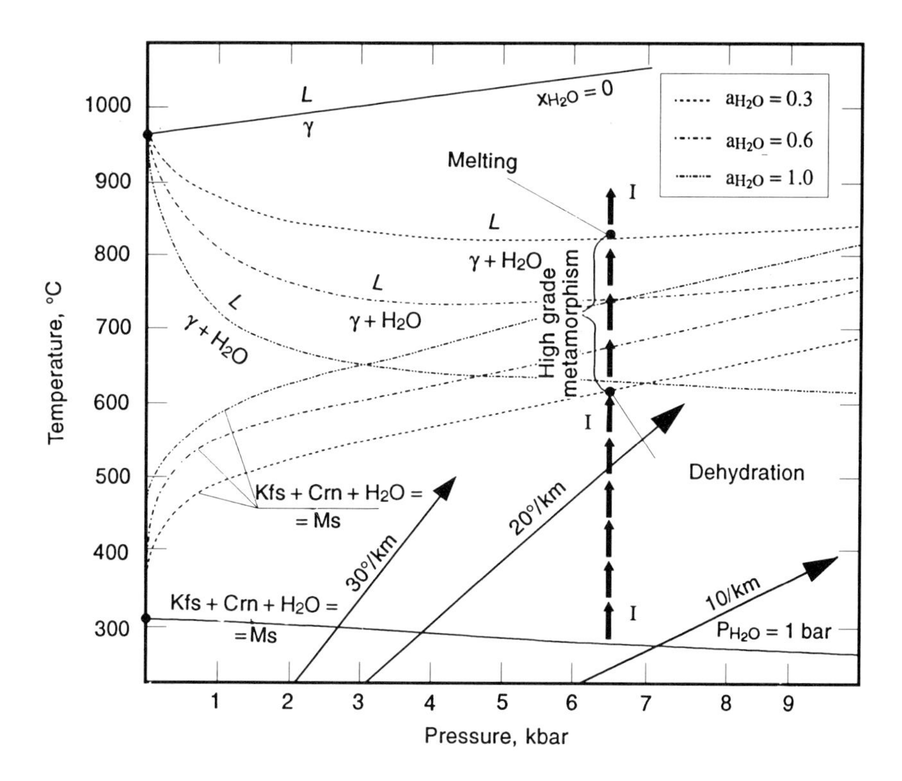

Figure 2.4. T–P diagram for muscovite dehydration (Ms ⇒ Kfs + Crn + H_2O) and for haplogranitic system melting, constructed from experimental and calculated data for conditions of a_{H_2O} = 1.0, 0.6, 0.3 (also P = 1 bar for the muscovite reaction). The diagram shows geothermal gradients of 10, 20 and 30 °C. Km^{-1}. I–I denotes the 6.5 kbar isothermal section, along with dehydration and melting reactions upon perfectly mobile behaviour of water (a_{H_2O} = 0.3).

dry system. The magnitude of the decrease is in proportion to the solubility of water in the melt.

Figure 2.4 includes melting curves (solidus) for the haplogranitic system at various values of a_{H_2O}, which illustrate this effect. They are based on the experimental data of Tuttle and Bowen (1958), Luth *et al.* (1964), Luth (1969), Wyllie (1977), Johannes (1984, 1985), and Epel'baum (1980) for $a_{H_2O} = 0$ and $a_{H_2O} = 1$ ($P = P_{H_2O}$). Also shown are solidus curves calculated for a_{H_2O} = 0.3 and 0.6 from experimental data.

Figure 2.4 can be used to predict the sequence of dehydration and melting reactions in systems in which water is perfectly mobile. In extreme cases (at $a_{H_2O} = 1$ and $P = P_{H_2O}$), partial melting of muscovite with quartz is possible at $P = P_{H_2O} > 3$ kbar, since the univariant equilibrium line Ms + Qtz = And (Sil) + Kfs + H_2O lies only 20–30 °C below the upper stability curve of muscovite shown here (see Zharikov *et al.*, 1972).

For normal conditions, some workers (e.g. Sevigny and Ghent, 1989; Perchuk *et al.*, 1985; Hansen *et al.*, 1984; Riciputi *et al.*, 1990; Aranovich *et al.*, 1987) have estimated the activity of water in the fluid as 0.5–0.7 for the amphibolite facies and as 0.2–0.3 for the granulite facies. For simplicity, we take values of 0.6 and 0.3, respectively. Then, with increasing temperature, the sequence of reactions is:

- muscovite dehydration (transition from greenschist and lower amphibolite facies to higher-temperature ones);
- amphibolite and/or granulite facies metamorphism in the presence of a fluid whose components show perfectly mobile behaviour;
- melting to produce granite at a eutectic near 750–800°C and $a_{H_2O} \approx 0.6$ (amphibolite facies) or 820–850°C and $a_{H_2O} \approx 0.3$ (granulite facies).

Figure 2.4 shows this sequence for P = 6.5 kbar, and also demonstrates that a typical crustal geothermal gradient of c.20°C.km^{-1} cannot result in extensive crustal melting; to do this requires an additional advective heat source, e.g. due to ascending magma or heated juvenile fluids, even in this, the most favourable case for melting.

(c) Water as an 'inert' component. In this case, an additional phase must be present to fix the content of water in the system. This may be a mineral phase that differs from other phases in the system in water content alone. For example, periclase (MgO) co-existing with brucite $Mg(OH)_2$: $MgO + H_2O = Mg(OH)_2$. For inert water behaviour, the periclase–brucite paragenesis (or any other hydration–dehydration reaction) should occupy a divariant field over a P–T range, rather than matching a univariant boundary between temperature facies, as is the case for perfectly mobile behaviour of water. In my view, divariant assemblages, corresponding to hydration–dehydration reactions, are not characteristic of natural conditions. For retrograde metamorphism, however, when these assemblages are observed, we interpret them as non-equilibrium, as supported by reaction relationships between minerals. Thus, paragenetic analysis constitutes convincing evidence for perfectly mobile behaviour of water and CO_2 during prograde metamorphism.

An alternative way of considering the immobile behaviour of water is to treat it as present as a separate isolated phase. In this case, it makes no difference whether the system is open or closed; what counts is the presence of a certain mass (volume) of water as a discrete phase (we do not consider the limiting case in which $m_{H_2O} \Rightarrow 0$ due to reaction, and the fluid phase is removed from the system). It is an easy matter to show that $\mu_{H_2O} = f(T, P, m_{H_2O})$ and

$$d\mu_{H_2O} = -S_{H_2O}\,dT + V_{H_2O}\,dP + \left(\frac{\partial \mu_{H_2O}}{\partial m_{H_2O}}\right)_{T,P} dm_{H_2O} \tag{2.9}$$

for inert water behaviour.

If the mass of water in the system remains unchanged, then:

$$\mathrm{d}\ln\gamma_{H_2O} \cdot \varphi_{H_2O} \cdot P_{H_2O} = \frac{S_{H_2O}\,\mathrm{d}T}{RT} + \frac{V_{H_2O}\,\mathrm{d}P}{RT}. \tag{2.10}$$

Because the aqueous phase is present in the system, then, in the case of complete equilibrium $P_{H_2O} = P_s = P$, phase reactions are the same as those occurring at $a_{H_2O} = 1$.

Other relationships are possible when $P_{H_2O} \neq P_s$: for example, P_{H_2O} (hydrostatic) $= \frac{1}{2.7} P_s$ (lithostatic). In this case, however, P_{H_2O} represents an independent parameter unassociated with the water content in the

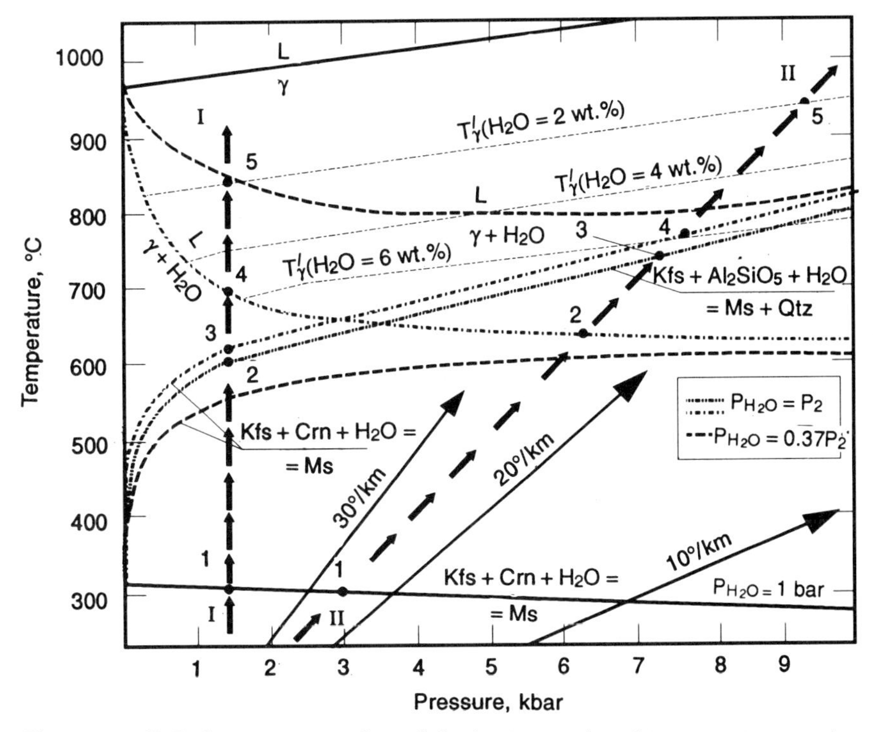

Figure 2.5 T–P diagram to analyse dehydration and melting reactions under inert water behaviour conditions. The diagram shows the muscovite dehydration reaction Ms⇒ Kfs + Crn + H_2O for conditions of $P_{H_2O} = 1$ bar, $P_{H_2O} = 0.37\,P_s$ and $P_{H_2O} = P_s$. Also plotted on the diagram are stable and metastable portions of the univariant line representing the reaction Qtz + Ms ⇒ Kfs + Al_2SiO_5 + H_2O for conditions of $P = P_{H_2O}$, using experimental data from Zharikov *et al.* (1972). To illustrate melting of the haplogranitic system, the diagram shows melting lines of the eutectic at $P_{H_2O} = P_s$, $P_{H_2O} = 0.37\,P_s$, $X_{H_2O} = 0$ and water contents of 2, 4, and 6 wt%. Also shown are lines representing geothermal gradients of 10, 20 and 30°C. km^{-1} and sections I–I (isobaric) and II–II (polybaric), for a geothermal gradient of 25°C. km^{-1}, for which phase reactions are analysed (see Fig. 2.6).

system, i.e. water possesses properties of a perfectly mobile component, though its fugacity is both T and P dependent. Figure 2.5 replicates muscovite dehydration curves and those for haplogranitic system melting for conditions of $P_{H_2O} = P_s$ and $P_{H_2O} = \frac{1}{2.7} P_s$. It can be seen that the sequence of phase reactions in the latter case is similar to that for perfectly mobile behaviour of water (see discussion above).

The differences in the sequence of reactions and mineral assemblages are also of significance for judging the extent of the conditions $P_{H_2O} = P$: dehydration and high-temperature metamorphism without melting are likely only at shallow depths. At higher pressures, muscovite-bearing parageneses partially melt to produce granitic liquid, which eliminates the fields of many naturally occurring mineral parageneses in both the granulite facies and the higher-temperature part of the amphibolite facies. In order for these assemblages to develop, it is necessary for P_{H_2O} to be substantially less than P. Finally, the question as to the 'accommodation' of the excess fluid phase in rocks remains an enigma. The mass of the fluid phase is liable to exceed the retrograde hydration reaction capacity of the rock.

We will now discuss the case of dehydration melting of a dry, muscovite-bearing rock. Water is in this case an inert (i.e. immobile) component, the amount present governing the amount of an isolated fluid phase or melt that can occur. Consider the resulting sequence of phase reactions in the KASH system, for which two alternative versions are described here (Figure 2.5).

(a) Version 1

At shallow depths (section I–I on Figure 2.5, P = 1.5 kbar), heating is possible due to conductive heat transport from magmatic bodies. This section is shown in Figure 2.6a with T–wt% H_2O coordinates.

At point 1, muscovite begins to undergo dehydration to be inhibited by the appearance of a free aqueous phase.

Between points 1–2, there exists a 'metastable' assemblage corresponding to a hydration reaction. As the temperature rises, the reaction shifts to muscovite if V_{H_2O} = constant or water is lost in order to hold $P_{H_2O} = P$.

At points 2–3, muscovite dehydrates:

$$Qtz + Ms \Rightarrow Kfs + Al_2SiO_5 + H_2O \text{ (point 2) or}$$

$$Ms \Rightarrow Kfs + Crn + H_2O \text{ (point 3).}$$

Between points 3–4, there exists a high-temperature assemblage, $Qtz + Kfs + Al_2SiO_5$ (or $Kfs + Al_2SiO_5 + Crn$) and a free volume of water, at $P_{H_2O} = P$ (excess P_{H_2O} may cause partial hydration due to heating if V_{H_2O} = constant or be lost).

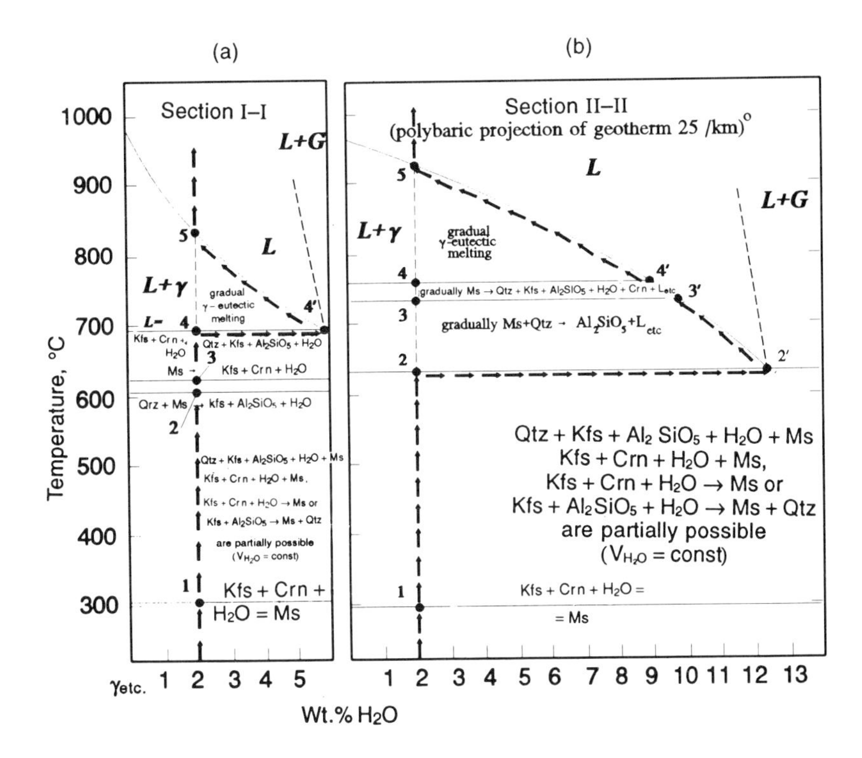

Figure 2.6 T–X section for the T–P diagram illustrating hydration and melting (Fig. 2.5). (a) Isobaric section I–I at P = 1.5 kbar. Arrows indicate the sequence of phase reactions, as well as variation in both the content of water and relative amount of eutectic melt. (b) Section II–II as a polybaric projection onto the T–X diagram for the section corresponding to a geothermal gradient of 25° C. km^{-1}. As in Fig. 2.6a, arrows indicate the sequence of phase reactions and change in both the content of water and relative amount of eutectic melt. For other explanations, see diagrams and text.

At point 4, quartz–Kfs eutectic melting commences under conditions of $P_{H_2O} = P$, and the amount of melt produced depends on the mass of water released by muscovite dehydration. For an H_2O content of muscovite of 4.5 wt%, a muscovite content in the metamorphic rock of 30–40%, and a total water content of about 2 wt%, around 30% of quartz + K–feldspar will melt to a water-saturated melt with 5.6–5.7 wt% H_2O, provided no water has been lost between point 1 and point 4.

Between points 4–5, melting of Qtz + Kfs continues, with the water content of the melt progressively decreasing.

Point 5 corresponds to the equilibrium eutectic for quartz + K–feldspar with 2.0 wt% H_2O at this pressure, and since this is the total H_2O content of the system, melting of quartz + K–feldspar will be complete at this point.

(b) Version 2

Section II–II (Figure 2.5) reflects phase relations along a steep geothermal gradient of 25°C. km^{-1}, and a T–wt% H_2O plot corresponding to this section is illustrated in Figure 2.6b.

At point 1, muscovite alone begins to undergo incipient dehydration, which is immediately inhibited by an isolated aqueous phase. Between points 1–2, there occurs a 'metastable' assemblage corresponding to the dehydration reaction. The increase in $P_{H_2O} = P$ (if $V_{H_2O} =$ constant) results in the partially reverse hydration and development of muscovite; water is lost in order to hold $P_{H_2O} = P_S$.

Between points 2–4, there is progressive melting of the eutectic composition, accompanied by gradual muscovite dehydration due to the dissolution of water in the melt.

By point 4, muscovite has completely disappeared; the amount of melt corresponds to the water content of the initial rock (2 wt% H_2O, assumed above). For a water content in saturated melt ≈ 9 wt%, the molten portion of the eutectic composition, i.e. Qtz + Kfs, will amount to around 22–23%.). As T and P rise, melting continues along the geotherm until point 5 is reached. This point corresponds to the solidus for this system with 2 wt% water.

Figure 2.6 shows the phase relations along sections I–I and II–II in terms of wt% H_2O versus T. For both these examples of inert water behaviour, the temperature ranges over which dehydration and eutectic melting occur extend for hundreds of degrees. On the T–P diagram, dehydration reactions occupy large divariant fields within which prograde hydration reactions are possible, as the temperature rises. Melting always commences at $P = P_{H_2O}$; the amount of melt is a function of water content and is invariably smaller than the total eutectic composition available. As the temperature increases, melting of the eutectic melt continues over a considerable temperature range. For a typical geothermal gradient of 25°C. km^{-1}, melting always commences under conditions where muscovite is stable.

In contrast, for perfectly mobile water behaviour, there are certain coordinates of dehydration and melting reactions that are fixed as a function of T, P, X_{H_2O} or a_{H_2O}. With these coordinates, dehydration and melting reactions proceed to completion, and the amount of the generated phases does not depend on the original water content of the initial rock. Rather, the quantity of melt is solely dependent on T, P, X_{H_2O}, as well as other possible intensive parameters of the system.

In the author's opinion, comparison of these predicted patterns of behaviour with the natural petrologic relations is powerful evidence in support of a perfectly mobile regime for water in magmatic and metamorphic processes.

2.2.3 Fluids under fine-pore medium conditions

The thermodynamic characteristics of the components of a fluid phase, be it aqueous or vapour, may be very different, if it is contained in a fine-pore medium (where pore sizes vary from c. 10^{-1} to c. 10^{-3} μm), from those in a 'free volume' (Zharikov, 1976b). Despite the fundamental importance of this to natural geochemical processes, there are remarkably few studies in this field. Shmonov *et al.* (1984) studied experimentally the partitioning of an H_2O–CO_2 mixture between zeolite and fluid as a function of pore size (0.25 to 250×10^{-2} μm) at temperatures of 400, 600, and 800 °C and pressures of 0.1, 0.5, 1 and 5 kbar. Belonozhko (1990) and Belonozhko and Shmulovich (1987a,b) carried out calculations of similar partitioning at P = 0.3, 1, 5, and 10 kbar, T = 400, 600, 800, and 1000 °C, $X_{CO_2} = 0.1$, 0.3, 0.5, 0.7, and 0.9, and pore sizes of 8, 10, and 25×10^{-4} μm, using molecular dynamics techniques. The results clearly indicate that the concentrations of water and CO_2 in the fluid contained in a porous medium and in free solution volume are different. Figure 2.7 presents some results obtained by Shmonov *et al.* (1984) and Belonozhko (1990). At high pressures, $X^{V}_{CO_2} > X^{pore}_{CO_2}$ and $X^{V}_{H_2O} < X^{pore}_{H_2O}$ by a factor of 1.5–4 (depending on T, P, and pore diameters).
Taking

$$d\mu_{CO_2} = RT d\ln \gamma_{CO_2} X_{CO_2} = 0, \tag{2.11}$$

i.e. $\mu^{V}_{CO_2} = \mu^{pore}_{CO_2}$, results in the relation

$$\gamma^{V}_{CO_2} \cdot X^{V}_{CO_2} = \gamma^{pore}_{CO_2} \cdot X^{pore}_{CO_2}. \tag{2.12}$$

For water, the relation is:

$$\gamma^{V}_{H_2O} \cdot X^{V}_{H_2O} = \gamma^{pore}_{H_2O} \cdot X^{pore}_{H_2O}, \tag{2.13}$$

where activity coefficients, $\gamma^{pore}_{CO_2}$ and $\gamma^{pore}_{H_2O}$, in pore solution reflect, as a first approximation, all the interactions of components of the fluid with its host under fine-pore medium conditions. From this, it follows that the activities of volatile components, notably, H_2O and CO_2, determined from mineral reactions, may be quite distinct from actual concentrations of these components in fluids in a fine-pore medium. In the cases above, $\gamma^{pore}_{CO_2} > 1 > \gamma^{pore}_{H_2O}$, and the mole fractions of CO_2 and water in pore solution will be, respectively, appreciably lower and higher than those estimated from mineral equilibria. This has important implications for the concept of 'dry' metamorphism and melting.

Measurements of diffusion coefficients of components through water-saturated rocks also provide direct evidence that the properties of components in a medium with fine-pores are distinct from those in bulk fluid. Balashov and co-workers (Balashov, 1992; Balashov *et al.*, 1983) conducted experiments on diffusion of chlorides of K, Na, Ca, Mg, Fe,

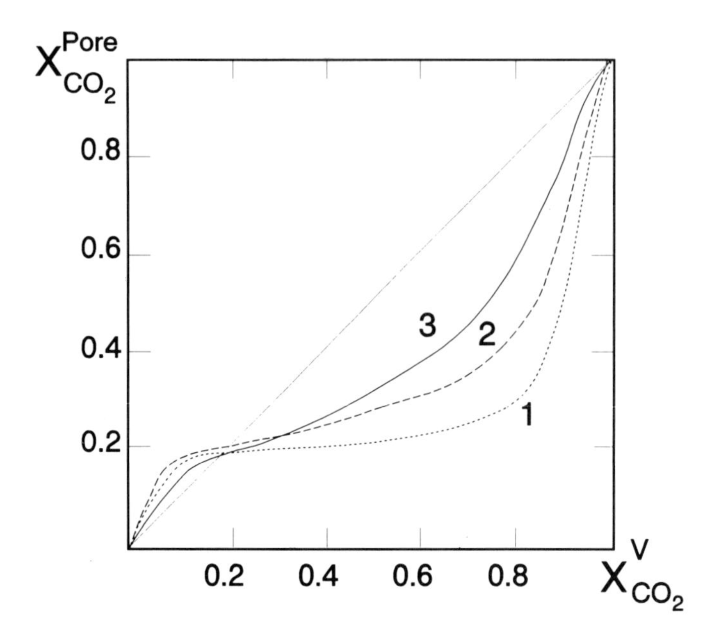

Figure 2.7. CO_2 concentration ratios (H_2O–CO_2 mixture) in pores ($X_{CO_2}^{pore}$) and in a free solution volume ($X_{CO_2}^{V}$). Symbols as follows: 1) 1000 bar, 400 °C, pores of size 0.25×10^{-2} μm in zeolites (Shmonov *et al.*, 1984); 2) 5000 bar, 600 °C, pores of size 0.26×10^{-2} μm in zeolites (Shmonov *et al.*, 1984); 3) 10000 bar, 800 °C, calculated pores of size 0.08×10^{-2} *μm* in quartz (Belonozhko, 1990).

and Al through artificial (sintered mullite, TiO_2 powder) and natural (granite, marble) membranes. The experiments showed that the difference between diffusion coefficients is a function of temperature, solution concentration and pore size. The diffusion coefficients in fine-pore media were invariably 0.5–5.5 times smaller than in a free solution volume because of the pronounced effects exerted by the surface and structure of porous media.

2.3 DYNAMIC PHENOMENA IN SYSTEMS WITH FLUID COMPONENTS

The effect of fluid components on the dynamics of geological processes is immense, and many aspects of this problem have been covered in the literature to a greater or lesser extent. It is well known (see Zharikov, 1976b; Fyfe *et al.*, 1978) that the diffusion rate of components in aqueous solutions is 10^5–10^7 faster than in a solid medium. The dissolution of water in melts disrupts the structure of melts and reduces considerably their viscosity (by 2–5 orders of magnitude; see Persikov, 1984). In addition, it affects the diffusion rates of components, the rates of cumulate deposition and convective processes in magmatic chambers. Experimental studies of rock permeability (see Shmonov and Vitovtova, 1992 and Chapter 11) reveal an intricate dependence of permeability on temperature, fluid pressure and total pressure. The wide range of permeability changes, including the appearance of impermeable zones in rocks at particular T and P conditions, implies a complex regime of infiltration of solutions in different geodynamic settings in the literature of diagenesis.

In this chapter, we will focus on one specific aspect of fluid behaviour in porous rocks: the differentiation of fluids upon their infiltration through fine-pore media, which is referred to here as the 'filtration effect' and is comparable to 'membrane filtration' in the literature of diagenesis.

2.3.1 The filtration effect and acid–base filtration differentiation

The concept of the filtration effect, i.e. a solute may move through a porous medium at a different rate to the solvent, was first introduced into the geological literature by Korzhinskiy (1947), although there had been some comment on the possible role of geological membranes in ore deposition processes (Mackay, 1946). More recently, there has been interest in such effects for their possible relevance to the origin of high salinities in oil field brines.

By employing Ershler's empirical equation

$$dm_i = f_i C_i dV, \tag{2.14}$$

whose other versions are

$$dm_i = f_i C_i dV = f_i C_i W dt = C_i W_i dt,$$

Korzhinskii (1947) presented the following equation for pure infiltration:

$$\frac{\partial C_i}{\partial t} + W f_i \frac{\partial C_i}{\partial x} + W C_i \frac{\partial f_i}{\partial x} = 0, \tag{2.15}$$

where C_i is the concentration of the ith component, f_i the coefficient of its filterability, W the infiltration rate of the solvent, and $W_i = f_i W$ the infiltration rate of the ith component (see above).

For a steady state, it follows from

$$W_{fi} \frac{\partial C_i}{\partial x} + W C_i \frac{\partial f_i}{\partial x} = 0; \frac{\partial (W f_i C_i)}{\partial x} = 0, \tag{2.16}$$

that $\varphi_i C_i = \text{constant}$, and $W_i C_i = \text{constant}$.

Alternatively, designating the concentration of the ith component as C_i^0 at $f_i = 1$, one obtains

$$f_i C_i = C_i^0; \; W_i C_i = W C_i^0, \tag{2.17}$$

the filterability coefficient is thus defined as

$$f_i = \frac{W_i}{W}; \; f_i = \frac{C_i^0}{C_i} \tag{2.18}$$

From Equations 2.17 and 2.18, it is evident, that if $f_i < 1$, then

$C_i > C_i^0$, and component concentration in the flux through a given medium increases.

The influence of membranes on infiltration of solutions (membrane effect) has received considerable attention in the geological literature, notably Hanshaw (1972), Hanshaw and Coplen (1973), Berry (1969), White (1955), Wyllie (1955), Sourirajan and Kennedy (1962). In these studies, much attention was given to the membrane effect as such, i.e. the semipermeability of rocks (e.g. clays, shales) on infiltration of solutions and brines. Note that in contrast to the membrane effect, the filtration effect describes phenomena occurring inside 'semipermeable' media rather than at their boundary, although these are two different aspects of the same phenomenon. From the 1960s, the author has conducted an extensive series of experimental and theoretical studies (Zharikov, 1965a, 1968; Zharikov and Alekhin, 1971, 1973; Zharikov *et al.*, 1962, 1973a; Alekhin *et al.*, 1983) and established the following:

1. Experiments on infiltration through fine-pore media show a regular change in the concentration of infiltrating solutions, which is too large to be simply the result of sorption phenomena and must be dependent on dynamic factors (see Figure 2.8).
2. The filtration effect is of an acid–base character, manifesting itself in the varying infiltration rate of anions and cations in electrolyte solutions. The conjugate change in the concentration of hydrogen and hydroxyl ions causes acid–base differentiation of solutions on infiltration. In rocks in which the mineral surfaces are negatively charged, the infiltration is faster, with acidic components concentrating at the head of the flux.
3. The experimental results suggest that the filtration effect is of an intricate electrokinetic nature complicated by processes of barodifferentiation (barodiffusion) and chemodiffusion, as well as by specific aspects of viscous flow in fine-pore media.

The total flux of component i may be expressed as:

$$I_i^S = I_i^P + I_i'^E + I_i^D = L_i^P X_i^P + L_i^E X_i^E + L_i^D X_i^D, \tag{2.19}$$

where I_i^S is the total flux of component, I_i^P, $I_i'^E$, and I_i^D the fluxes in response to pressure gradients, electrical potential, and chemical potential, respectively, X_i^P, X_i^E, X_i^D the operating forces, and L_i^P, L_i^E and L_i^D the kinetic coefficients for flux constituents.

Zharikov (1965b) expanded the expression for the operating forces to apply to conditions under which the force imposed on the system is a pressure gradient. Then, with terms grouped together, Equation 2.19 becomes

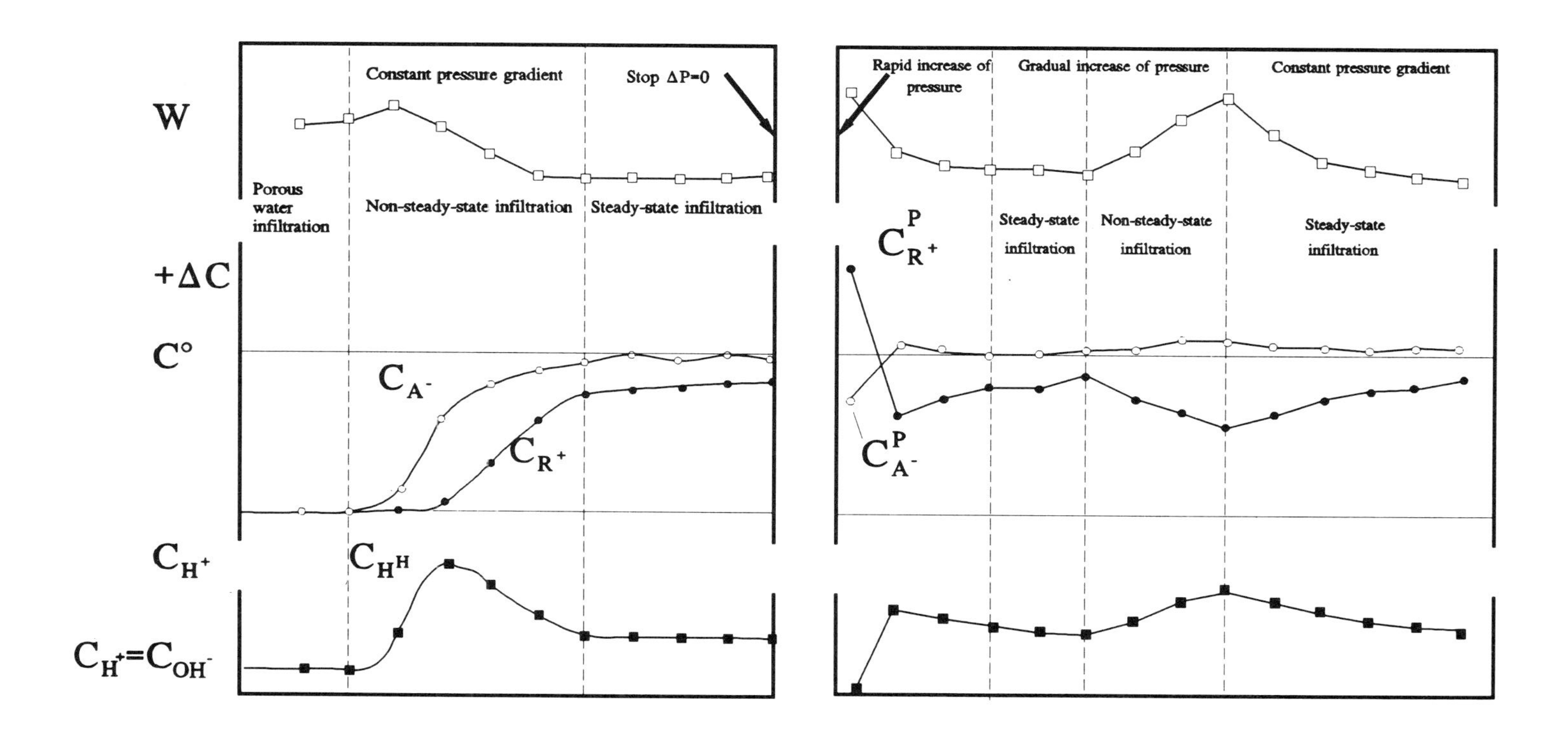

Figure 2.8. Schematic generalized diagram illustrating infiltration of electrolyte solutions through fine-pore filters. The diagram also shows variations in cation (C_R)–anion (C_A) concentrations, infiltration (W′) rate, and in hydrogen ion activities. A resumption of infiltration in response to a sharp pressure increase causes a pore solution to squeeze out. Note that ion concentrations (C_R^{pore}, C_A^{pore}) in the solution are the reverse of the filtrate.

$$-I_i^S = -\frac{dm_i}{dt} = \left(L_i^P - L_i^{DP}\overline{V}_i - \frac{L_i^{EP}\overline{V}_i}{z_iF}\right)\text{gradP}$$

$$-\left(L_i^E - L_i^{DE} z_iF\right)\text{grad}\psi - \left(L_i^D - \frac{L_i^{ED}}{Z_iF}\right)\text{grad}\mu_i. \quad (2.20)$$

Analysing Equation 2.20 and the resulting steady-state equations, one can gain a basic impression of the complex mechanism of acid–base differentiation, as follows.

For component infiltration involving the flux of solution (L^PgradP), when effective pore sizes become comparable to free path lengths for hydrated ions or molecules, transport parameters are different for different components. Varying solution density between adjacent pores of different size (see Belonozhko, 1990) should cause differentiation of solutions in response to viscous frictional forces, depending on the partial viscosity of components and the forces that bind the components to the solvent.

Infiltration of components will be subjected to an electric field (electric double layer), which will be imposed on it, at the solid phase–fluid phase boundary:

$$\left(\frac{L^{ED}V_i}{z_iF}\text{gradP}, L^{DE} z_iF\ \text{grad}\psi\right).$$

Being governed by the surface charge ψ, the resulting flux will reduce the concentration of ions that have the same charge as the surface (co-ions), but raise the concentration of oppositely charged (counter) ions. The effect of separation becomes greater with pressure, depending on the partial volume of the component concerned.

Infiltration of solution components will be complicated by barodifferentiation (barodiffusion) phenomena due to pressure gradients ($L^{DP}V_i$ grad P). The direction and intensity of barodifferentiation depend on the sign and magnitude of the partial volume of components as well.

The processes of separation of solution components are retarded by their diffusion (L^D gradμ_i), which is also subjected to the electric field set up by the charged surface. Naturally, the variation in the concentration of ions in aqueous solution is invariably compensated by the variation in the concentration of hydrogen and hydroxyl ions, thereby causing the acid–base differentiation.

The empirical filterability coefficient must take into account all these relationships. Adding up forced fluxes (i.e. infiltration-induced fluxes such as electrokinetic differentiation, barodiffusion and chemodiffusion in the field of pressure gradients) and imposing the steady-state condition, we have:

$$-\left(I_i^{DE} + I_i^{DP} + I_i^D\right) = 0. \quad (2.21)$$

Uncovering the expression for fluxes and passing on immediately to a unidimensional model (expressed in terms of complete differentials), we obtain the equation:

$$\mathrm{dln}\, C_i = -\frac{L_i^{DE}\, z_i F}{L_i^{D}\, RT} \cdot d\psi - \frac{L_i^{DP}\, V_i}{L_i^{D}\, RT} \cdot dP, \tag{2.22}$$

which can be solved as follows:

$$C_i'' = C_i' \cdot \exp \frac{L_i^{DE}\, z_i F}{L_i^{D}\, RT}(\psi' - \psi'') \times \exp \frac{L_i^{DP}\, V_i}{L_i^{D}\, RT}(P' - P''). \tag{2.23}$$

By comparing Equation 2.23 with the phenomenological determination of a steady-state, $f_i' C_i' = f_i'' C_i''$ = constant (Equations 2.16 and 2.17), one can see that

$$f_i = \exp \frac{L_i^{DE}\, z_i\, F}{L_i^{D} \cdot RT} \cdot \psi \cdot \exp \frac{L_i^{DP}\, V_i}{L_i^{D} \cdot RT} \cdot P \tag{2.24}$$

and

$$W_i = W \cdot \exp\left(\frac{L_i^{DE}\, z_i\, F_\psi + L_i^{DP}\, V_i\, P}{L_i^{P} \cdot RT}\right). \tag{2.25}$$

Equations 2.24 and 2.25 describe the mechanism of filtration effect. Specific physico-chemical and kinetic characteristics for individual solution components (z_i, V_i, L_i^{DE}, L_i^{DP}, L_i^{D}) lead to different rates of infiltration, and hence the acid–base infiltration effect. Although experimental determination of kinetic characteristics is possible, they are difficult to determine because of the superimposition of different processes. However, the different physico-chemical characteristics of components (z_i, V_i) do allow a basic pattern of differentiation to be envisaged. Figure 2.9 presents the characteristics of a range of ions. If the surface of most minerals is negatively charged, ions with a negative charge and positive partial molar volume will have large filterability coefficients and will move to the head of the solution flux. This is characteristic of anions of many acids. Conversely, ions with a positive charge and negative partial volume will be slowed down, leading to an increased concentration in the rear flux portion, possibly to the point of causing precipitation. Typical representatives of this group will be di- and trivalent cations like Ca^{+2}, Mg^{+2}, Zn^{+2}, Pb^{+2}, Cu^{+2}, Fe^{+2}, Fe^{+3}, etc.

Clearly, the nature of the filtration effect will also depend on a number of other variables (e.g. isotropic and anisotropic media, constant pressure gradient or constant solution flux); however, here we will consider only one, the possibility of two-phase fluids in porous rocks.

For porous sedimentary rocks, this problem has been addressed in connection with oil production; for fine-pore media, investigations are sparse. Most recently, Koshemchuk (1993) has conducted a series of

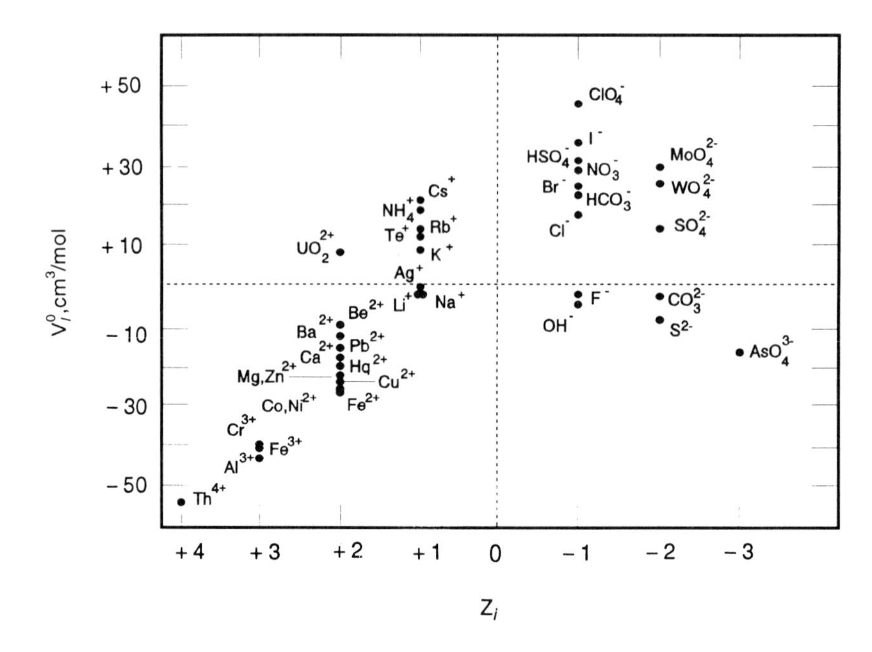

Figure 2.9. Charge (z_i) and partial molar volumes V^0 of some ions at infinite dilution and $\bar{V}^+_H = 0$. For more information on the nature of filtration effect, see text.

experiments on infiltration of water–gas mixtures (e.g. $H_2O–CO_2$, $H_2O–Ar$, $H_2O–CO_2–Ar$) through different fine-pore filters at various pressure gradients. The experiments revealed the major dependence of the infiltration regime on permeability: in high-permeability materials, there is a 'bubbling regime', where the gas phase leaves the liquid one behind at $K_D = 3.0–10.0$ md; a regime of homogeneous flow exists at $K_D = 3.0 \times 10^{-2}–3.0$ md; while at low permeabilities, the liquid phase leaves the gas one behind at $K_D = 1.0 \times 10^{-6}–3.0 \times 10^{-4}$ md. The fractionation of components between gas and liquid phases produces chemical differentiation of material in a heterogeneous fluid flux. For example, acid gases such as CO_2 or H_2S may be fractionated into the gas phase. Decrease in the pressure gradient causes the degree of component separation onto the fine-pore filters to become greater.

In conclusion, it should be stressed that the study of the nature and systematics of filtration differentiation processes are in their infancy and undeniably require further detailed physico-chemical investigations, particularly in the hydrothermal range. Yet acid–base filtration differentiation is clearly demonstrated, and its leading role in the evolution of hydrothermal processes is known from geological observations.

ACKNOWLEDGEMENTS

I am grateful to O.P. Rumyantseva, V.N. Orlov, and D.A. Varlamov for their assistance in preparing the manuscript and figures for publication. The research described in this chapter was supported by the Russian Fund of Fundamental Investigations under Grant No. 93–05–8190.

REFERENCES

Alekhin,Yu.V., Lakshtanov, L.Z., and Zharikov, V.A. (1983). Transport phenomena in pore solution under hydrothermal conditions. *4th Int. Symp. on Water–Rock Interaction, Extend. Abstracts*, Misasa, Japan, pp. 5–8.

Aranovich, L.Ya., Shmulovich, K.I., and Fed'kin, V.V. (1987). Distinctive features of the regime of H_2O and CO_2 in regional metamorphism, in *Contributions to Physico-chemical Petrology*, XIV (eds V.A. Zharikov and V.V. Fed'kin), pp. 96–118 (in Russian).

Balashov, V.N. (1992). Diffusion mass transport in hydrothermal systems. Doctoral dissertation, Institute of Experimental Mineralogy, Chernogolovka (in Russian).

Balashov, V.N., Zaraisky, G.P., Tikhomirova, V.I., and Postnova, L.E. (1983). Diffusion of rock-forming components in pore solutions at 200°C and 100 MPa. *Geochem. Int.*, **20** (1), 28–39.

Belonozhko, A.B. (1990). Calculation of the properties of supercritical fluid in fine pores. Ph.D. thesis, Institute for Chemical Physics, Moscow (in Russian).

Belonozhko, A.B. and Shmulovich, K.I. (1987a). A molecular-dynamics study of a dense fluid in micropores. *Geochem. Int.*, **24** (6), 1–12.

Belonozhko, A.V. and Shmulovich, K.I. (1987b). Fluid phase in a fine-pore medium at high pressures. *Doklady Akademii Nauk SSSR*, **295** (3), 625–9.

Berry, F.A.F. (1969). Relative factors influence on membrane filtration effect in geological environments. *Chemical Geology*, **4**, 295–301.

Bodnar, R.J. and Sterner, S.M. (1985). Fluid inclusions in natural quartz. II: Application to P–T–V studies. *Geochim. Cosmochim. Acta*, **49** (9), 1855–9.

Boettcher, A.L. (1984). The system SiO_2–H_2O–CO_2 melting solubility mechanism of carbon and liquid structure to high pressures. *Am. Mineral.*, **69**, 823–33.

Bowers, T.S. and Helgeson, H.C. (1983). Calculation of the thermodynamic and geochemical consequences of non–ideal mixing in the system H_2O–CO_2–NaCl on phase relations in geologic systems: Equation of state for H_2O–CO_2–NaCl fluids at high pressures and temperatures. *Geochim. Cosmochim. Acta*, **47**, 1247–75.

Carmichael, L. and Eugster, H. (eds) (1987). *Thermodynamic Modelling of Geological Materials: Minerals, Fluids and Melts. Reviews in Mineralogy*, **17**.

Chukhrov, F.V. (1974) (ed.) *Minerals. Diagrams of phase equilibrium, part 1*, Nauka Press, Moscow (in Russian).

Crawford, M.L., Kraus, D.W., and Hollister, L.S. (1979). Petrologic and fluid inclusion study of calc–silicate rocks, Prince Rupert, British Columbia, *Am. J. Sci.*, **279**, 1135–59.

Epel'baum, M.B.(1980). *Silicate Melts Involving Volatile Components*, Nauka Press, Moscow (in Russian).

Franck, E.U.(1985). Aqueous mixtures to supercritical temperatures and at high pressures. *Pure Appl. Chem.*, **57** (8), 1065–70.

Fyfe, W.S., Price, N.L., and Thompson, A.B. (1978). *Fluids in the Earth's Crust. Developments in Geochemistry*, **1**, Elsevier, Amsterdam.

Fyfe, W.S., Turner, F., and Verhoogen, J.(1958). Metamorphic Reactions and Metamorphic Facies. *Geol. Soc. Am. Mem.*, **73**.

Gehrig, M. (1980). Phasengleichegmichte und P–V–T Daten ternarer Mixhungen aus Wassen Kohlendioxide und Natrium-Chlorid bis 3 kbar und 550°C. Doktor Dissertation Univer. Karlsruhe Fveiburg. Verlag.

Grant, G.A. (1985). Phase equilibria in partial melting of pelitic rocks, in *Migmatites* (ed. J.R. Ashworth), Blackie, Glasgow, pp. 86–140.

Hansen, E.C., Newton, R.C., and Janardhan, A.S. (1984). Fluid inclusions in rocks from the amphibolite–facies gneiss to charnockite progression in southern Karnataka, India: direct evidence concerning the fluids of granulite metamorphism. *J. Met. Geology*, **2**, 294–364.

Hanshaw, B.B. (1972). Natural-membrane phenomena and subsurface waste emplacement, in *Underground Waste Management and Environmental Implications, The AAPG Memoir*, **18**, 308–15.

Hanshaw, B.B. and Coplen, T.B. (1973). Ultrafiltration by compact clay membrane. II. Sodium ion exclusion at various ionic strengths. *Geochim. Cosmochim. Acta*, **37**, 2311–27.

Jacobs, G.M. and Kerrick, D.M. (1981). Methane: an equation of state with application to the ternary system H_2O–CO_2–CH_4. *Geochim. Cosmochim. Acta*, **45** (5), 607–14.

Johannes, W. (1984). Beginning of melting in the granite system Qtz–Or–Ab–An–H_2O. *Contrib. Mineral. Petrol.*, **86**, 264–73.

Johannes, W. (1985). The significance of experimental studies for the formation of migmatites, in *Migmatites* (ed. J.R. Ashworth), Blackie, Glasgow, pp. 36–82.

Johnson, E.L. (1991). Experimentally determined limits for H_2O–CO_2–NaCl immiscibility in granulites. *Geology*, **19**, 925–8.

Korzhinskii, D.S. (1940). The mineral equilibrium factors and abyssal mineralogical facies. *Proceedings of the USSR's Academy of Sciences Institute of Geological Sciences*, **12**, Academy of Sciences Press, Moscow (in Russian).

Korzhinskii, D.S. (1942). The concept of geochemical mobility of elements. *Zapiski Vsesoyuznogo Mineralogicheskogo Obshchestva*, **71** (3–4), 160–76 (in Russian).

Korzhinskii, D.S. (1945). Regularities of mineral assemblages in Archaean rock, eastern Siberia. *Proceedings of the USSR's Academy of Sciences Institute of Geosciences*, **61**, 111 pp. Academy of Sciences Press (in Russian).

Korzhinskii, D.S. (1947). Filtration effect in solutions and its implication for geology. *Izvestia Akademii Nauk SSSR, ser. Geol.*, **2**, 35–48 (in Russian).

Korzhinskii, D.S. (1949). Open systems with perfectly mobile components and the phase rule. *Izvestia Akademii Nauk SSSR, ser. Geol.*, **2**, 3–14.

Korzhinskii, D.S. (1959). *Physico-chemical Basis for the Analysis of the Paragenesis of Minerals*. Translated from Russian. N.Y:, Consult. Bur.INC: Chapman & Hall, London.

Korzhinskii, D.S. (1966). On thermodynamics of open systems and the phase rule (A reply to D.F. Weill and W.S. Fyfe). *Geochim. Cosmochim. Acta*, **30**, 829–35.

Korzhinskii, D.S.(1967). On thermodynamics of open systems and the phase rule

(A reply to the second critical paper of D.F. Weill and W.S. Fyfe). *Geochim. Cosmochim. Acta*, **31**, 1177–80.

Kotel'nikov, A.R. and Kotel'nikova, Z.A. (1990). The phase state of the H_2O–CO_2–NaCl system examined from synthetic fluid inclusions in quartz. *Geochem. Int.*, **27** (11), 55–65.

Koshemchuk, S.K. (1993). Study of the regularities of two-phase infiltration through natural fine membranes in the water–gas system. Ph.D. thesis, Institute of Experimental Mineralogy, Chernogolovka (in Russian).

Krauskopf, K.B. (1957). Heavy metal content of magmatic vapor. *Econ. Geol.*, **52**, 786–807.

Luth, W.C. (1969). The systems $NaAlSi_3O_8$–SiO_2 and $KAlSi_3O_8$–SiO_2 and relationship between H_2O content, P_{H_2O} and P in granitic magmas. *Am. J. Sci.*, **276**–A, 325–41.

Luth, W.C., Johns, R.H., and Tuttle, O.F. (1964). The granite system at pressure of 4 to 10 kilobars. *J. Geophys. Soc.*, **69**, 759–73.

Mackay, R.A. (1946). The control of impounding structures in ore deposition. *Econ. Geol.*, **41** (1), 13–46.

Ming–Chou, I. (1986). Redetermination of phase equilibrium properties in the system NaCl–H_2O to 1000 °C and 1500 bars. *14th Gen. Int. Miner. Assoc., Standford, California, 13–14 July. Abst. program.*

Niggli, P. (1937). *Das magma und Seine Produkte*, 261 S. Leipzig.

Pampura, V.D., Karpov, I.K., and Kuz'min, L.A. (1981). A physico-chemical model for the equilibrium composition in the system CO_2–NaCl–H_2O in the pressure–temperature ranges 49–1000 bar and 25–300 °C. *Doklady Akademii Nauk SSSR*, **258** (4), 989–92 (in Russian).

Perchuk, L.L., Aranovich, L.Ya., Podlesskii, K.K., *et al.* (1985). Precambrian granulites of the Aldan shield, eastern Siberia. *J. Met. Geol.*, **3**, 265–310.

Persikov, E.S. (1984). *Viscosity of Magmatic Melts*, Nauka Press, Moscow (in Russian).

Plyasunova, N.V. (1993). Experimental studies of equilibria in the systems H_2O–CO_2–NaCl and H_2O–CO_2–$CaCl_2$ between 500–700 °C at P = 5 kbar. Ph.D. thesis, Institute of Experimental Mineralogy, Chernogolovka (in Russian).

Plyasunov, A.V. and Zakirov, I.V. (1991). Evaluation of the thermodynamic properties of H_2O–CO_2 homogeneous mixtures at high *T* and *P*, in *Contributions to Physico-chemical Petrology*, XVII, (eds V.A. Zharikov and V.V. Fed'kin), Nauka Press, Moscow (in Russian), pp. 71–88.

Robie, R.A., Hemingway, B.S., and Fisher, J.R. (1978). Thermodynamic properties of minerals and related substances at 298.15 K and 1 Bar (10^5 Pascals) pressure and at higher temperatures. *U.S. Geol. Survey Bull.*, **1452**, 456 pp.

Riciputi, L.R., Valley, J.W. and McGregor, V.R. (1990). Conditions of Archaean granulite metamorphism in the Godthab–FisKenaesset region, southern West Greenland. *J. Met. Geol.*, **8**, 171–90.

Ryabchikov, I.D. (1975). *Thermodynamics of a Fluid Phase of Granitoid Magmas*, Nauka Press, Moscow (in Russian).

Sevigny, J.H. and Ghent, E.D. (1989). Pressure, temperature and fluid composition during amphibolite facies metamorphism of graphitic metapelites, Howard Ridge, British Columbia, *J. Met. Geol*, **7**, 497–505.

Shmonov, V.M. and Vitovtova, V.M. (1992). Rock permeability for solution of the fluid transport problem. *Experiment in Geosciences*, **1** (1), 1–49.

Shmonov, V.M., Vostroknutova, Z.N., and Vitovtova, V.M. (1984). On the possible influence of adsorption on fluid concentration in pores and gas–liquid inclusions, in *Contributions to Physico-chemical Petrology*, **XII**, (eds V.A. Zharikov and V.V. Fed'kin), Nauka Press, Moscow (in Russian), pp. 78–84.

Shmulovich, K.I. (1975). A mineral equilibria diagram in the system

$CaO–Al_2O_3–SiO_2–CO_2$, in *Contributions to Physico-chemical Petrology*, V, (eds V.A. Zharikov and V.V. Fed'kin), Nauka Press, Moscow (in Russian), pp. 258–66.

Shmulovich, K.I., Shmonov, V.M., and Zharikov, V.A. (1982). The thermodynamics of supercritical fluid systems, in *Advances in Physical Geochemistry* (ed. Saxena, S.K.), **v. 2**, pp. 173–190. Springer–Verlag.

Shmulovich, K.I., Mazur, V.A., Kalinichev, A.G., and Khodarevskaya, L.I. (1980). P–V–T and component activity–concentration relations for the systems of H_2O–non-polar gas type. *Geochem. Int.*, **17** (6), 18–30.

Shmulovich, K.I. (1988). *Carbon Dioxide in High–temperature Mineral Formation Processes*, Nauka Press, Moscow (in Russian).

Sourirajan, S. and Kennedy, G.C. (1962). The system H_2O–NaCl at elevated temperatures and pressures. *Am. J. Sci.*, **260**, 115–41.

Sterner, S.M. and Bodnar, R.I. (1991). Synthetic fluid inclusions. X. Experimental determination of P–V–T–X properties in the CO_2–H_2O system to 6 kb and 700°C. *Am. J. Sci.*, **291** (1), 1–54.

Susarla, V.R., Ever, A. and Franck, E.U. (1987). A new method of determining solubility of salts and simple salt mixtures at higher temperatures (0–200°C) and up to 100 bars pressure. *Proc. Indian Acad. Sci., Chem. Sci.*, **99** (3), 195–202.

Thompson, A.B. (1982). Dehydration melting of pelitic rocks and the generation of H_2O–undersaturated granitic liquids. *Am. J. Sci.*, **282**, 1567–95.

Tkachenko, S.I. and Shmulovich, K.I. (1992). Liquid–vapour equilibria in the systems water–salt (NaCl, KCl, $CaCl_2$, or $MgCl_2$) at 400–600°C. *Doklady Rossiiskoi Akademii Nauk*, **326** (6), 1055–9.

Trommsdorff, V. and Skippen, G. (1986). Vapour loss (boiling) as a mechanism for fluid evolution in metamorphic rocks. *Contrib. Mineral. Petrol.*, **94**, 317–22.

Tuttle, O.F. and Bowen, N.L. (1958). *Origin of Granite in the Light of Experimental studies in the System $NaAlSi_3O_8$–$KAlSi_3O_8$–SiO_2–H_2O. Geol. Soc. Am. Memoir.*, 74.

Valyashko, V.M., Kravchuk, K.G. and Korotayev, M.Yu. (1984). Phase states of binary aqueous solutions of inorganic substances at high *T* and *P*. *Obzory Teplophisicheskikh Svoistv Veshchestv*, **5** (49), 57–126 (in Russian).

Voigt, W. and Fanghaenel, T. (1985). Determination of the solid–liquid equilibrium in highly concentrated salt-water systems up to 250°C. *Z. Phys. Chem.*, **266** (3), 522–8.

Weill, D.F. and Fyfe, W.S. (1964). A discussion of the Korzhinskii and Thompson treatment of thermodynamic equilibrium in open systems. *Geochim Cosmochim. Acta*, **28**, 565–76.

Weill, D.F. and Fyfe, W.S. (1967). On equilibrium thermodynamics of open systems and the phase rule (A reply to D.S. Korzhinskii). *Geochim. Cosmochim. Acta*, **31**, 1167–76.

White, D.E. (1955). Magmatic, connate and metamorphic waters. *Bull. Geol. Soc. Am.*, **68**, 1657–1706.

Wyllie, P.J. (1977). Crustal anatexis: an experimental review. *Tectonophysics*, **43**, 41–71.

Wyllie, M.R.I. (1955). Role of clay in well–log interpretation. *Proceed. First Nat. Conf. on Clays and Clay Tech. Div. of Mines, State California, Bull*, **169**, 282–305.

Yardley, B.W.D. and Bottrell, S.H. (1988). Immiscible fluids in metamorphism: implications of two-phase flow for reaction history. *Geology*, **16**, 199–202.

Zavaritskii, A.N. (1926). *Physico-chemical Foundations of the Petrology of Igneous Rocks*, Nauchnoye Khimiko-tekhnicheskoye Izdatel'stvo, St. Petersburg (in Russian).

Zhang, Y. and Frantz, J.D. (1989). Experimental determination of the compositional limits of immiscibility in the system $CaCl_2$–H_2O–CO_2 at high temperatures and pressures using synthetic fluid inclusions. *Chemical Geology*, **74**, 289–308.

Zharikov, V.A., Dyuzhikova, T.N., and Maksakova, E.M. (1962). Experimental and theoretical studies of the infiltration effect. The varying rate of infiltration of anions and cations. *Izvestia Akademii Nauk SSSR, ser. Geol.*, **1**, 41–63 (in Russian).

Zharikov, V.A. (1965a). On the possible geochemical role of electrokinetic phenomena, in *Problems of Geochemistry*, volume dedicated to the 70th Anniversary of A.P.Vinogradov's birth, Nauka Press, Moscow (in Russian), pp. 278–285.

Zharikov, V.A. (1965b). Thermodynamic characteristic of irreversible natural processes. *Geochem. Int.*, **2** (5), 873–84.

Zharikov, V.A. (1968). Theoretical and experimental investigations of filtration effect. III: Electrokinetic mechanisms and a possible geochemical role, in *Metasomatism and Other Issues of Physico-chemical Petrology*, Nauka Press, Moscow (in Russian), pp. 9–29.

Zharikov, V.A. (1976a). *The Foundations of Physico-chemical Petrology*, Moscow University Press (in Russian).

Zharikov, V.A. (1976b). Some burning problems of experimental mineralogy. *Zapiski Vsesoyuznogo Mineralogicheskogo Obshchestva*, **105** (5), 543–70 (in Russian).

Zharikov, V.A. and Alekhin, Yu.V. (1971). Runs on infiltration of solutions through rock media. *Doklady Akademii Nauk SSSR*, **198** (2), 433–6 (in Russian).

Zharikov, V.A., Ivanov, I.P, and Fonarev, V.I. (1972). *Mineral Equilibria in the System K_2O–Al_2O_3–H_2O*, Nauka Press, Moscow (in Russian).

Zharikov, V.A. and Alekhin, Yu.V. (1973). A filtration effect as a reason for the evolution of hydrothermal solutions. *Proceedings of the International Geochemical Meeting, V. II: Hydrothermal processes*, Nauka Press, Moscow (in Russian), pp. 346–61.

Zharikov, V.A., Alekhin, Yu.V., and Rysikova, V.T. (1973a). Some physico–chemical regularities of infiltration of solutions through rocks, in *The Role of Physico–chemical Properties of Rocks in the Localization of Endogenic Deposits*, Nauka Press, Moscow (in Russian), pp. 7–25.

Zharikov, V.A., Alekhin, Yu.V. and Zakirov, I.V. (1973b). The system H_2O–CO_2 and planetary atmosphere, in *Results of Scientific Research and Engineering, ser. Geochemistry, Mineralogy, and Petrography*, **7**, All-Union Institute of Scientific and Technical Information (VINITI) (in Russian), pp. 5–79.

Zharikov, V.A., Shmulovich, K.I., and Bulatov, V.K. (1977). Experimental study of the CuO–MgO–Al_2O_3–SiO_2–H_2O–CO_2 system and conditions of contact metamorphism. *Tectonophysics*, **43**, 145–62.

GLOSSARY OF SYMBOLS

G	Gibbs free energy
G_m	Korzhinskii's free energy for systems involving perfectly mobile components
ΔG^R	free energy change on reaction
μ_j	chemical potential of the jth component
m_j, Δm_j	mass and change in the mass of the jth component, respectively

$S, \Delta S$	entropy and entropy change, respectively
$V, \Delta V$	volume and volume change, respectively
T	temperature
T^l	melting temperature
P	pressure
n	the number of degrees of freedom
k_i	the number of inert components
r	the number of phases
γ_j	activity coefficient of the jth component
φ_j	fugacity coefficient of the jth component
α_j	activity of the jth component
X_j	mole fraction of the jth component
m_j^l, m_j^s	mass of the jth component in liquid and solid phases, respectively
R	gas constant
$H^l, \Delta H^l$	enthalpy and enthalpy change of melting, respectively
f_i	filterability coefficient of the ith component
C_i	concentration of the ith component
W, W_i	infiltration rate and infiltration rate of the ith component, respectively
I_i^S	the total flux of the ith component
I_i^P, I_i^E, I_i^D	fluxes in response to pressure gradients, electrical potential, and chemical potential, respectively
X_i^P, X_i^E, X_i^D	operating forces of fluxes
L_i^P, L_i^E, L_i^D	kinetic coefficients for fluxes
ψ	electric field potential
z_i	charge on the ith component
V_i	partial volume of the ith component
F	Faraday constant

CHAPTER THREE

Hydrothermal experimental techniques used at the Institute of Experimental Mineralogy, Russian Academy of Sciences

Kirill I. Shmulovich, Vladimir I. Sorokin and Georgiy P. Zaraisky

3.1 INTRODUCTION

The aim of this chapter is to provide an outline of the main experimental techniques used to obtain the data presented in the rest of the book, and to indicate the sort of uncertainties that arise in the work. In addition to the equipment described here, some specialized devices to measure permeability and related parameters at elevated temperatures and pressures are introduced in the relevant chapters (10 and 11) because of the distinctive nature of those experiments. The essential principles of hydrothermal experimental techniques are universal, and have been described extensively in the literature (for example, Ulmer and Barnes, 1987; Tziklis, 1976; Holloway and Wood, 1988). They will not be repeated here, but our own designs for dealing with specific problems are outlined in detail.

The main problems in the study of hydrothermal fluids centre around sampling under experimental conditions, protection of reactors from corrosion, and conservation of solution purity between sampling and analysis.

Fluids in the Crust: Equilibrium and transport properties.
Edited by K.I. Shmulovich, B.W.D. Yardley and G.G. Gonchar.
Published in 1994 by Chapman & Hall, London. ISBN 0 412 563207

Two main types of high-pressure apparatus are used for routine hydrothermal experiments: isothermal autoclaves in which the pressure vessel is heated externally and cold-seal exoclaves (Tuttle bombs) in which the portion of the vessel containing the sample is heated, while the sealing nut is cold.

An autoclave is placed in its entirety in a furnace; not infrequently, up to 12 autoclaves are loaded into one furnace. The temperature in the furnace is measured with a chromel–alumel thermocouple. The pressure in the autoclaves is controlled internally by adding a measured quantity of water, so that, after allowing for the internal volume of the autoclave and the volume of the capsule, the water will exert the desired pressure at the temperature of the run. There is no external pressure control. Calibration experiments using a manometer showed that the error of pressure estimates does not exceed 10%. The design of the furnace allows the temperature gradient along the outside of the autoclave to be reduced to 10–20 °C; the metal body of the autoclave reduces this gradient further, so that the conditions inside the autoclave are essentially isothermal.

In exoclaves (Tuttle bombs), the pressure and temperature are controlled and monitored on a continuous basis; each pressure vessel is fitted with its own manometer, as well as with controlling and measuring thermocouples. The error in measured pressure is less than 60 bar, the error in measured temperature is ± 10 °C. Whereas the performance of autoclaves depends on the effectiveness of the locking nuts, the performance of Tuttle bombs with their strong temperature gradients is governed by the properties of the Ni 'superalloys' from which they are made. For reactors in use at the Institute of Experimental Mineralogy (IEM), Russian Academy of Sciences, a special EI–455 alloy has been developed and the optimal mode of heat treatment investigated, to enhance long-term strength which permits high pressures to be maintained at high temperatures. Investigations of nickel-base alloys revealed EI–455A to possess the best characteristics ($\sigma_b = 125$ kg. mm^{-2} at 750 °C). In principle, bombs of this alloy may be used to conduct experiments up to 9 kbar at 700 °C, but routine experiments with Tuttle bombs are normally run to 8 kbar and 750 °C or to 5 kbar and 800 °C. At more extreme P–T conditions, internally-heated gas pressure vessels (furnace inside a bomb, bomb walls are cold) or a piston–cylinder type of apparatus are employed.

When dealing with solutions whose composition must be controlled or measured, large sample volumes are required, and Tuttle bombs prove unsuitable. In this case, either large autoclaves, 200 to 300 cm^3 in internal volume, or special reactors (below) are used. The large autoclaves are heated uniformly to the desired temperature, thereby eliminating the temperature gradients that are inevitably present in a Tuttle bomb. The working limits of such apparatus commonly do not exceed 2 kbar and 700 °C; however, for some specific problems, particularly heat-resistant alloys (to 800 °C) or thick-walled vessels (to 8 kbar and 700 °C) were employed. For example, an externally heated vessel was used for runs to 700 °C at 8 kbar in the measurement of the P–V–T properties of CO_2 (Shmonov, 1977).

In addition to the experimental work described in this volume, experimental studies on crystal growth are also carried out at the Institute of Experimental Mineralogy, using purpose built apparatus. These include Ni–'superalloy' autoclaves, 0.5 l in capacity, to work up to 2 kbar and 700 °C, 5 l capacity Ti–alloy autoclaves whose operating limits are up to 450 °C and 2 kbar, 24 l capacity copper-lined autoclaves of heat-resistant stainless steel to work up to 350 bar and 400 °C, and a crystallizer of 250 l capacity to work up to 100 bar and 400 °C. The reader is welcome to apply to the institute for details of this equipment.

3.2 PARTS OF THE APPARATUS FOR STUDYING HYDROTHERMAL PROCESSES

A number of experimental techniques have been developed for solubility measurements and studies of hydrothermal processes.

3.2.1 Thermal wedge lock

Traditionally, autoclaves are sealed on the unsupported area principle, using gaskets of soft metal (Cu, Au) between stainless steel rings. Sorokin and Kapustin (1982) have suggested a more straightforward lock based on the locking force built up due to the different thermal expansion of different autoclave parts. A schematic sketch of the autoclave is shown in Figure 3.1. The design, configuration and fabrication of this vessel are extremely simple. The key principle of the lock is that the thermal expansion coefficient (α) of the ring material should be higher than that of the autoclave body. For instance, a titanium body ($\alpha = 8.5 \times 10^{-6}$ deg^{-1}) coupled with a stainless steel ring ($\alpha = 2 \times 10^{-3}$ deg^{-1}) is ideally suited for this purpose. A Ni–'superalloy' body with a stainless steel ring is also used. The autoclaves are so designed that they are easy to assemble and disassemble; this procedure does not require much effort because the thermal expansion is of course reversible. Autoclaves of this design are used to 1 kbar and 450 °C (titanium) or to 650 °C (Ni–'superalloy'), have a capacity up to 100 ml, and an inner diameter from 6 to 25 mm.

3.2.2 Tensoresistive pressure transducer

An improved transducer is shown schematically in Figure 3.2. It is used to record pressure in a gold-lined extraction-quench hydrothermal

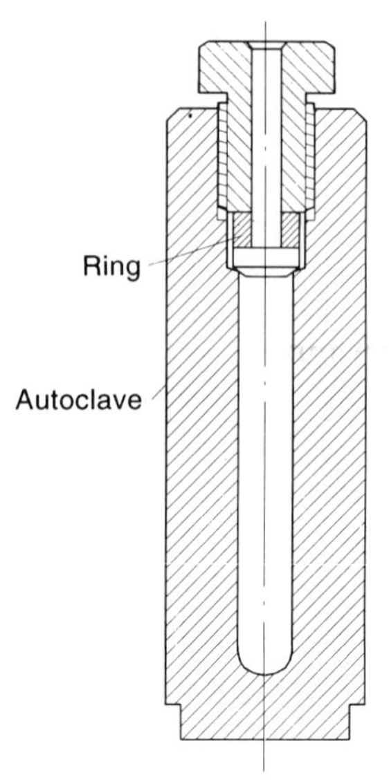

Figure 3.1. Autoclave with a 'thermal wedge' lock. The ring has higher thermal expansion than the other parts.

apparatus. The principle of operation of this transducer consists in transmitting deformation of the metallic membrane to a sapphire single crystal soldered to the former and provided with a bridge circuit. The bridge imbalance is then measured, which is linearly pressure dependent. To improve the accuracy of measurements, the transducers are first 'aged'

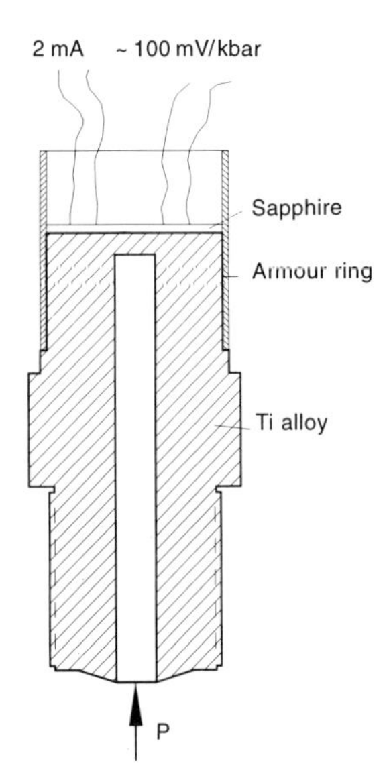

Figure 3.2. Pressure transducer for 5 kbar. Its inside volume may be filled by a metal cylinder.

at pressure to 120% of the maximum working pressure, and calibrated using a dead-weight pressure-gauge tester. The inert titanium body of the transducer allows operation in almost all hydrothermal media. Two types of pressure transducers are available : D–100 (to 1 kbar) and D–500 (to 5 kbar). The D–500 transducer is produced jointly with the *Teplopribor* Research Institute (Figure 3.2). Apart from the high sensitivity ($\approx$ 100 mV.kbar^{-1}), the pressure transducers have appreciably small internal capacities. With an inner Ti cylinder, the capacity of the D–500 pressure transducer, for example, does not exceed 10–15 mm^3.

Conventional transducers can be used without calibration and thermostatic control; the errors are around 1% of the upper measurement limit. The precision may be enhanced by at least an order of magnitude provided that it is pre-aged, thermostatically controlled and calibrated. Measurements of coexisting phase compositions in the systems H_2O–salt (Chapter 8, this volume) to 2 kbar were made with an accuracy of $\pm$ 1.5 bar.

3.2.3 Lining

There are several ways of protecting solutions from corrosion products of reactor walls. For example, the use of Teflon inserts at T < 200°C, of Au and Pt 'floating' inserts with aggressive solutions inside and pure water outside, and the use of 'rigid' liners. Rigid lining, i.e. coating the inner surface of the reactor with a noble metal layer (Au, Pt) presents the problem of obtaining a reliable bond between the lining and the inner surface of the reactor vessel. Two methods have been devised to overcome this. In the first method, the linegar and reactor bodies are bonded under shock wave. A small charge of condensed explosive (standard detonator is used) and a soft filler (polymer) are placed in an Au piston inserted into the working volume of the reactor. After detonating the explosion, the Au piston is bonded to the reactor and becomes a liner. In the second method, gas-saturated water is frozen into a gold capsule, loaded into the reactor. Both methods yield almost the same results.

Experience in operating Au or Pt lined reactors showed the rigid lining to be short lived. Heating–cooling cycles produce a gap between the liner and reactor body due to the difference in thermal expansion of metal liner and Ni 'superalloy'. Water or diffusing hydrogen may find their way into the gap to form a gas bubble after 10–20 cycles. The sealing of the liner is monitored by chemical analysis of solution samples from within the liner; the appearance of Ni or Fe from the reactor in solution is indicative of deterioration in sealing.

3.2.4 Capillary tubing

Special-purpose capillary tubes have been developed to offer further scope to apparatus designers. Titanium capillaries with external diameters of 3 and 4 mm are widely used to work in acidic and neutral media at 1–1.5 kbar. Most of the conventional hydrothermal apparatus uses water as a pressure medium, with stainless steel capillaries, 4.0 and 0.8 mm in external and internal diameter, respectively. For very high liquid and gas pressures, heat-resistant capillaries, 5 mm in diameter, are employed, which are capable of withstanding pressures to 20 kbar. At pressures below 8 kbar, welded and thread joints are used. Such joints are weaker than capillaries themselves, hence a collet clamp is used for higher pressures, in which case the limiting strength is that of the capillary itself.

3.3 SAMPLING TECHNIQUES

Studies of phase equilibria and component fractionation require samples to be extracted from the working volume of the reactor. Two types of sampling are carried out at IEM: maintaining constant P and T and allowing P, T, and X of the system to change in response to sampling.

3.3.1 Sorokin–Kapustin sampling device

The measurements whose results are presented in Chapter 4 have been primarily made using the device illustrated in Figure 3.3, which is designed to study poorly soluble substances to temperatures of 500°C and pressures of 1.5 kbar. The arrangement permits repeated sampling at constant pressure in the course of one experimental series, and pressure and temperature may be changed over the course of the run. The pressure vessel is fitted with a detachable sampler and valve connected directly to the internal sample volume, and contains a rotating magnetic stirrer to eliminate temperature gradients and accelerate reaction.

The pressure vessel with sampler is connected to a high-pressure hydraulic press filled with a fluid. The set-up permits repeated isobaric or isothermal solubility measurements without any need to dismantle it until the investigated substance is completely consumed, which may be replenished during the operation process. The fluid-supply apparatus is also connected to a vacuum pump and argon source, making it possible to evacuate the system or purge it with inert gas.

Those parts of the apparatus that are in contact with the fluid are all made from a heat-resistant titanium alloy. The locking part of the vessel is designed to seal as it heats up, due to the differential expansion of the materials, as in the autoclave described above. The reactor capacity is 50 cm^3, the sampler capacity, 5–10 cm^3. This design is particularly useful for studying reactions that proceed relatively fast (hours to a few weeks).

The sequence of operations is as follows. After adding the investigated substance to the container (Figure 3.3), the pressure vessel is sealed and,

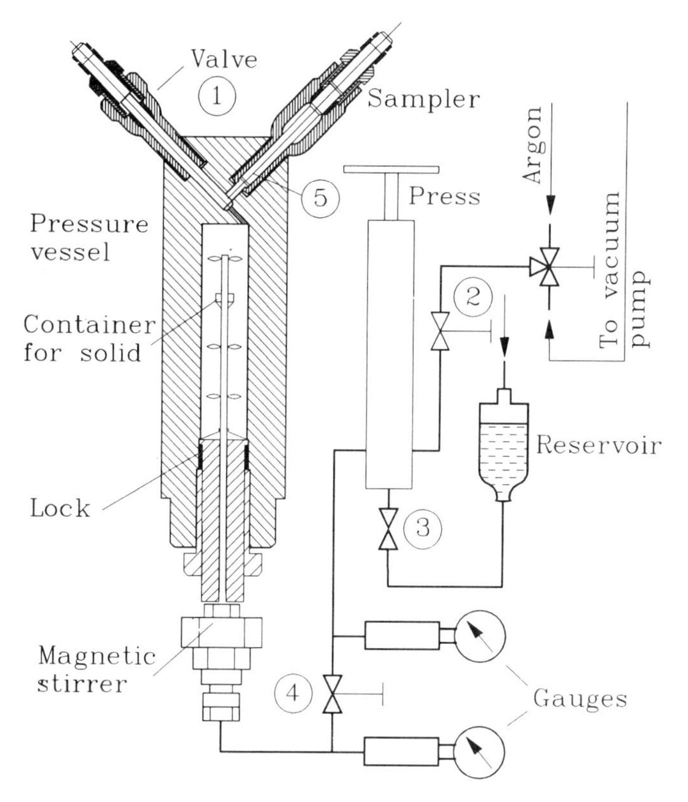

Figure 3.3. Device for sampling under isothermal and isobaric conditions.

if necessary, evacuated or blown with an inert gas to clean the system. Keeping valves 1,2,3 closed, the reactor is filled with working solution from the cold hydraulic press and heated. After the working P and T are attained, the reactor is isolated from the feeding system by closing valve 4. In this valve position, the press may be refilled with working solution from a reservoir. To sample the fluid from the run, the sampler with its valve 5 closed is attached to the reactor. The pressures in the vessel and press are equalized and sampling is then carried out with valves 1 and 4 open, by slowly opening the sampler valve 5. The pressure drop in the system is compensated for by increasing the pressure at the press. The movable membrane disc prevents fresh solution from entering the sampling zone. After the pressure ceases to drop, as evidenced by manometer readings, valve 1 and the sampler valve 5 are both closed, and the sampler removed for analysis. After new working P and T are achieved, valve 4 is closed and the cycle repeated.

This device has been used to study Hg and S solubilities in water, along with the systems SnO_2–H_2O, SnO_2–HCl–H_2O, and SnO_2 + HNO_3–H_2O described in Chapter 4.

3.3.2 Zakirov and Sretenskaya device

An alternative type of the isothermal sampler, located inside the working volume of the reactor, has been developed by Zakirov and Sretenskaya (in press). They suggested that graphite be used to seal the valve needle in the hot zone. This allows valve temperatures of up to 500°C without deterioration in the sealing. In this apparatus, part of the space in the autoclave may be isolated from the remainder at the T and P of the run by a valve or needle controlled externally. In this way, fluid may be isolated from reacting solids under run conditions, although only one sample may be collected during each run.

3.3.3 Non-isothermal sampling

Samples may be easily withdrawn from a reactor through a valve into an external sampler. Inevitably, part of the sampler must be in a cold zone within the pressure vessel, but the technique is viable provided the sample volume:cold zone ratio is 20:1, i.e. the error of composition measurement does not exceed 5%. Figure 3.4 is a sketch of such a design with two-way sampling. The reactor and shutter parts are made from EI–437 Ni 'superalloy'. The working capacity of the reactor is 120 cm^3, the volume of samples, 1–5 cm^3, while that of the cold zone (capillary and valve unit) is less than 0.05 cm^3.

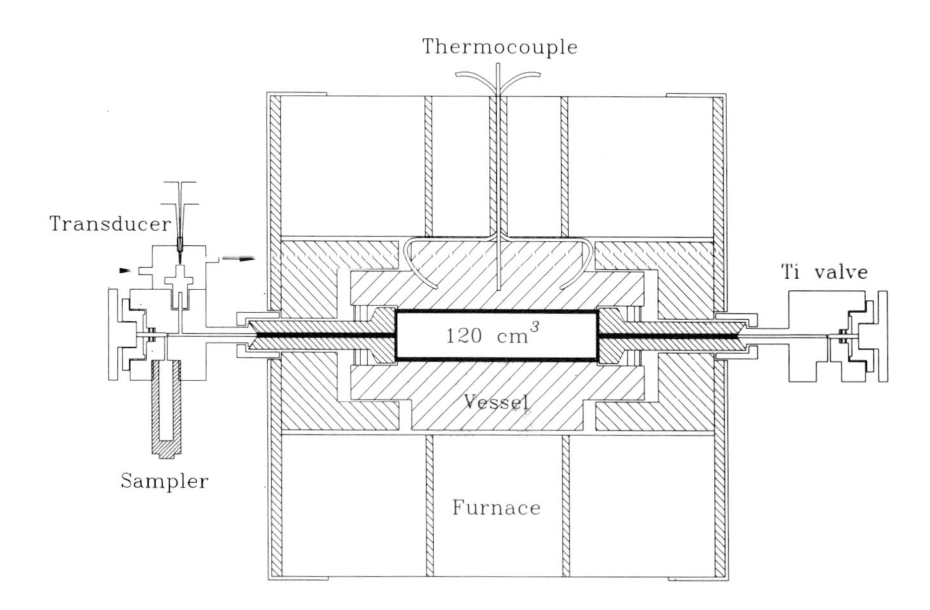

Figure 3.4. Device for collecting external (cold) samples. The dark solid line represents a gold coat, the arrows denote the flow of water to and from the thermostat.

The reactor and two shutters are lined with gold throughout the volume. Only outside the furnace, at the shutter–capillary junction, does solution come into contact with titanium, from which the capillary tubing, valve units, and bodies of receiving samplers are made. The entire apparatus may be rotated around a horizontal axis to provide the maximum surface for exchange between coexisting fluid phases when the reactor is positioned horizontally or to permit the extraction of either vapour or liquid from the top or bottom, when it is positioned vertically. The temperature gradient along the working zone of the reactor does not exceed 0.5°C in the horizontal position. Checking for the accuracy of measurements along the liquid–vapour equilibrium line of pure water yielded values that differed from tabulated data by less than 0.5°C for pressures less than 220 bar.

This device has been employed to study 'L–G' equilibria in the systems H_2O–NaCl, H_2O–KCl, H_2O–$MgCl_2$, and H_2O–$CaCl_2$ to 700°C and 2 kbar (see Chapter 8, this volume).

3.4 APPARATUS AND TECHNIQUES FOR EXPERIMENTS ON DIFFUSION AND INFILTRATION METASOMATISM

Experimental modelling of diffusion metasomatism can be carried out easily without employing sophisticated equipment and without permanent control over the run progress. It suffices to produce a concentration gradient in the pore solution in a synthetic rock for the process to evolve spontaneously until the gradient disappears as a result of diffusion.

3.4.1 Autoclave technique

Most of our experiments (Zaraisky, 1989) have been done employing an autoclave technique that uses large solution volumes. Rocks are crushed to < 0.07 mm grain size and loaded as compact powder into small gold or platinum capsules left open at one end (d = 5 mm, l = 50 mm; rarely, d = 3 mm, l = 25 mm). The capsule with rock is mounted vertically in a 150 cm^3 capacity sealed insert of BT–8 corrosion-stable titanium alloy, which is placed in an autoclave (Figure 3.5). The reacting solution is poured into the insert to which dry ice (weighed amount), powdered quartz, elemental sulphur or other poorly soluble components are added depending on the conditions of the runs. If need be, oxygen or other buffers are also added. Ti–alloy insert walls, passivated with a stable TiO_2 film, are very inert and have no effect on the composition and properties of the solution except under conditions of high alkalinity or when fluoride is present, in which cases, gold-lined inserts with a thermally-sealed lock are used.

Over the course of the run, the solution reacts with the rock by

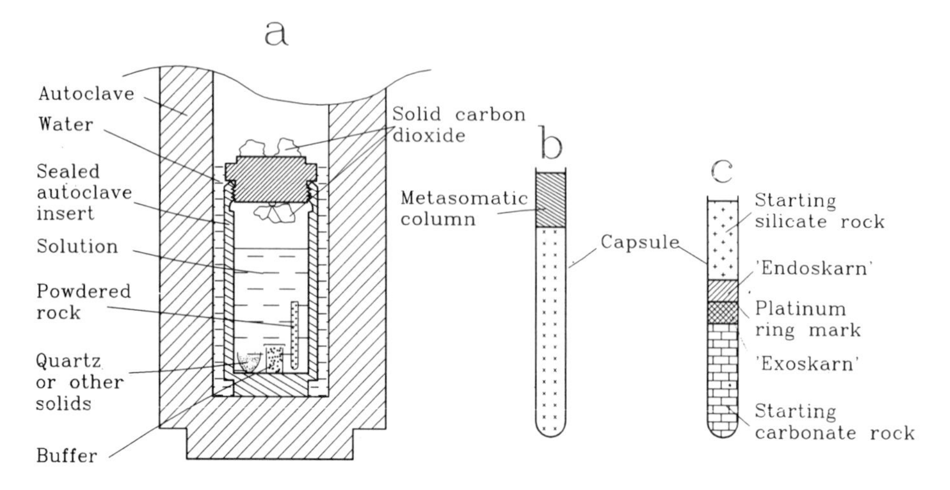

Figure 3.5. Schematic representation of the autoclave experiment on modelling of diffusion metasomatism. (a) Autoclave equipment before the experiment. (b) Capsule after the experiment on modelling of simple diffusion metasomatism. (c) Capsule after the experiments on bimetasomatism modelling.

diffusion through the open capsule end. Because of the high solution:rock ratio (from 50 to 100 and above), the variation in the composition of the solution is negligible.

After a run of appropriate duration, the initially homogeneous rock has been replaced by a column of metasomatic zones that are strongly transformed at the open capsule end, but remain similar in composition to the initial rock at its bottom.

In bimetasomatic interaction experiments (skarn formation modelling), the bottom of the capsule was loaded with one rock (e.g. carbonate), and the top with another (aluminosilicate). In those experiments, bimetasomatic zones develop outwards from the rock contact (Figure 3.5c). Apparatus of this type permits diffusion metasomatism to be modelled to temperatures and pressures of 600°C and 3 kbar. The optimum run duration is 2 weeks at 400–600°C, 4 weeks at 250–350°C. Lower-temperature experiments prove ineffective.

3.4.2 Diffusion studies

The composition of the external solution in the vessel will remain constant during the run only for components with concentrations in excess of $10^{-2}-10^{-1}$ molar. Maintaining lower concentrations in the external solution requires that a more elaborate diffusion arrangement be used (Figure 3.6). A cold buffer reservoir is located outside the furnace and contains an external solution. In practice, this buffer reservoir has a limited volume (in our case, 100 cm^3), and the solution is refreshed as it is used up. This

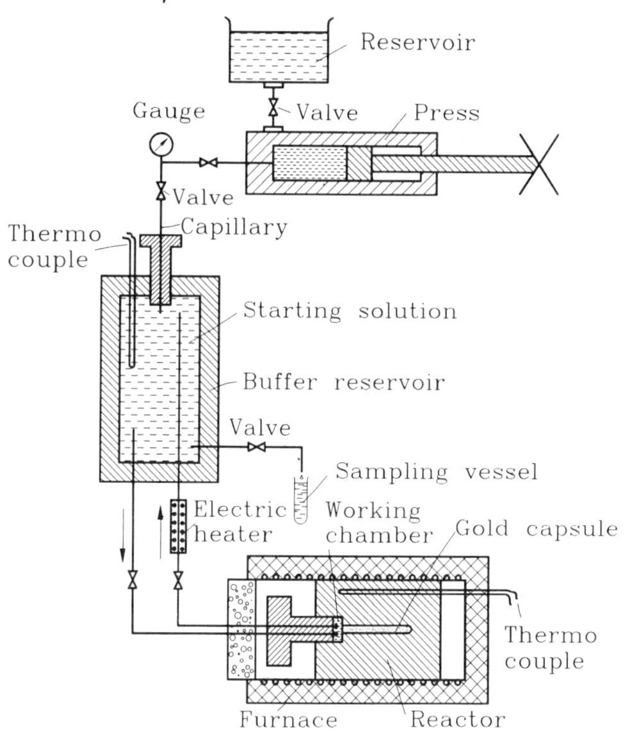

Figure 3.6 Schematic diagram of the convection apparatus for modelling of diffusion metasomatism.

arrangement permits solution to be replenished from the reservoir near the open capsule end by thermal convection. The reactor with capsule may be isolated by a locking valve, either to replenish the reservoir or to sample solutions at intervals during the experiment. The apparatus is made from titanium alloy and allows long-term experiments to 450 °C and 2 kbar. If need be, the inner surfaces of all passageways in the hot zone may be lined with gold tubes. Due to the 'spontaneous' thermal convection, the apparatus operates automatically and requires the attention of an operator only during sampling or replenishment of the buffer reservoir.

3.4.3 Infiltration metasomatism

Infiltration metasomatism experiments (Chapter 6) require more sophisticated equipment and regular control over the reaction progress (Figure 3.7). The same device was used to measure rock permeabilities at elevated temperatures and pressures (Chapter 10).

In carrying out experiments on infiltration metasomatism, both massive and powdered rocks have been used. The massive rocks are used as cylindrical core samples, 9.6 mm in diameter and 25 mm long, with well polished lateral surfaces and strictly parallel ends. The sample is

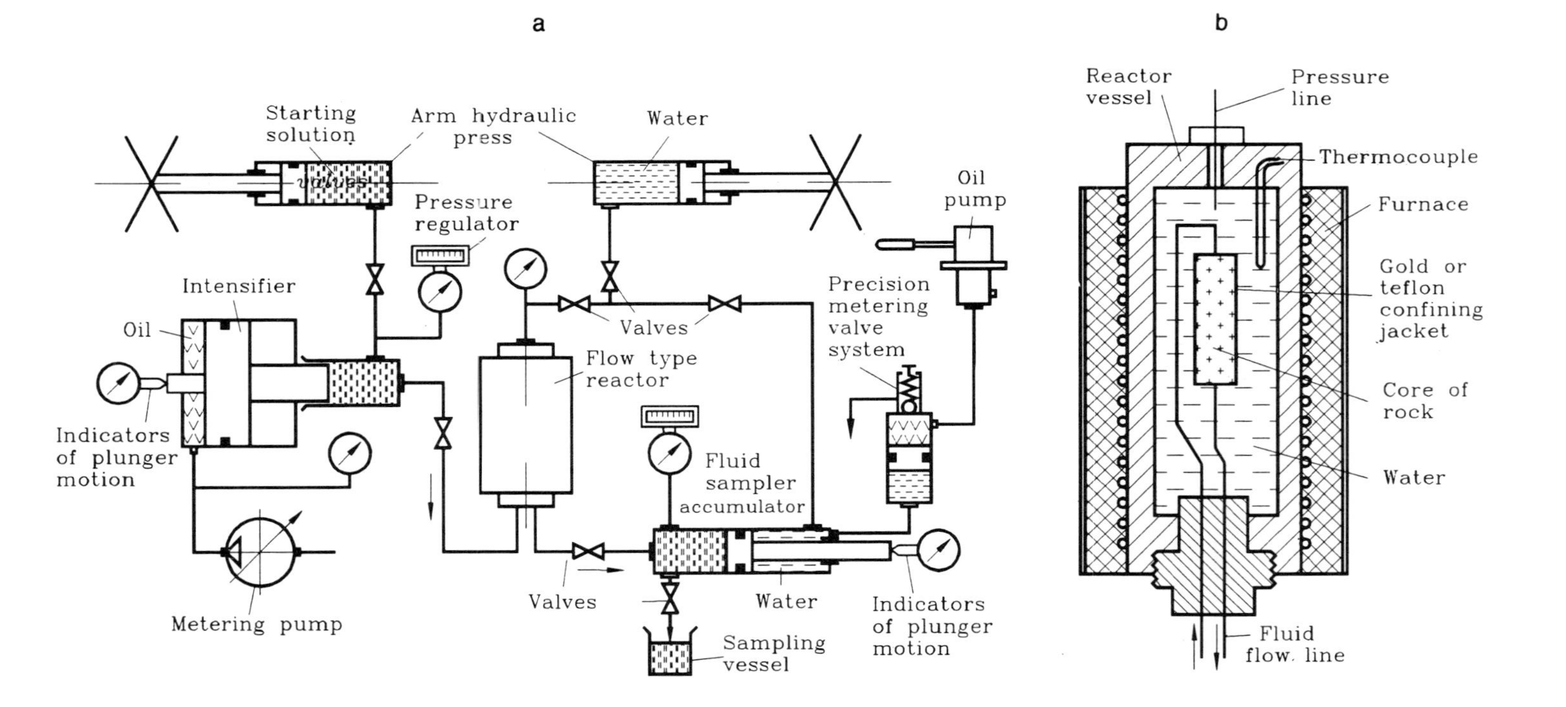

Figure 3.7. Schematic diagram of the apparatus for modelling of infiltration metasomatism. (a) Schematic of the apparatus. (b) Schematic diagram of the flow-type reactor.

loaded in a thin-walled (0.2 mm) gold tube, which is sealed by applying hydrostatic confining pressure during the run. The reacting solution is fed and removed through the sample ends. Fluid pressure at the inlet and outlet of the sample is controlled to provide a constant pressure differential, which is > 10 bars for a total fluid pressure of 1–1.5 kbar. The maximum operating temperature is 600°C. The volume of fluid which passes through the sample is determined from the movement of a piston coupled to the receiving chamber piston and led out through a Teflon gland and is recorded by a micrometer with an accuracy of 1 μm.

Inlet pressure is monitored by a precision metering pump combined with an electrocontact manometer. Outlet pressure is held at a given level by an automatic metering valve system that responds to signals from the electrocontact manometer. The receiving chamber, in which the solution passing through the rock sample is collected, also acts as a sampling device. Because the metering pump allows continuous variation of fluid flow rates over a wide range of 2×10^{-2} to $10\ cm^3.hr^{-1}$, highly permeable rock or even rock powders can be investigated using this arrangement; however, such materials are investigated under constant flow rate conditions rather than under a fixed pressure gradient.

3.5 SUMMARY AND CONCLUSIONS

In this brief review, we have focused on the main apparatus in use in 1993 or used in the experiments described in the following chapters. The apparatus available at IEM allows the following experimental investigations with fluid systems to be carried out:

1. Measurements of mineral solubilities up to 1 kbar and 600°C using Ti– and Ni–alloy autoclaves, with or without a noble metal coat.
2. Measurements of P–T–X properties of systems up to 2 kbar and 700°C, using fluid sampling techniques, and those of fluids in synthetic fluid inclusions up to 7 kbar and 750°C.
3. Routine experiments, including solubility measurements, in Tuttle bombs to 5 kbar and 750°C. For specific problems, experiments can be run to 8 kbar and 700°C or to 5 kbar and 800°C; however, pressure vessels have only a short working life under such extreme conditions.
4. Specialist measurements (e.g. H_2 fugacity, electrical conductivity, viscosity) using internally-heated gas bombs to 10 kbar and 1300°C. Special cells of piston–cylinder type are available, which are located inside the gas bomb to permit working with a partial hydrogen pressure up to 2 kbar.
5. Experimental modelling of zoning and of diffusion and infiltration metasomatism processes to 2 kbar and 600°C.
6. Permeability measurements on natural and synthetic rock samples and materials to 600°C and 2 kbar.

Cells for optical, X-ray, IR and Raman spectroscopy studies of solutions are also in use at the institute, and have been discussed elsewhere (Bondarenko and Gorbaty, 1984; Gorbaty and Bondarenko, 1993; Gorbaty and Okhulkov, in press).

REFERENCES

Bondarenko, G.V. and Gorbaty, Yu.E. (1984) High-temperature cell for Raman studies at high pressures. *Pribory i Tekhnika Experimenta*, **1**, 208–10 (in Russian).

Gorbaty, Yu.E. and Bondarenko, G.V. (1993) High-pressure high-temperature two-chamber cell with changeable path length for accurate measurements of absorption coefficients. *Rev. Sci. Instr.*, **64** (8), 2346–49.

Gorbaty, Yu.E. and Okhulkov, A.V. (in press) High-pressure X-ray cell for studying the structure of fluids with the energy-dispersive technique. *Rev. Sci. Instr.*

Holloway, J.R. and Wood, D.A. (1988) *Simulating the Earth – Experimental Geochemistry*. Unwin Hyman, Winchester, MA.

Shmonov, V.M. (1977) Apparatus for measuring P–V–T properties of gases to 1000 K and 8000 bar, in *Contributions to Physico-chemical Petrology*, **VI**, (ed. V.A. Zharikov), Nauka Press, Moscow (in Russian), pp. 236–45.

Sorokin, V.I. and Kapustin, N.V (1982) Device for repeated isobaric sampling in an isolated reservoir, in *Experimental Problems with High-pressure Solid-media and Hydrothermal Apparatus* (eds I.P. Ivanov and Yu.A. Litvin), Nauka Press, Moscow (in Russian), pp. 132–4.

Tziklis, D.S. (1976) *The Physico-chemical Investigation Technique at High and Very High Pressures*, Khimia, Moscow (in Russian).

Ulmer, G.C. and Barnes, H.L. (eds) (1987) *Hydrothermal Experimental Techniques*, Wiley-Interscience, New York.

Zakirov, I.V. and Sretenskaya, N.G. (in press) Experimental technique of phase compositions determination under heterogeneous conditions, in *Experimental Problems of Geology*, Nauka Press, Moscow (in Russian).

Zaraisky, G.P. (1989) *Zoning and Formation Conditions of Metasomatic Rocks*. Nauka Press, Moscow (in Russian).

CHAPTER FOUR

Solubility and complex formation in the systems Hg–H_2O, S–H_2O, SiO_2–H_2O and SnO_2–H_2O

Vladimir I. Sorokin and Tat'yana P. Dadze

4.1 INTRODUCTION

Elements which occupy closely related positions in the periodic table may exhibit contrasting styles of behaviour in geochemical processes. A well known example is the difference in the behaviour of chlorine and fluorine during crystallization of magma, where chlorine is accommodated in fluid, and plays a major role in mass transport of ore metals, while fluorine is concentrated in the residual melt, playing an important role in the formation of pegmatities and greisens.

The geochemical behaviour of an element is governed not only by its own chemical properties, but by the physical properties of the element and its compounds also. In order to calculate mass transport by hydrothermal solution, it is necessary to understand which components in solutions behave as elements or ions, and which are dominated by interaction with water to form complexes.

The present paper considers the behaviour of four components in hydrothermal solutions, Hg, S, SiO_2, and SnO_2, which illustrate this point. Mercury and sulphur are very different in their chemical properties, but have one feature in common: high vapour pressure at geological temperatures. We have therefore chosen to study the behaviour of this

Fluids in the Crust: Equilibrium and transport properties.
Edited by K.I. Shmulovich, B.W.D. Yardley and G.G. Gonchar.
Published in 1994 by Chapman & Hall, London. ISBN 0 412 563207

pair of elements in aqueous fluids to investigate similarities and differences in their geochemical behaviour.

Conversely, SiO_2 and SnO_2 are chemically related, since they are oxides of elements belonging to the same subgroup. These elements are also similar in geochemical behaviour, although not in abundance, and cassiterite is typically associated with quartz in veins, which are of major economic importance. We have investigated similarities and differences in the hydrothermal behaviour of these chemically related oxides in the simple model systems, SiO_2–H_2O and SnO_2–H_2O, as a contrast to the study of Hg and S.

Determining solubility of minerals or elements as a method of investigating their behaviour over a wide range of P–T conditions is an established approach in geochemistry. We have chosen this method because it allows us to follow the behaviour of components over a range of temperature and/or pressure, using simple binary systems.

The experimental studies reported here were initiated in 1970 and are still in progress, and our ideas have inevitably evolved. In this chapter, we have attempted to summarize data obtained at different times using different experimental facilities and synthesize our current views on complexation and transport in hydrothermal fluids.

4.2 MERCURY SOLUBILITY IN WATER

The high vapour pressure of mercury and its remarkable chemical inertness suggest that dissolved elementary species may be present in solutions. That the Hg^o_{aq} species exists in aqueous solution has been demonstrated spectroscopically (Reichardt and Bonhoeffer, 1931; Gushchina *et al.*, 1989). Sorokin (1973) showed that it is the predominant form of dissolved Hg in the Hg–H_2O system if no oxidation occurs. This chapter considers the dissolved elementary species Hg^o_{aq} only. It is shown below that this species is likely to be the dominant mercury species in hydrothermal ore-forming solutions, except in the presence of very high concentrations of complexing ligands. Other aquo complexes of mercury are also stable under hydrothermal conditions, subject to ligand availability, and some of them are discussed in detail by Zotov *et al.* (Chapter 5, this volume).

4.2.1 Previous work

The solubility of mercury in water as Hg^o_{aq} has received considerable attention at low temperatures (< 100°C), (Stock *et al.*, 1934; Choi and Tuck, 1962; Glew and Hames, 1971; Sanemasa, 1975) and many others (Clever, 1987), but studies at higher temperatures are fewer (Reichardt and Bonhoeffer, 1931; Sorokin, 1973; Sorokin and Gruzdev, 1975; Sorokin *et al.*, 1978, 1986, 1988; Varekamp and Busek, 1984; Gushchina *et al.*, 1989). Not all the experimental results from these studies are in agreement.

4.2.2 Experimental

Experiments were conducted over the temperature–pressure ranges 100–500 °C and 405–1033 bar using the Sorokin–Kapustin sampling device described in detail in the foregoing chapter (see Chapter 3, this volume). Starting materials were mercury, purified twice by distillation, and triply distilled water, freed of oxygen by boiling and blowing high purity Ar through it. Sampling and a preliminary analytical procedure were carried out under an Ar atmosphere. Depending on the Hg content of water, the solution was analysed by one of three methods: the amalgamation method, colorimetry or the method of Aidin'yan (1969), which involves precipitating mercury together with lead by H_2S, evaporating the mercury, and determining it turbometrically.

4.2.3 Results and discussion

Results of experiments from our work and from the literature are presented in Table 4.1 and Figures 4.1 and 4.2. Note that not all low T data (< 120 °C) from the literature have been omitted (Sorokin *et al.*, 1988; Clever, 1987). The data sets fall into two groups; most studies yielded relatively low Hg concentrations, but the studies of Sanemasa (1975) and Gushchina *et al.* (1989) yielded higher Hg concentrations for the same T.

Table 4.1. Experimental data for mercury solubility in deoxygenated water

T, °C	*P*, bar	*– log m*	*Reference*	*T*, °C	*P*, *bar*	*– log m*	*Reference*
1	2	3	4	5	6	7	8
120	2	5.277	Reichardt, 1931	68.19	1	6.071	Glew, 1971
25	1	7.001	Stock, 1934	72.44	1	6.013	Glew, 1971
30	1	6.825	Stock, 1934	300	507	2.840	Sorokin, 1973
85	1	5.825	Stock, 1934	300	648	2.922	Sorokin, 1973
100	1	5.524	Stock, 1934	298	912	3.024	Sorokin, 1973
25	1	6.502	Choi, 1962	400	405	1.775	Sorokin, 1973
35	1	6.265	Choi, 1962	400	507	1.861	Sorokin, 1973
50	1	6.054	Choi, 1962	400	502	1.794	Sorokin, 1973
65	1	5.966	Choi, 1962	400	709	1.910	Sorokin, 1973
80	1	5.886	Choi, 1962	400	709	1.855	Sorokin, 1973
90	1	5.779	Choi, 1962	398	932	1.954	Sorokin, 1973
4.3	1	6.633	Glew, 1971	401	922	1.974	Sorokin, 1973
4.3	1	6.655	Glew, 1971	500	507	0.92	Sorokin, 1973
11.41	1	6.624	Glew, 1971	502	517	0.927	Sorokin, 1973
11.41	1	6.635	Glew, 1971	500	527	0.997	Sorokin, 1973

Table 4.1 (Contd)

T, °C	*P*, bar	*– log m*	*Reference*	*T*, °C	*P*, *bar*	*– log m*	*Reference*
1	2	3	4	5	6	7	8
11.41	1	6.652	Glew, 1971	507	709	1.004	Sorokin, 1973
14.97	1	6.669	Glew, 1971	503	973	1.175	Sorokin, 1973
19.80	1	6.621	Glew, 1971	495	765	1.036	Sorokin, 1973
19.80	1	6.603	Glew, 1971	498	1003	1.089	Sorokin, 1973
25.02	1	6.524	Glew, 1971	5	1	7.019	Sanemasa, 1975
30.02	1	6.445	Glew, 1971	10	1	6.865	Sanemasa, 1975
30.09	1	6.463	Glew, 1971	20	1	6.648	Sanemasa, 1975
30.09	1	6.457	Glew, 1971	30	1	6.390	Sanemasa, 1975
39.16	1	6.358	Glew, 1971	40	1	6.162	Sanemasa, 1975
45.07	1	6.335	Glew, 1971	50	1	5.958	Sanemasa, 1975
45.07	1	6.352	Glew, 1971	60	1	5.736	Sanemasa, 1975
49.80	1	6.261	Glew, 1971	105	1013	5.639	Sorokin, 1978
50.24	1	6.269	Glew, 1971	202	1023	4.216	Sorokin, 1978
50.24	1	6.341	Glew, 1971	101	1021	5.942	Sorokin, 1988
50.24	1	6.309	Glew, 1971	110	1013	5.464	Sorokin, 1988
53.71	1	6.269	Glew, 1971	202	1033	4.209	Sorokin, 1988
59.98	1	6.162	Glew, 1971	202	1016	4.219	Sorokin, 1988
59.98	1	6.210	Glew, 1971	150	sat	3.710	Gushchina, 1991
62.28	1	6.132	Glew, 1971	200	sat	3.090	Gushchina, 1991
68.19	1	6.049	Glew, 1971	250	sat	2.470	Gushchina, 1991

Note: First author only of publication given.

The temperature dependence of mercury solubility in water for the first group above 100°C may be expressed in the form:

$$\log m = a + b \cdot T^{-1} + c \cdot T^{-1} + d \cdot \log T + f \cdot T^{-2} \qquad (4.1)$$

for which the numerical coefficients are listed in Table 4.2 for pressures of 1000 bar, 500 bar and saturated vapour pressure.

Table 4.2. Parameters for the equation describing the temperature dependence of Hg solubility in water

Pressure parameters	*Saturated vapour*	*500* bar	*1000* bar
a	– 1.7684	20.2134	– 8.58718
b	– 4988.61	– 12114.8	– 1751.74
c	1.16456×10^{-2}	-1.84202×10^{-2}	1.33024×10^{-2}
d	0.364418	0.409684	0.776497
e	-6.34888×10^{-6}	8.27514×10^{-6}	-5.2235×10^{-6}
f	7.22607×10^{5}	1.57386×10^{6}	186859

To describe the solubility of Hg_{liq} within the entire range of experimental conditions, and to permit extrapolation, the following model was employed:

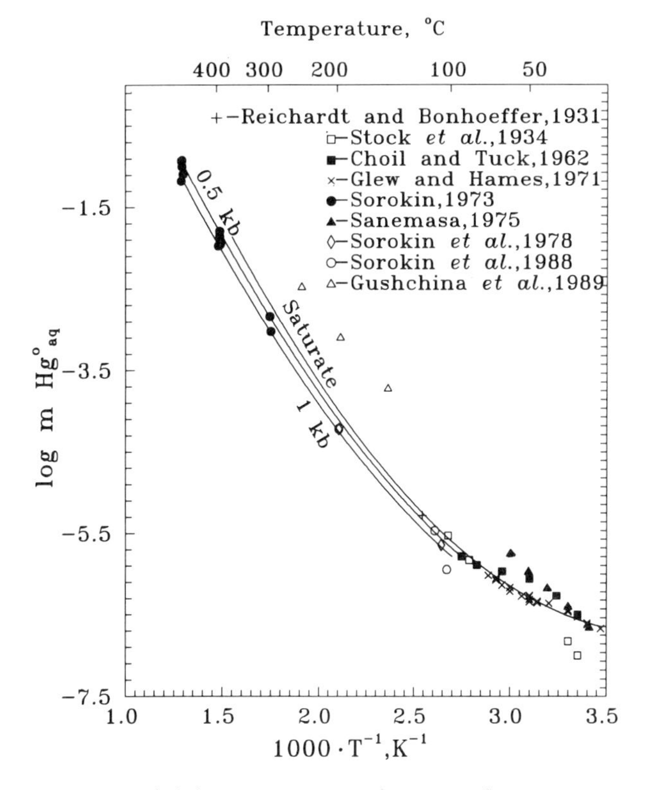

Figure 4.1. Mercury solubility in water as a function of temperature.

$$RT.\ln K^{\circ}(T, P_2) = RT.\ln K^{\circ}(T, P_1) - \int_{P_1}^{P_2} \Delta V^0 \, dP, \tag{4.2}$$

where K^o is the equilibrium constant for the reaction

$$Hg_{liq} = Hg^o_{aq} \tag{4.3}$$

determined from the experiments, $\ln K^0(T,P) = \ln m_{Hg^0_{aq}}$ and $K^{\circ}(T,P_l)$ is the value of this constant at T and a reference pressure P_1. Below 300 °C, the saturated vapour pressure of water was used as the reference pressure P_1, while above 300 °C, the reference pressure was 500 bars, $\Delta V^0 = V^0_{Hg^0_{aq}} - V^0_{Hg_{liq}}$ is the molar volume of Reaction 4.3. ΔV^0 was calculated with Equation 4.2 from the experimental data at different pressures (Table 4.1), while $V^0_{Hg_{liq}}$ was calculated from Vukalovich *et al.* (1971) and Sorokin *et al.* (1988). To 350 °C, $V^0_{Hg^0_{aq}}$ is independent of pressure and adequately described by the equation:

$$V^0_{Hg^0_{aq}}(T) = 13.494 + 0.026256T - 1.2952 \times 10^{-4} T^2 + 3.7774 \times 10^{-7} T^3. \tag{4.4}$$

up to 1 kbar at least. For experiments above 350 °C, the estimated $V_{Hg^0_{aq}}$

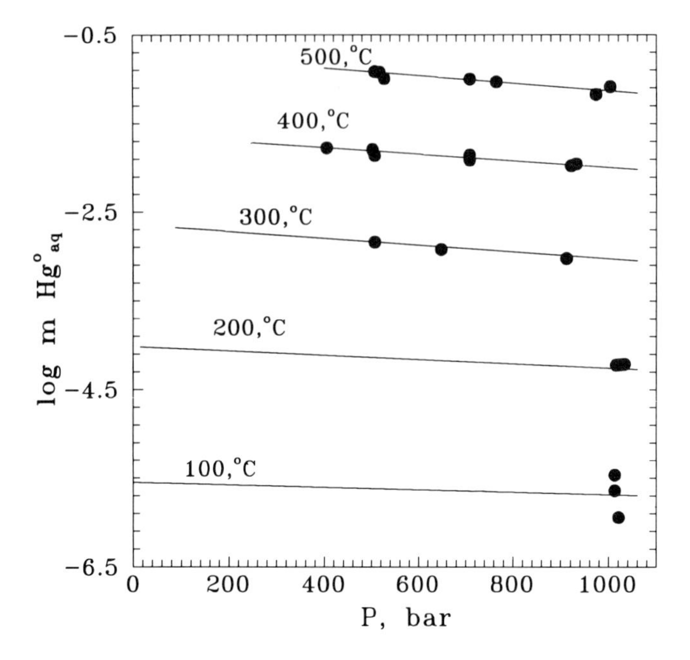

Figure 4.2. Mercury solubility in water as a function of pressure.

values do not fit Equation 4.2; they are 65.3, 69.1, 75.3 $cm^3.mol^{-1}$ at 400, 450 and 500°C, respectively, for the same pressures. This may be attributed to the fact that in low-density water, $V^0_{Hg^0_{aq}}$ is dependent on both temperature and pressure. There are insufficient experimental data to take this dependence into account.

The Henry's law constant was computed from the experimental data and fugacity of Hg_{liq} via the equations:

$$K^0_H\,(P,T) = \frac{m_{Hg^0_{aq}}}{f_{Hg}{}^{(P,T)}} \tag{4.5}$$

and

$$RT.\ln f_{Hg}\,(P) = RT \cdot \ln f^0_{Hg}\,(P_{sat}) + \int_{P_{sat}}^{P} V^0_{Hg}\,dP, \tag{4.6}$$

where f^0_{Hg} (P_{sat}) stands for the fugacity of mercury on the saturation curve, $f_{Hg}(P)$ denotes the mercury fugacity at $P >> P_{sat}$, and V^0_{Hg} refers to the molar volume of liquid mercury at P and T. The apparent Gibbs free energy of formation, $\Delta G_{f(T)}$, $\Delta H_{f(T)}$, $\log K^0_H$, were calculated directly from the experimental data and are shown in Table 4.3, apparent ΔG_{Hgliq} and ΔH_{Hgliq} were taken from modified version of Johnson *et al.*'s (1992) HKF model.

Table 4.3. Thermodynamic properties of Hg^0_{aq}

T,K	log K^0_H	$\Delta G^0_f(T)^*$ kcal. mol^{-1}	$\Delta H^0_f(T)^*$ kcal. mol^{-1}	V^0 cm^3. mol^{-1}
		Saturation pressure		
298.15	-0.983	8.951	4.063	19.8
373.15	-2.122	8.053	10.271	24.9
473.15	-2.387	5.276	16.118	36.9
573	-2.208	1.401	20.099	57.1
		500 bar		
298	-1.156			19.8
373	-2.296	8.347	10.185	24.9
473	-2.585	5.664	15.691	36.9
573	-2.431	1.978	19.330	57.1
673	-2.183	-3.215	36.203	
773	-1.896	-10.081	39.257	
		1000 bar		
298	-1.330			19.8
373	-2.470	8.642	10.100	24.9
473	-2.789	6.100	15.271	36.9
573	-2.684	2.634	19.005	57.1
673	-2.436	-2.450	36.381	
773	-2.151	-9.196	38.718	

* Apparent properties.

4.2.4 Conclusion

The thermodynamic properties of Hg^0_{aq} presented in Table 4.3 have been used, together with data on $Hg_{(liq)}$, $HgS_{(c)}$, and other aqueous Hg species taken from Khodakovsky *et al.* (1977) and Sorokin *et al.* (1988) to compute the pH–Eh diagram shown in Figure 4.3. This demonstrates the important role of Hg^0_{aq} in natural processes. Since it is the dominant species over a wide range of geochemically important conditions, spanning both the sulphate and sulphide regions, its role becomes increasingly important with increased temperature.

4.3 SULPHUR SOLUBILITY IN WATER AND THE THERMODYNAMICS OF THE AQUEOUS SPECIES S^0_{aq}, H_2S_{aq}, SO_{2aq}, $H_2S_2O_3$, HSO_4^- TO 440°C

Sulphur–water interaction provides a classic example of the importance of hydrolysis reactions in hydrothermal processes.

Aqueous fluids in the system S–H_2O have received extensive experimental study to 360°C; however, this has mostly concentrated on H_2S and H_2SO_4 (Ellis and Giggenbach, 1971; Ohmoto and Lasaga, 1982; Robinson, 1973; Rafalskii *et al.*, 1983; Alekseyev *et al.*, 1985; Dadze and Sorokin, 1993). S^0_{aq} has been studied to 250°C (Laptev *et al.*, 1987), but

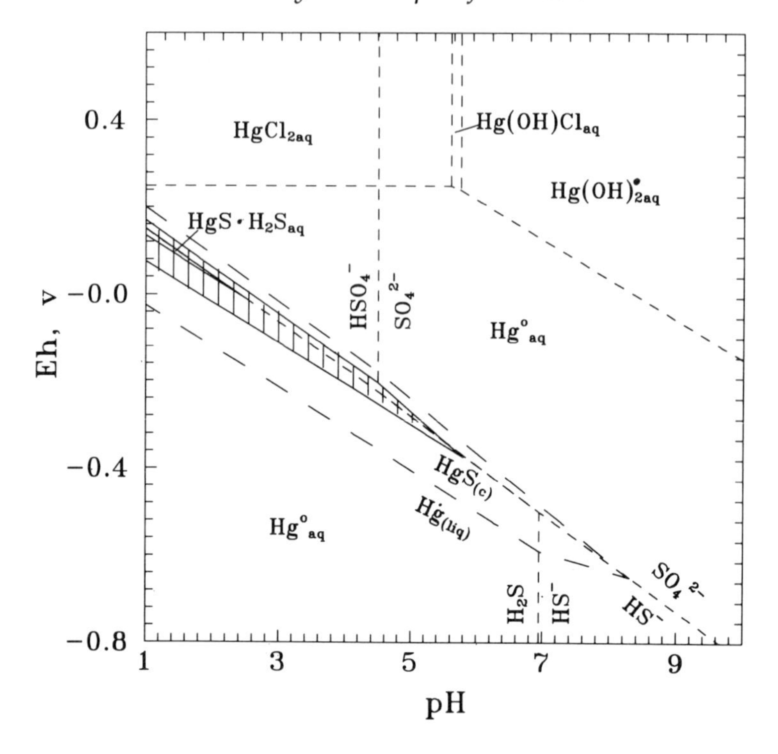

Figure 4.3. Eh vs pH diagram for the system Hg–S–Cl–O–H; T = 200°C; P = 1 atm; total S = 10^{-3}; total Cl^- = 10^{-2}; total Hg = 10^{-6} mol. kg H_2O^{-1}; dashed area – cinnabar.

there are few high-T studies of other aqueous sulphur species in S-saturated water.

4.3.1 Experimental

Experiments in the range 200–374 °C were largely carried out at pressures close to the saturated vapour pressure of water. At higher temperatures, experiments were conducted along the isochore with a density of water = 0.35 g.cm^{-3}. Sulphur solubility in water was investigated using a Ti-alloy Sorokin–Kapustin sampling device described in Chapter 3. Reagents were high purity sulphur and deoxygenated water. To purify the water, triply distilled water was boiled for 2 hours, cooled, and flushed with high purity Ar. The sulphur was loaded in a container located on the magnetic mixer shaft inside the reactor (Figure 3.3). The entire sampling device was evacuated, flushed with Ar and filled with H_2O under Ar pressure before the start of the run. The time required for the system to reach equilibrium was estimated from the data of Alekseyev *et al.* (1985).

The system is complex, with a large set of possible disproportionation reactions due to sulphur–water interaction, where S may have valencies of 2^-, 0, 4^+, 6^+, as well as possible recombination reactions during sampling and preliminary analytical procedures. To avoid or diminish the effect of recombination reactions, we used a method based on 'H_2S capture', i.e. the precipitation of H_2S as CdS at the time of sampling. Four to five cm^3 of 10% $Cd(CH_3COO)_2$ solution was preloaded into the sampler before the sample was transferred into it from the reactor. After cooling, the sample (with the CdS precipitate) was transferred from the sampler into a flask, to which 10 cm^3 of glycerin had been added to inhibit oxidation of SO_{2aq} and $H_2S_2O_3$; 25–30 cm^3 of hexane was also added to the flask to extract S^0_{aq}. After one day, the hexane fraction was separated and analysed for S^0_{aq} on a Specord-UV–VIS spectrophotometer. The remaining solution containing the CdS precipitate was filtered into a volumetric flask.

The H_2S removed as a CdS precipitate earlier, as well as $H_2S_2O_3$ and SO_{2aq} in the filtered solution, were determined separately by iodometric titration. HSO_4^- was determined by titration with $BaCl_2$ using a separate aliquot of the filtered solution.

To measure S_{tot}, we repeated the experiment under the same conditions, and all the sulphur present was oxidized for determination as HSO_4^- by adding bromine water or H_2O_2 and heating. The final measurement of S_{tot} in the total sample was made by weighing and titration with $BaCl_2$.

4.3.2 Results

In the range 220–440 °C, the sulphur component concentrations in the system S–H_2O may be represented as:

$$\log m_{S_{tot}} = 5.33548 - \frac{6383.76}{T} + \frac{1599.83 \times 10^3}{T^2} \pm 0.0062 \tag{4.7}$$

$$\log m_{H_2S} = 5.55242 - \frac{6508}{T} + \frac{1511.57 \times 10^3}{T^2} \pm 0.0039 \tag{4.8}$$

$$\log m_{SO_{2aq}} = 5.13812 - \frac{6141.31}{T} + \frac{1234.05 \times 10^3}{T^2} \pm 0.087 \tag{4.9}$$

$$\log m_{H_2S_2O_3} = -68.4834 + \frac{109895}{T} - \frac{59673.4 \times 10^3}{T^2} + \frac{10601.7 \times 10^6}{T^3} \pm 0.085 \tag{4.10}$$

$$\log m_{HSO_4^-} = -60.9029 + \frac{90189.9}{T} - \frac{45414.4 \times 10^3}{T^2} + \frac{77503.99 \times 10^6}{T^3} \pm 0.0058, \tag{4.11}$$

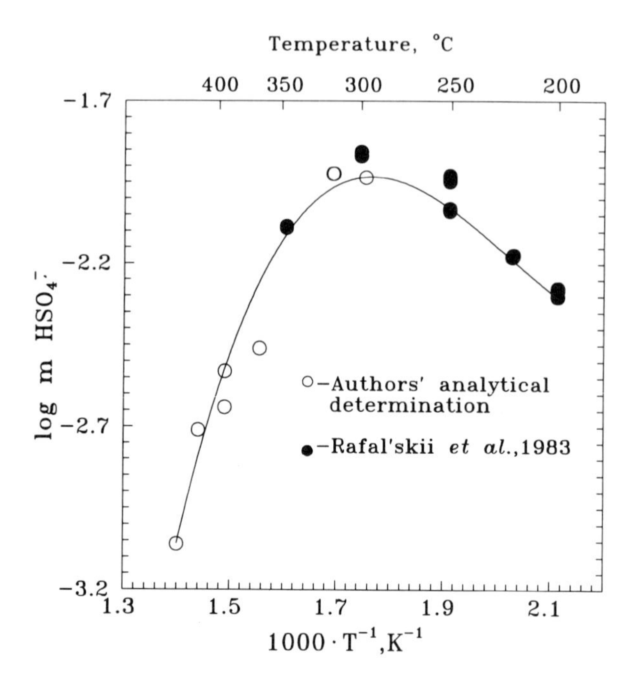

Figure 4.4. The system S–H_2O. Temperature dependencies of log $m_{HSO_4^-}$.

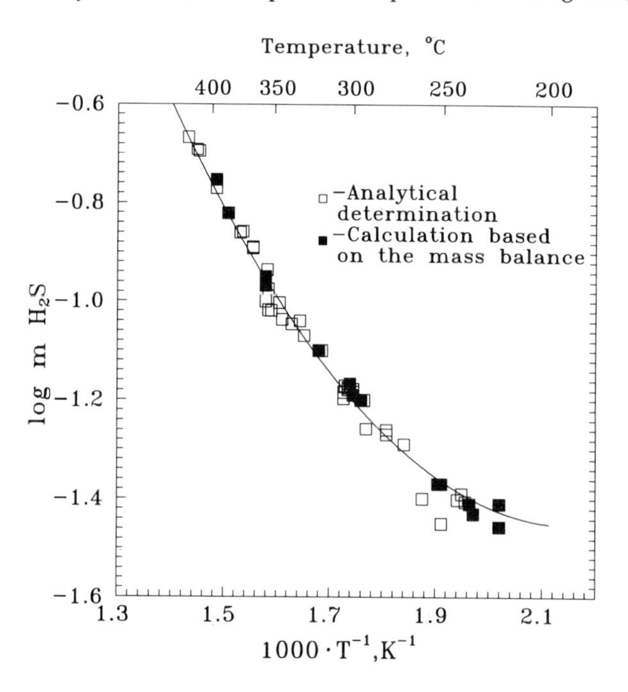

Figure 4.5. The system S–H_2O. Temperature dependencies of log m_{H_2S}.

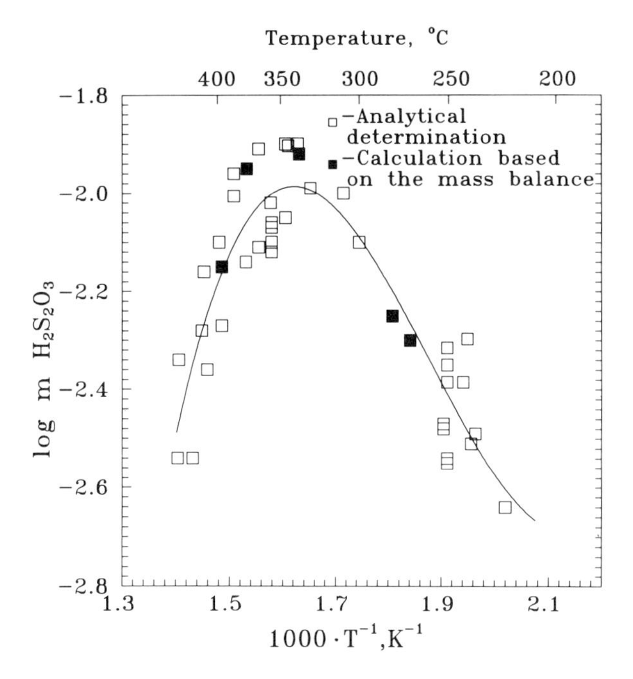

Figure 4.6. The system S–H_2O. Temperature dependencies of log $m_{H_2S_2O_3}$.

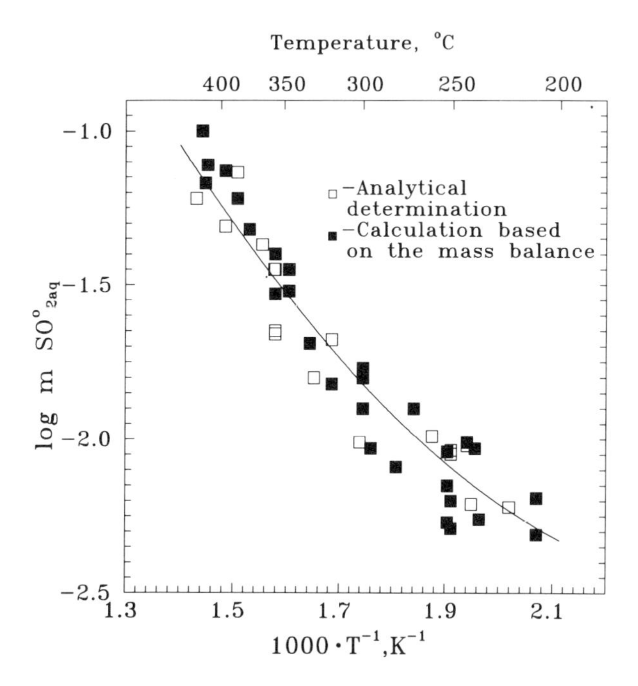

Figure 4.7. The system S–H_2O. Temperature dependencies of log m_{SO_2}.

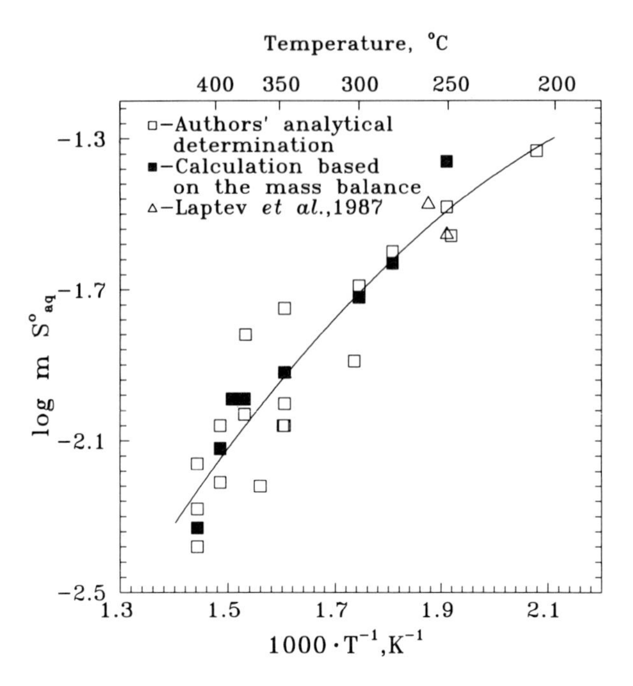

Figure 4.8. The system S–H_2O. Temperature dependencies of log $m_{S^\circ_{aq}}$.

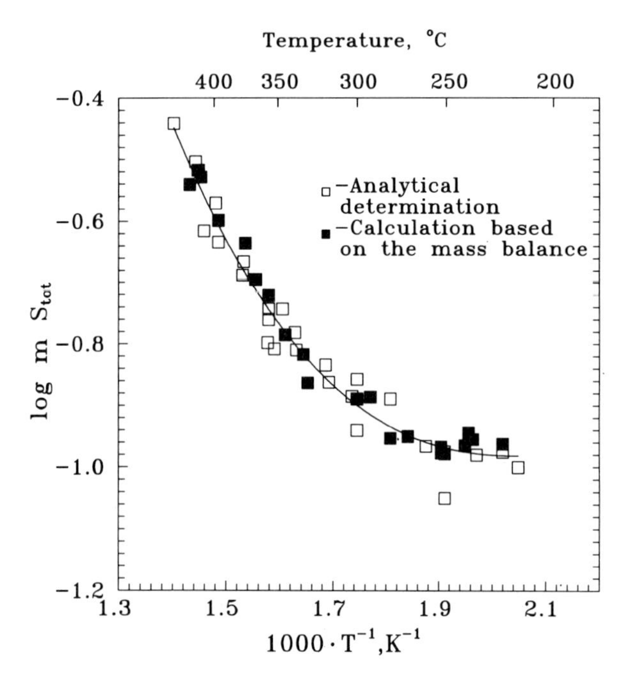

Figure 4.9. The system S–H_2O. Temperature dependencies of log $m_{S_{total}}$.

$$\log m_{S^0_{aq}} = -0.106359 - \frac{7458.63}{T} + \frac{6064.84 \times 10^3}{T^2}$$

$$- \frac{11327.72 \times 10^6}{T^3} \pm 0.131. \tag{4.12}$$

The results of the experimental measurements of sulphur species concentrations in equilibrium with water and S are shown in Figures 4.4, 4.5, 4.6, 4.7, 4.8. Note that the mass balance of all analysed species gives a good fit to the direct measurement of S_{tot} (Figure 4.9).

4.3.3 Discussion of results

Sulphur dissolution in the system S–H_2O under the conditions investigated may be represented as the sum of the following reactions:

$$4S_{liq} + 4H_2O = 3H_2S + H^+ + HSO_4^-, \tag{4.13}$$

$$4S_{liq} + 3H_2O = 2H_2S + H_2S_2O_3, \tag{4.14}$$

$$3S_{liq} + 2H_2O = 2H_2S + SO_{2aq}, \tag{4.15}$$

$$S_{liq} = S^0_{aq} \tag{4.16}$$

The data on HSO_4^- and H_2S^0 concentrations above 270°C are in good agreement with published data and thermodynamic extrapolations (Figure 4.10a). At temperatures below 250°C, the content of hydrogen sulphide is anomalously high, probably due to a contribution from the sulphane component H_2S_n, determined in combination with the monomeric form. A similar phenomenon is also observed below 300°C for SO_{2aq} (Figure 4.10b) and can again be assigned to a contribution from polymeric forms (polythionates). That this effect is not a random one is evidenced by the agreement between total sulphur concentrations measured directly with those obtained by summing the species concentrations.

Measured concentrations of $H_2S_2O_3$ were higher than expected, with the maximum concentration occurring at a surprisingly high temperature, around 330°C. Note, however, that this species is never predominant.

The measurements of elemental sulphur concentration proved the most unexpected. Results are consistent with published data and with theoretical predictions of the sulphur vapour pressure to 250°C. At higher temperatures, however, the concentrations of S^0_{aq} decrease drastically (Figure 4.8). Without additional experimental data it is difficult to offer a sound explanation for this effect.

4.3.4 Thermodynamic properties

The thermodynamic properties of the aqueous sulphur species are not well understood at temperatures above 100–150°C. This is particularly

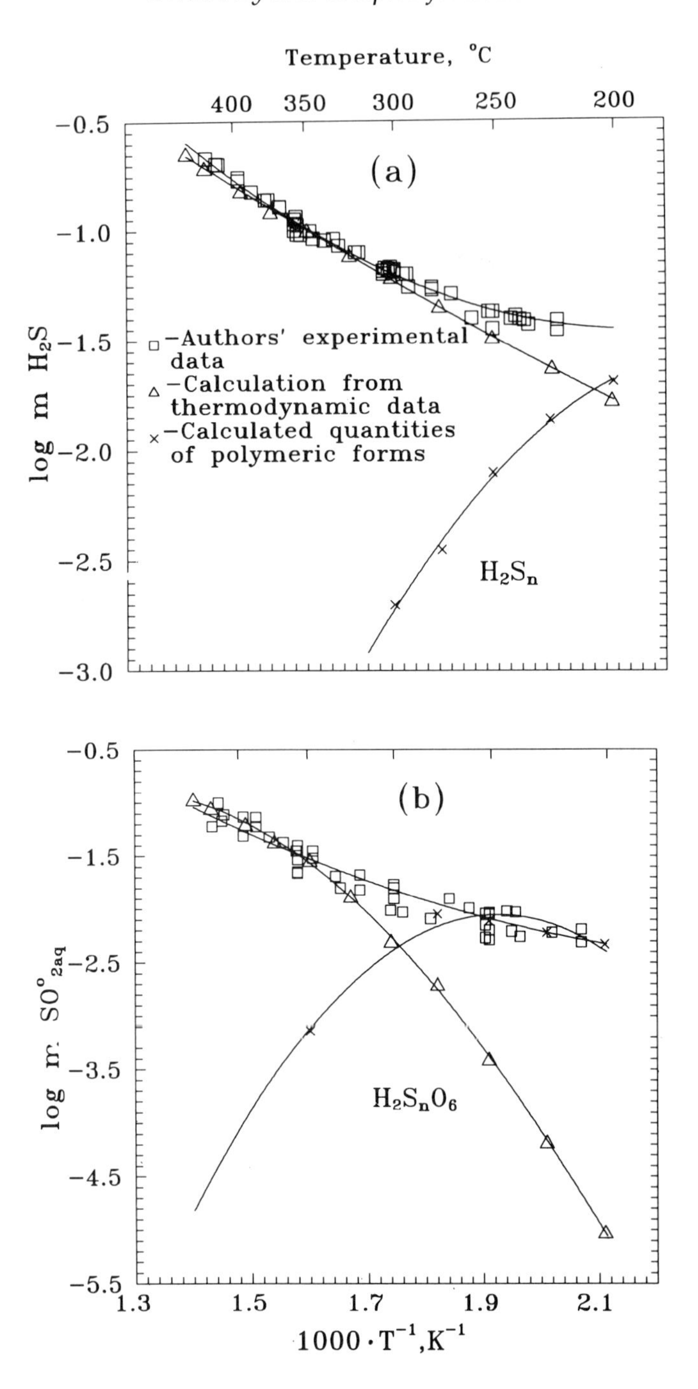

Figure 4.10. Predicted quantities (molality) of sulphanes (a) and polythionates (b) in the system S–H_2O.

true for the intermediate oxidation levels, because they are relatively unstable in water, and disproportionation reactions occur more readily. In contrast, the thermodynamics of the end members of the oxidation–reduction series, S^{2-} and S^{6+} (sulphide and sulphate species), in the system $S–H_2O$ is reasonably well understood to 360°C.

The data on the thermodynamics of sulphate seem the most preferable because both its aqueous species and products of their dissociation display high temperature stability in water-bearing systems.

The sources of basic thermodynamic data we chose were a modified version of Johnson *et al.*'s (1991) HKF model and HSO^-_{4aq} data. The results of the calculations are presented in Figures 4.11, 4.12, 4.13, 4.14.

The Gibbs free energy ($\Delta G^0_{f,T}$) for $H_2S^0_{aq}$ was calculated from reaction 4.13 via the equation:

$$\Delta G^0_{f,(T)H_2S^0_{aq}} = \frac{1}{3}\,[4\Delta G^0_{f,(T)S_{liq}} + 4\Delta G^0_{f,(T)H_2O_{liq}} - \Delta G^0_{f,(T)HSO_4^-} + \Delta G^0_{13\,(T)}], \tag{4.17}$$

where $\Delta G^0_{13\,(T)}$ is the apparent conventional Gibbs free energy of reaction 4.13, defined as:

$$\Delta G^0_{13\,(T)} = -\,RT.\ln K_{13}. \tag{4.18}$$

K_{13} was retrieved from experimental data, and activity coefficients for ions were calculated with the Debye–Hückel equation of third approach (Helgeson *et al.*, 1981). The coefficients A and B were taken from Helgeson and Kirkham (1974), å and b_γ from Helgeson *et al.* (1981), while $\Delta G^0_{f,(T)S_{liq}}$ is from IVTANTERMO (1984).

Table 4.4. Apparent Gibbs free energy of aqueous sulphur species, retrieved from experiments ($kcal.\ mol^{-1}$)

T (°C)	*P*(*bar*)	H_2S_{aq}	SO_{2aq}	$H_2S_2O_3$	S^0_{aq}
200	99	−14.2	−86.1	−147.4	1.12
225	105	−15.2	−86.2	−148.4	1.10
250	115	−16.3	−86.3	−149.6	1.17
275	132	−17.4	−86.6	−151.0	1.29
300	159	−18.5	−87.1	−152.5	1.43
325	196	−19.7	−87.7	−154.0	1.60
350	247	−20.8	−88.5	−155.4	1.78
375	300	−21.9	−89.6	−157.0	1.95
400	402	−22.9	−91.0	−158.65	2.12
425	512	−23.8	−92.7	−160.4	2.28
440	589	−24.2	−94.0	−161.5	2.37

Values of $\Delta G^0_{f,(T)H_2S}$ are given in Figure 4.11 and Table 4.4. Also included in the figure are P–T estimates based on the data of other workers (e.g.

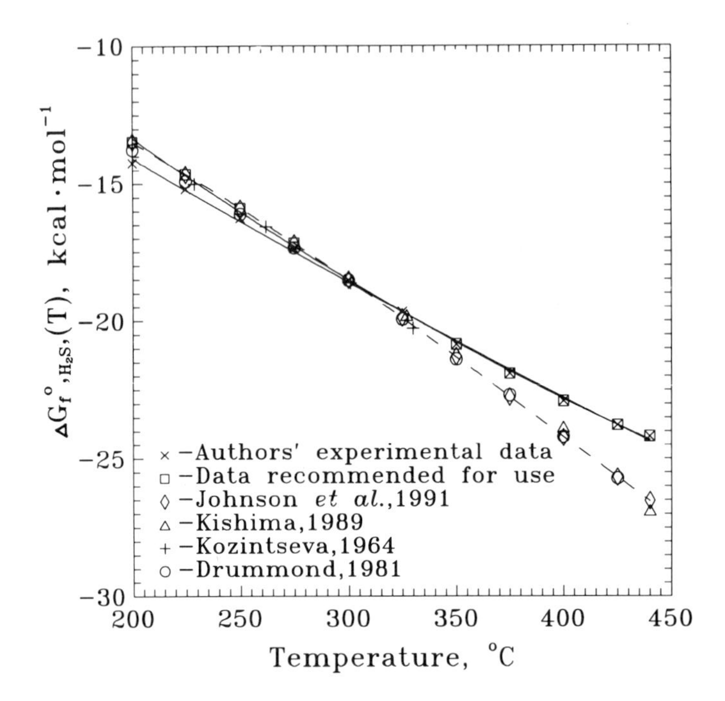

Figure 4.11. The temperature dependence of the Gibbs free energy for H_2S.

Kozintseva, 1964; Drummond, 1981; Kishima, 1989), reduced to our experimental curve.

Similarly, values of $\Delta G^0_{(T)H_2S}$ of formation retrieved from reaction 4.13 were used to retrieve $\Delta G^0_{(T)SO_{2aq}}$ of formation and $\Delta G^0_{(T)H_2S_2O_3}$ of formation from reactions 4.14 and 4.15, respectively (Figures 4.12, 4.13, Table 4.4).

The results of calculations based on our experiments (Table 4.4) yield similar patterns for the behaviour of H_2S_{aq} and SO_{2aq} to the available thermodynamic data (Johnson *et al.*, 1992). In the region 250–320°C, the data are in good agreement, but there are marked differences between published thermodynamic values and those calculated from our experimental data at higher or lower temperatures (Figures 4.11, 4.12). Below 250°C, the peculiarities of the solubility of sulphur in water (discussed above) allow us to suggest that in this instance sulphanes (H_2S_n) and polythionates ($H_2S_nO_6$) have contributed to the experimental sulphide and sulphate analyses. By combining the data of Johnson *et al.* (1992) for temperatures below 280°C with our experimental results for higher temperatures, we have obtained a set of values for H_2S_{aq} and SO_{2aq}, which we recommend for use (Figures 4.11, 4.12).

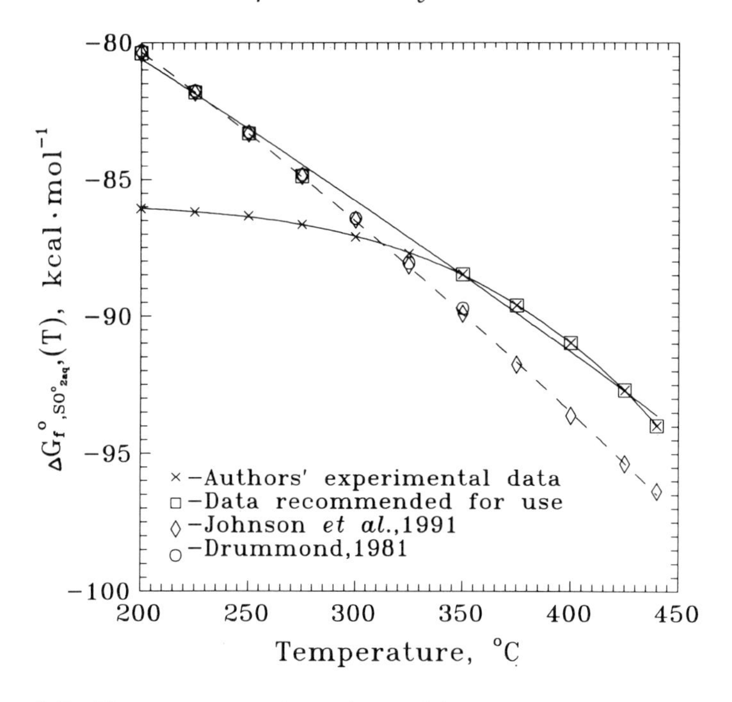

Figure 4.12. The temperature dependence of the Gibbs free energy for SO°_{2aq}.

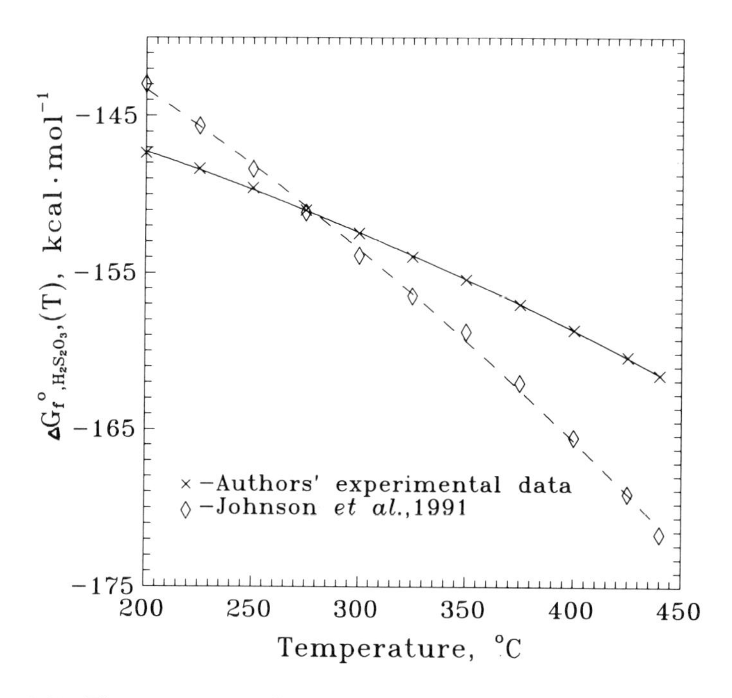

Figure 4.13. The temperature dependence of the Gibbs free energy for $H_2S_2O_3$.

Table 4.5. Apparent Gibbs free energy of H_2S_{aq} and SO_{2aq} (kcal. mol^{-1})

P(bar)	*Saturated H_2O*		*500*	*500*	*1000*	*1000*
T(°C)	*H_2S*	*SO_2*	*H_2S*	*SO_2*	*H_2S*	*SO_2*
200	−14.3	−86.2	−13.9	−85.7	−13.5	−85.2
225	−15.3	−86.3	−14.8	−85.8	−14.4	−85.3
250	−16.3	−86.4	−15.9	−85.9	−15.5	−85.4
275	−17.4	−86.7	−17.0	−86.2	−16.6	−85.7
300	−18.6	−87.2	−18.2	−86.7	−17.8	−86.2
325	−19.8	−87.9	−19.4	−87.3	−18.9	−86.8
350	−21.1	−88.9	−20.6	−88.1	−20.1	−87.6
375			−21.6	−89.2	−21.2	−88.7
400			−22.7	−90.6	−22.3	−90.1
425			−23.8	−92.8	−23.2	−91.9

Table 4.5 lists data for the Gibbs free energies of formation of aqueous sulphur species at P_{sat} and P = 0.5 and 1kbar, retrieved with the equation:

$$\Delta G^0_{(T,P)} = \Delta G^0_{(T,P)ex} + \int_{P_{ex}}^{P} VdP, \tag{4.19}$$

where P_{ex} is the pressure of the run (Table 4.4), and $\Delta G^0_{(T,P)ex}$ is the apparent Gibbs free energy of formation at the experimental pressure. The molar volumes (V) of the species are from Johnson *et al.* (1991). The calculation of the composition of the system on the basis of the apparent free energy data obtained from Shvarov's 'Gibbs' programme provides some insight into the predicted amounts of sulphanes and polythionates in the system (Figure 4.10a, b).

The $H_2S_2O_3$ values were calculated from the data for $S_2O_3^{2-}$ (Johnson *et al.*, 1991) and from the dissociation constants (K_1 and K_2) of $H_2S_2O_3$ along the saturation curve of water (Naumov *et al.*, 1971) using the equation:

$$\Delta G^0_{(T)H_2S_2O_3} = RT.\ln K_1 + RT.\ln K_2 + \Delta G^0_{(T)S_2O_3^{2-}}. \tag{4.20}$$

The Henry's law constant for S^0_{aq}, designed as:

$$K_H(P,T) = \frac{m_{S^0_{aq}}}{f_{S_g}} \tag{4.21}$$

was calculated from the equation:

$$RT.\ln f_{S_g}(P) = RT.\ln f_{S_g}(P_{sat}) + \int_{P_{sat}}^{P} V_{S_{liq}}\, dP, \tag{4.22}$$

where $f_{S_g}(P_{sat})$ is the sulphur fugacity on the saturation curve, and $V_{S_{liq}}$ the molar volume of liquid sulphur (West, 1950). Values obtained for K_H were 0.291, − 0.533, − 1.234 and − 1.896 at 250, 300, 350 and 400°C, respectively.

The Gibbs free energy of formation of S^0_{aq} was calculated with the equation:

$$\Delta G^0_{(T)S^0_{aq}} = \Delta G^0_{(T)S_{liq}} - RT.\ln m_{S^0_{aq}} \quad (4.23)$$

and the resulting values are given in Table 4.4 and plotted in Figure 4.14.

The temperature dependence of $m_{S^0_{aq}}$ is the most unexpected, passing through a maximum despite the progressive increase in the sulphur vapour pressure (Figure 4.8). This leads to the conclusions that the mechanisms by which dissolution occurs vary with temperature, and that dispersed polymeric species can occur in solution over wide temperature ranges.

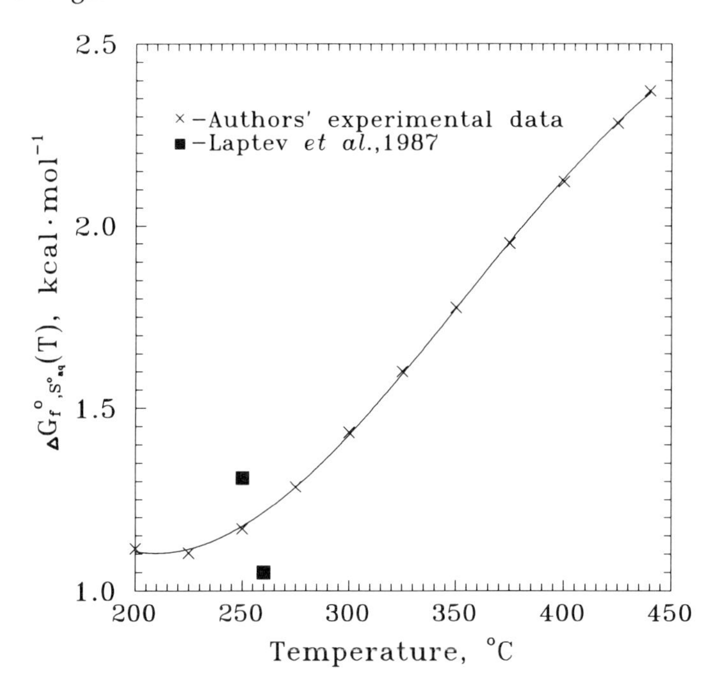

Figure 4.14. The temperature dependence of the Gibbs free energy for S^0_{aq}.

4.4 SiO_2 SOLUBILITY IN WATER AND IN ACID SOLUTIONS (HCl, HNO_3) AT 100–400 °C AND 1013 BAR

Silica is a major constituent of the Earth's crust, quartz constitutes the most important vein mineral, and hence silica behaviour in hydrothermal process, including ore formation, is of prime importance. All forms of silica are poorly soluble, even in strong solvents, and the pattern of its behaviour in hydrothermal solutions is distinctive. For example, it displays an anomalous volume change of dissolution, and silica solubility has an unusual relation to acidity and to the chloride ion. This study was particularly concerned with the aqueous forms of silica, and so we have investigated the solubility of the most soluble, amorphous, form of silica.

4.4.1 Previous work

Although much has been published on the solubility of silica in water and in electrolyte aqueous solutions (Anderson and Burnham, 1965, 1967; Walther and Helgeson, 1977), fewer studies have investigated its behaviour in acidic solutions, including acidic chloride ones. The possibility that silica may combine in solution with hydrochloric acid was indicated by Sadek (1952), and Alexander *et al.* (1954) observed a slight increase in the SiO_2 solubility in the pH range 2–6 and suggested that the complex $Si(OH)_{4aq}$ might react with acids at low pH values. Elmer and Nordberg (1958), however, reported a decrease in SiO_2 in the presence of nitric acid over a wide range of concentrations at 36–95°C. Kitahara (1960) recorded an increase in the solubility of SiO_2 in NaCl solutions relative to pure water. It is generally accepted that over a wide range of neutral to acid pH values, the solubility of silica in neutal and acidic electrolyte solutions is independent of acidity and governed by the complex $H_4SiO_4^0$ or $SiO_2.nH_2O$ (Anderson and Burnham, 1965, 1967; Busey and Mesmer, 1977; Fournier and Rowe, 1977; Walther and Helgeson, 1977; Sorokin and Dadze, 1980).

4.4.2 Experimental

Experiments were conducted in Ti-alloy autoclaves with an internal capacity of 20 cm^3 at 100–400°C and 1013 bar. Pressure was controlled by filling the autoclave with a measured mass of fluid, determined from P–V–T data for water (Kennedy, 1950) and for aqueous chloride solutions (Egorov and Ikornikova, 1973). Quenching was carried out in water within 0.5–1 min. In addition, a set of test runs was performed in Pt capsules, yielding results similar to those obtained using the Ti-alloy autoclaves.

Finely divided amorphous SiO_2 was used as a starting material, which was placed in a container in the top of the autoclave. X-ray analysis demonstrated that this material remained amorphous at all run temperatures. The time required for equilibrium was determined experimentally as 2, 3, 4 and 6 weeks for 400, 300, 200 and 100°C, respectively. The concentration of silicic acid in quenched solutions was measured photocolorimetrically as monosilicic acid, using ammonium molybdate (blue complex). In order to ensure all dissolved silica was in this form, an aliquot of the quenched solution was immediately taken into 47% NaOH solution of high purity. Note that for some runs solutions were analysed without preliminary treatment with alkali. If the time between quench and the sampling of the aliquot was less than 15 minutes, then there was no difference from results on solutions which were treated with alkali. This allows us to conclude that at the temperature and pressure of the run, SiO_2 in solutions was monomeric. If solutions were analysed at longer time intervals, then lower SiO_2 concentrations were obtained, possibly due to losses upon cooling.

4.4.3 Results

The results of the experiments are given in Table 4.6 and Figures 4.15 and 4.16. Table 4.6 presents the average of the results from between 3 to 7 runs at each T. It can be seen that neither HCl nor HNO_3 has any appreciable effect on silica solubility (Figure 4.15), and that within the compositional range studied there is no noticeable salting-out effect. Figure 4.16 shows our measured solubility of amorphous silica in water at 1013 bar as a function of T, with the data of Fournier and Rowe (1977) for comparison. Also shown are results for silica solubility in seawater from Wildly (1974) and Jones and Pytkowicz (1977). A least-squares fit to our experimental data gives the following equation for the solubility of amorphous silica:

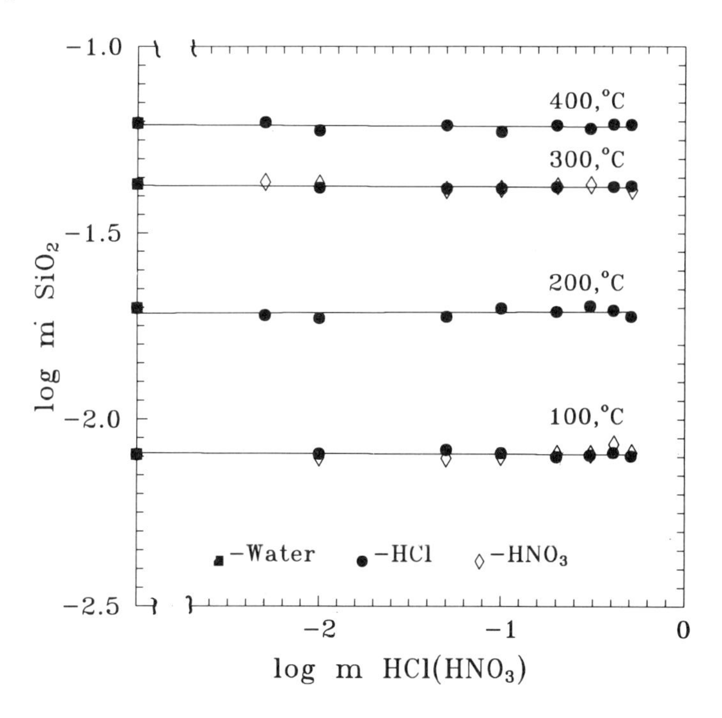

Figure 4.15. The solubility of amorphous SiO_2 at P = 1013 bar.

Table 4.6. Solubility of amorphous SiO_2 in water and in HCl and HNO_3 solutions, P=1013 bar

HCl	$m_{SiO_2} \times 10^2$ *mol. kg* H_2O^{-1}				HNO_3	$3m_{SiO_2} \times 10^2$ *mol.kg* H_2O^{-1}	
molality	*100°C*	*200°C*	*300°C*	*400°C*	*molality*	*100°C*	*300°C*
0.00	0.805	0.199	4.28	6.24	–	–	–
0.005	–	1.91	–	6.26	0.005	–	4.33
0.010	0.809	1.87	4.19	5.95	0.010	0.789	4.29
0.050	0.830	1.89	4.16	6.15	0.050	0.786	4.13
0.100	0.813	1.99	4.17	5.92	0.100	0.795	4.16
0.200	0.795	1.95	4.21	6.15	0.202	0.809	4.24
0.303	0.801	2.02	4.21	6.01	0.306	0.809	4.27
0.406	0.814	1.97	4.22	6.20	0.410	0.858	–
0.509	0.798	1.89	4.24	6.18	0.516	0.815	4.12

$$\log m_{SiO_2} = \frac{-769.5}{T} - 0.055 \pm 0.002, \tag{4.24}$$

where m_{SiO_2} is the silica molality, T is temperature (in K). This is closely comparable to the equation in Fournier and Rowe (1977). Extrapolation to low temperatures (< 100°C) is in reasonable agreement with the compilation of Walther and Helgeson (1977) (Figure 4.15).

4.4.4 Discussion

The independence of silica solubility from pH or ligand concentration (Cl^- or NO_3^{2-}) indicates that aside from $SiO_2.2H_2O$, no additional silica complexes form (Figure 4.16). Furthermore, even at very low pH (< 1), $SiO_2 . 2H_2O$ does not undergo a base-type dissociation with the formation of positively charged ions. Both the experimental results themselves, and the similar monosilicic acid concentrations found irrespective of whether solutions had been stabilized with alkali (above), indicates that to, at least, 200°C, the solubility of silica and its concentration in both acid and neutral hydrothermal solutions are determined by interaction of SiO_2 with water with the formation of $SiO_2.nH_2O$. Above 200°C, this reaction leads to the formation of $SiO_2.2H_2O$, or, $H_4SiO_4^0$.

4.4.5 Conclusions

The solubility of silica in hydrothermal solutions (except those having significant fluoride concentrations, see Chapter 6, this volume) is governed by its reaction with water, accompanied by the formation of the complex $H_4SiO_4^0$. Under alkaline conditions, this complex undergoes an acid-type stepwise dissociation, i.e. $H_nSiO_4^{n-4}$. Hydrogen ions may be replaced by alkaline metal ions, as proposed by Anderson and Burnham (1965) for alkaline solutions.

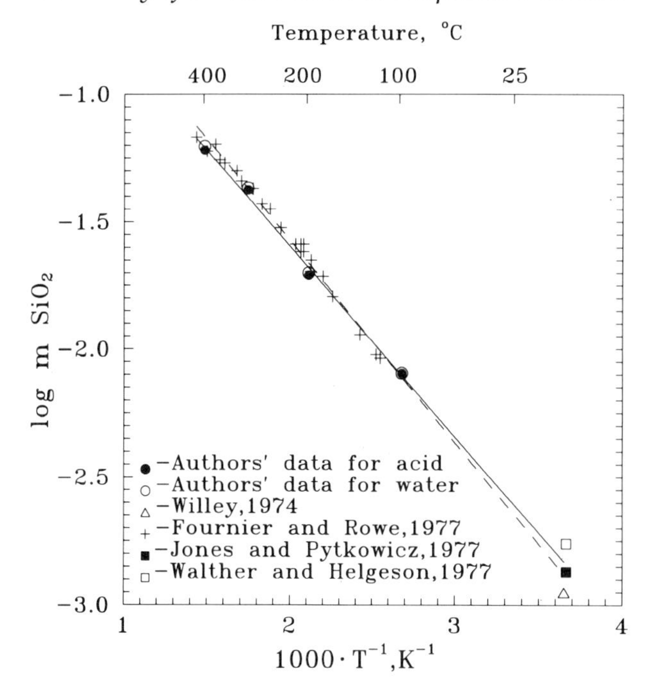

Figure 4.16. The temperature dependence of the solubility of amorphous SiO_2. Solid line represents our data; dashed line, data of Fournier and Rowe (1977).

4.5 SOLUBILITY OF SnO_2 IN WATER AND AQUEOUS ELECTROLYTE SOLUTIONS AT 200–400°C AND 16–1500 BAR

Chemically, tin is a silicon analogue and falls into the same subgroup of the periodic system as carbon, germanium and lead. Tin and silicon dioxides occur as cassiterite and quartz in nature, and cassiterite generally does not occur without silica. Geochemically, however, Si and Sn diverge considerably. For instance, divalent Si is probably not present in nature, whereas divalent Sn is quite common. Therefore, a comparative experimental assessment of their behaviour under hydrothermal conditions is of interest.

4.5.1 Previous work

Although there are some earlier publications in the literature, concerned with SnO_2 solubility in aqueous solutions, only during the past 20 years have investigations in this field acquired a systematic and quantitative character (Klintsova and Barsukov, 1970, 1973; Klintsova *et al.*, 1975; Dadze *et al.*, 1981; Dadze and Sorokin, 1986; Kovalenko *et al.*, 1986, 1992; Wilson and Eugster, 1990; Barsukov *et al.*, 1991). As in the case of silica, tetravalent tin forms a neutral complex $SnO_2.2H_2O$ interpreted as $Sn(OH)_4^0$

over a wide range of conditions (Dadze *et al.*, 1981) The consensus among different workers is that at 200°C, the complex $Sn(OH)_4^0$ dominates in solutions at a HNO_3 concentration up to 0.1 m (e.g. Klintsova and Barsukov, 1973). A series of complexes $Sn(OH)_{4+n}^{n-}$ are also established for tetravalent tin (Klintsova and Barsukov, 1973), along with its chloro and hydroxychloro complexes (Kovalenko *et al.*, 1986; Barsukov *et al.*, 1991). Many workers have reported data on divalent Sn in complexes of the type $Sn(OH)_n^{2-n}$, which are probably of major importance for tin transport in nature at relatively low redox values (Jackson and Helgeson, 1985; Kovalenko *et al.*, 1986; Wilson and Eugster, 1990; Barsukov *et al.*, 1991).

4.5.2 Experimental

As in the case of SiO_2, experiments were conducted in Ti-alloy autoclaves, using a similar experimental approach. Experiments in HCl solutions at low pH (1.3–1.6) were performed in Pt capsules to prevent interaction with the autoclave material. The time required for equilibrium was determined experimentally as 10, 14 and 20 days at 400, 300 and 200°C, respectively. High-purity crystalline cassiterite (SnO_2) was used as a starting material. X-ray analysis revealed the material to be identical before and after the runs. The concentration of tin in quenched solutions was determined photocolorimetrically using phenylfluoron. At least three experiments were run under each set of conditions. For each experiment, 2–3 independent measurements of tin concentration were made, and the results averaged. The solubility of SnO_2 was measured in water and in HCl, HCl + KCl (I = 0.029 and 0.038) and HNO_3 solutions at temperatures from 200 to 400°C, 1013 bar pressure and total concentrations in the range 0.001–0.5 m. Over the same temperature range, measurements of SnO_2 solubility in pure water were also obtained at pressures between 16 and 1500 bar.

4.5.3 Results

The results of the experiments are presented in Tables 4.7, 4.8 and Figures 4.17, 4.18 and 4.19.

The experiments in pure water demonstrate that unlike SiO_2, the solubility of SnO_2 in water decreases with increasing pressure.

Table 4.7. Experimental data on SnO_2 solubility in water as a function of pressure

P(bar)	*$m_{Sn} \times 10^6$ mol. kg H_2O^{-1}* 200°C	300°C	400°C
16	0.920	–	–
50	0.733		
90		2.30	
250	0.860		

Table 4.7 (Contd)

P(bar)	*$m_{Sn} \times 10^6$ mol. kg H_2O^{-1}* 200°C	300°C	400°C
300		2.29	
400			4.38
550	0.780	1.84	3.91
600			3.42
1000	0.576	1.35	2.50
1250		1.20	–
1500	0.495	–	–

Table 4.8. Solubility of SnO_2 in HCl, HCl + KCl, and HNO_3 solutions; P=1013 bar

Acid + KCl or acid molality	*$m_{Sn} \times 10^6$ mol. kg H_2O^{-1}*				
	200°C		*300°C*		*400°C*
	HCl	*HNO_3*	*HCl*	*HNO_3*	*HCl*
0.001	0.572	–	1.36	–	2.5
0.001+0.0029	–	–	1.38	–	–
0.001+0.0029	–	–	1.52	–	–
0.005	0.995	0.700	2.63	–	3.25
0.005+0.025	–	–	2.63	–	–
0.005+0.025	–	–	2.00	–	–
0.008+0.032	–	–	2.24	–	–
0.010	1.86	0.842	2.95	2.40	5.10
0.010+0.020	–	–	2.95	–	–
0.010+0.030	–	–	2.91	–	–
0.015+0.015	–	–	5.14	–	–
0.015+0.025	–	–	4.10	–	–
0.015+0.025	–	–	4.47	–	–
0.020	2.51	–	7.40	–	–
0.020+0.010	–	–	6.00	–	–
0.030+0.010	–	–	8.80	–	–
0.03	3.55	–	10.40	–	–
0.03	–	–	9.20		–
0.03	–	–	10.40	–	–
0.04	5.06	–	12.00	–	–
0.04	–	–	9.60	–	–
0.04	–	–	13.00	–	–
0.05	5.50	2.27	–	8.6	–
0.05	–	2.44	–	–	–
0.08	7.41	–	–	–	–
0.1	–	3.79	31.0	13.9	–
0.2	–	–	64.0	–	–
0.202	–	6.32	–	30.4	–
0.202	–	8.00	–	–	–
0.306	–	–	–	38.9	–
0.410	–	15.70	–	52.8	–
0.516	–	–	–	78.3	–

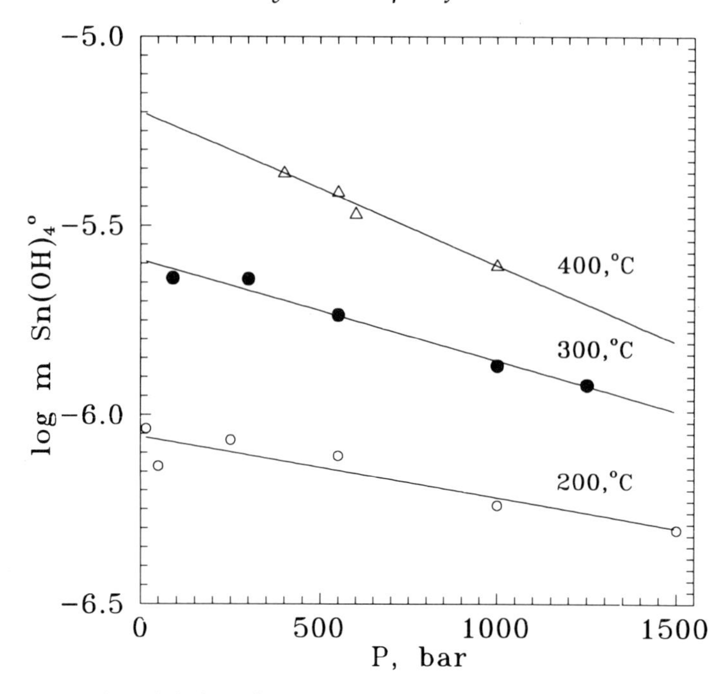

Figure 4.17. The solubility of cassiterite in water.

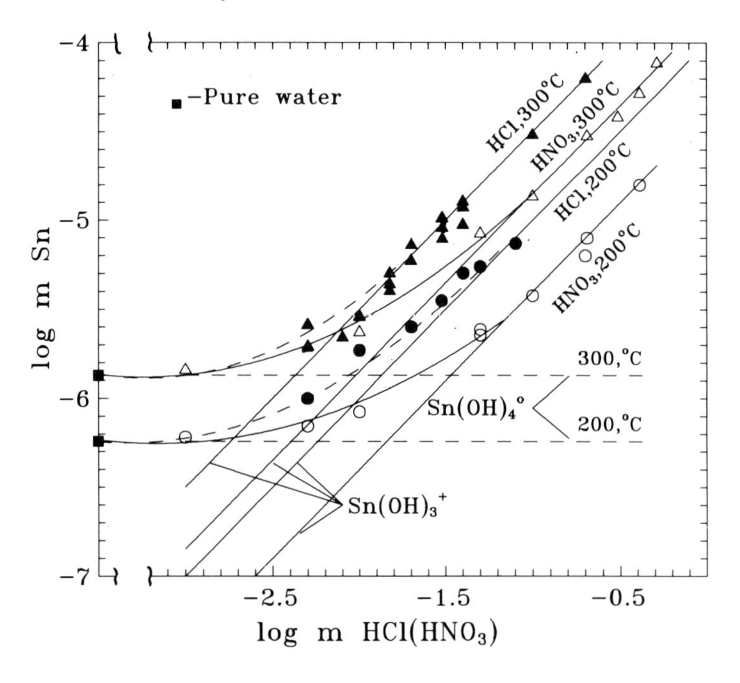

Figure 4.18. SnO_2 solubility in HCl and in HNO_3 solutions. Solid symbols HCl, open symbols HNO_3; P = 1013 bar.

Experiments using acid solutions yield a similar dependence of solubility on pH for both acids, demonstrating that tin(IV) concentrations are a function of the solution acidity rather than the available ligands. The increased concentration of the chloride ion in HCl + KCl solution has no appreciable effect on SnO_2 solubility, slightly decreasing it in a few cases, due to the salting-out effect perhaps. In Ti-alloy autoclaves, solution interacts with the reactor walls when HCl concentrations are larger than 0.03–0.05 m. Therefore, for measurements in strongly acid solutions, either Pt capsules were employed or HCl was replaced by HNO_3. Corrosion of the reactor in HCl solutions was accompanied by a decrease in the solution redox potential measured after the run, as well as by a sharp increase in the SnO_2 solubility of 2–3 orders of magnitude. This may be due to the appearance of more stable Sn(II) complexes, possibly chloro complexes. Model calculations of experimental values of redox potential support this suggestion, but are not presented here. There is reason to believe that in all other cases we dealt with Sn(IV) compounds only.

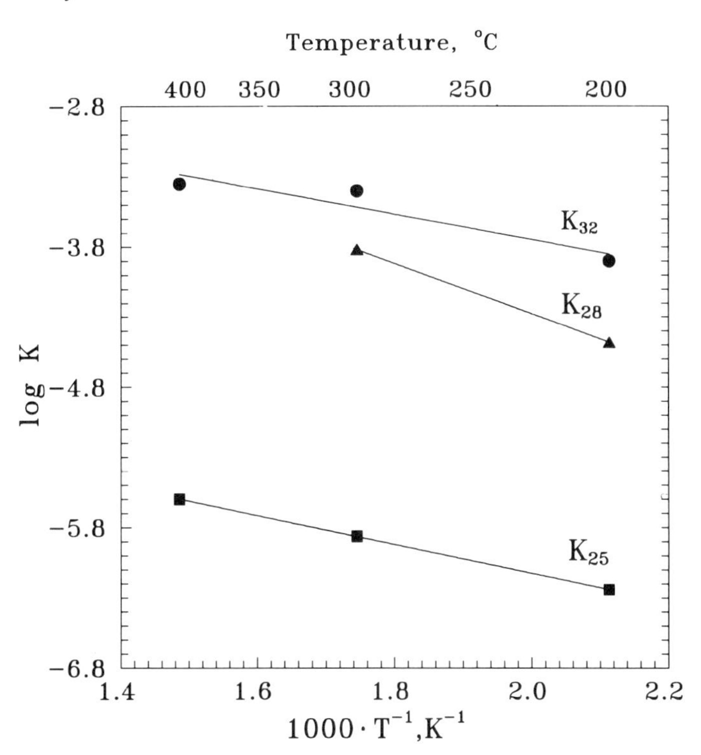

Figure 4.19. The temperature dependence of SnO_2 solubility at P = 1013 bar · K_{25} = constant for the reaction $SnO_2 + H_2O = Sn(OH)_4^0$, K_{28} = constant for the reaction $SnO_2 + H_2O + H^+ = SnOH)_3^+$. K_{32} = constant for the reaction $SnO_2 + H_2O + HCl = Sn(OH)_3Cl$.

4.5.4 Discussion and thermodynamics

The data presented here give evidence that, as in the case with silica, a neutral complex is widespread in dilute solutions, which forms according to the reaction:

$$SnO_{2\,(c)} + H_2O = Sn(OH)^0_{4aq}. \qquad (4.25)$$

At P = 1013 bar, the temperature dependence of the constant K_{25} for reaction 4.25 is adequately described by the rigorously linear equation (Figure 4.19):

$$\log K_{25} = -\frac{1023.02}{T} - 4.007733. \qquad (4.26)$$

Based on data for the pressure–temperature dependence of solubility (Table 4.7), the apparent Gibbs free energies of formation of $Sn(OH)^0_{4aq}$ have been calculated as follows:

$$\Delta G^0_{f(P,\,T)Sn(OH)^0_4} = \Delta G^0_{f\,(P_1T)Sn(OH)^0_4} + \int_{P_1}^{P_2} V^0_{Sn(OH)_4}\, dP \qquad (4.27)$$

where

$$\Delta G^0_{f(P_1,\,T)Sn(OH)^0_4} = \Delta G^0_{f(P_1,\,T)SnO_2} + 2\Delta G^0_{f(P_1,\,T)H_2O} - RT.\ln K_{25},$$

$P_1 = 1$ kbar, $V_{Sn(OH)^0_4}$ is calculated from experimental data using the equation $\frac{[(\ln K_{25})]}{P}{}_T = -\frac{\Delta V^0_T}{RT}$, where ΔV^0_T is the volume change of reaction (4.25), $V^0_{SnO_2}$ and $V^0_{H_2O}$ are taken from Johnson *et al.* (1991). $V^0_{Sn(OH)^0_4}$ is practically independent of pressure, varying with temperature according to the equation:

$$V^0_{Sn(OH)^0_4} = 49.696 - 0.17776.T + 3.275.10^{-4}.T^2\ \text{cm}^3\ \text{mol}^{-1}. \qquad (4.27a)$$

The calculated values of the apparent Gibbs free energies of formation of $Sn(OH)^0_4$ are listed in Table 4.9.

Table 4.9. Apparent Gibbs free energy of formation, ΔG^0_f (T, P) of $Sn(OH)^0_4$ (kcal. mol^{-1})

P, bar	*200°C*	*300°C*	*400°C*
satur.	–234.453	–240.214	–
500	–233.690	–239.382	–246.094
1000	–232.903	–238.378	–244.872
2000	–232.117	–237.374	–243.651

The 1:1 correlation between m_{Sn} in solution and acid concentration (Figure 4.18), together with the lack of any influence of chloride ion (as KCl) concentration on solubility, suggests that the difference in SnO_2 solubility between HCl and HNO_3 solutions is the result of a difference

between the dissociation constants of these acids. The solubility data and slopes on Figure 4.18 are consistent with the dissolution reaction:

$$SnO_{2\,(c)} + H_2O + H^+ = Sn(OH)_3^+ \qquad (4.28)$$

for all HNO_3, HCl, and HCl + KCl solutions (pH < 2.5).

Table 4.10. Constants for reactions: 4.28 and 4.32

HNO_3, HCl	*KCl*	log K_{28}		log K_{32}			
		200°C	*300°C*	*200°C*	*300°C*		*400°C*
molality		*HNO_3*	*HNO_3*	*HCl*	*HCl + KCl*	*HCl*	*HCl*
0.005	0.025	–4.608	–	–4.034	–3.651	–3.528	–3.660
0.010	0.020	–4.578	–3.970	–3.829	–3.714	– 3.708	–3.353
0.015	0.015	–	–	–	–3.563	–	–
0.020	0.010	–	–	–3.932	–3.514	–3.400	–
0.030	0.010	–	–	–3.906	–3.463	–3.399	–
0.040	–	–	–	–3.842	–	–3.435	–
0.050	–	–4.462	–3.800	–3.889	–	–	–
0.080	–	–	–	–3.929	–	–	–
0.100	–	–4.478	–3.844	–	–	–3.301	–
0.200	–	–4.521	–	–	–	–3.215	–
0.202	–	–4.416	–3.759	–	–	–	–
0.306	–	–	–3.805	–	–	–	–
0.410	–	–4.399	–3.773	–	–	–	–
0.516	–	–	–3.675	–	–	–	–

The equilibrium constants for reaction 4.28 (see Table 4.10, Figure 4.19) were calculated assuming that activity coefficients for equally charged ions cancel:

$$\log K_{28} = \log m_{Sn(OH)_3^+} - \log m_{H^+} - \log a_{H_2O} \qquad (4.29)$$

where $\log m_{H^+} = \frac{1}{2}(\log K_{diss} - \log m_{acid})$, K_{diss} is the dissociation constant of an acid, and m_{acid} refers to the concentration of the undissociated acid portion. The temperature dependence of K_{28} is:

$$\log K_{28} = \frac{-1793.48}{T} - 0.6904. \qquad (4.30)$$

If reaction 4.28 describes cassiterite dissolution in both HCl and HNO_3 solutions, then the following relation should be true:

$$\frac{[m_{Sn\,(OH)_3^+}]_{HCl}}{[m_{H^+}]_{HCl}} = \frac{[m_{Sn\,(OH)_3^+}]_{HNO_3}}{[m_{H^+}]_{HNO_3}}. \qquad (4.31)$$

In fact this is not the case. We infer that dissolution in HCl involves an additional mechanism which enhances solubility, while reaction 4.28 is taken as the major dissolution mechanism for SnO_2 in HNO_3 solutions. For HCl solutions, the following dissolution reaction of SnO_2 may be suggested:

$$SnO_{2\,(c)} + H_2O + HCl = Sn\,(OH)_3Cl_{aq}, \tag{4.32}$$

which accounts for both the slope of the solubility curve and lack of any effect of chloride ion added as KCl.

The constant for reaction 4.32 may be calculated as follows:

$$\log K_{32} = \log m_{Sn\,(OH)_3Cl} - \log \gamma_{st_{HCl}} - \log m_{HCl}, \tag{4.33}$$

where γ_{st} is the stoichiometric activity coefficient of $HCl = \alpha.\gamma^{\pm}$, where $\gamma^{\pm}$ denotes the average ionic activity coefficient calculated from the Debye–Hückel equation, and α stands for the degree of dissociation of HCl (Plyasunov, 1989).

The temperature dependence of K_{32} may be expressed by:

$$\log K_{32} = \frac{-909.619}{T} - 1.9297. \tag{4.34}$$

Based on the above data, the apparent Gibbs free energies of formation of $Sn(OH)_3^+$ and $Sn(OH)_3Cl$ at 1013 bar have been calculated from free energy of reactions 4.28 and 4.32 as follows:

$$\Delta G^{O}_{P(T)} = -RT.\ln K. \tag{4.35}$$

Values for $Sn(OH)_3^+$ are –176.7 and –181.0 at 200 and 300°C, respectively, and –209.3, –215.7 and –223.2 kcal for $Sn(OH)_3Cl$ at 200, 300 and 400°C, respectively.

4.5.5 Conclusions

The solubility of SnO_2 shows some similarity to that of SiO_2. That is, the occurrence of a neutral complex, the same pattern of behaviour under alkaline conditions. There is also a tendency to form aquo complexes with F, which may gain great importance (Klintsova *et al.*, 1975; Jackson and Helgeson, 1985; Barsukov *et al.*, 1989, 1991; Kovalenko *et al.*, 1992). Aside from these similarities between SnO_2 and SiO_2 solubilities, there are a lot of essential differences:

1. The solubility of SnO_2 as $Sn(OH)_4^0$ is several orders of magnitude lower than that of SiO_2 as $H_4SiO_4^0$.
2. The volume changes for dissolution reactions of cassiterite and silica in water are opposite in sign.
3. For Sn(IV), the positively charged complex $Sn(OH)_3^+$ (to which there is no analogue for silica) is believed to play an important role in acidic hydrothermal solutions. The neutral $Sn(OH)_3\,Cl^0$ complex may be important in HCl solutions.

The present study has not considered the effect of redox on SnO_2 solubility, although existing work suggests that Sn is predominantly divalent at f_{O_2} values below the HM buffer (Wilson and Eugster, 1990; Jackson and Helgeson, 1985). This type of behaviour has no analogue for SiO_2. Sn(II) occurs in solution as chloride and mixed hydroxychloride complexes (Wilson and Eguster, 1990).

Thus, unlike SiO_2, tin acts as a typical amphoter under hydrothermal conditions, playing the role of either an acid or a base, depending on the solution acidity (Figure 4.20). The metallic properties of tin are most pronounced when it is divalent, and tin, like many other typical metals, may form sulphide and sulphide–oxide ore deposits. Precipitation of cassiterite from many hydrothermal solutions probably involves oxidation of aqueous Sn(II) to Sn(IV).

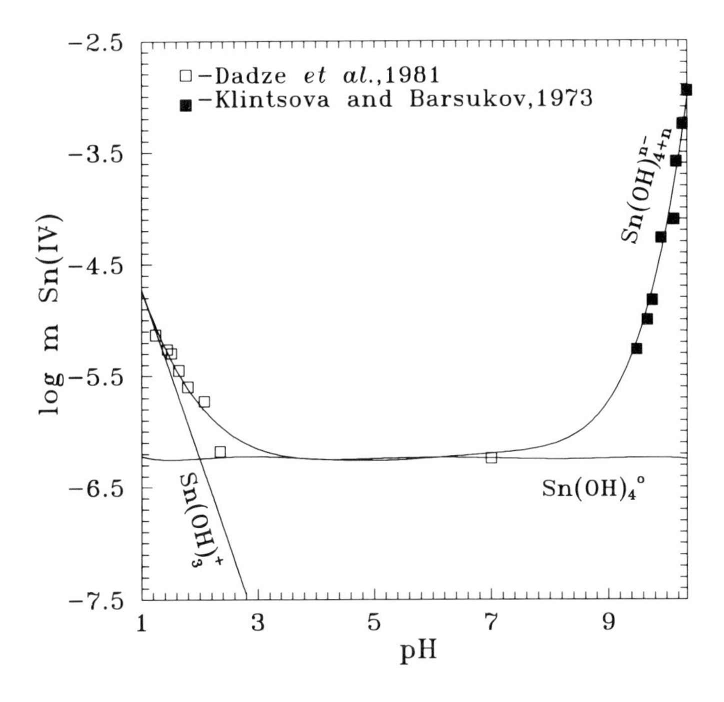

Figure 4.20. The dependence of SnO_2 solubility on the solution pH at T = 200°C.

4.6 SUMMARY

The experimental data presented here on the solubility of four minerals in pure water under hydrothermal conditions provide insight into the specific character of their geochemical behaviour. Mercury and sulphur are markedly distinct in chemical properties, but both are stable in water and have very high vapour pressures at elevated temperatures. As a consequence, they may migrate as Hg^0_{aq} and S^0_{aq}. For Hg, mass transport in neutral media largely occurs by means of elemental species. S is polyvalent, and mass transport thus involves more complex species at elevated P and T. From field observations and experimental data it is known that in addition to S^0_{aq}, sulphite, sulphate and sulphide species may also play an important role in the transport of sulphur. However, even in the simple model system S–H_2O, complex types of interaction are observed, especially below 300°C. Examples are the formation of aquo complexes, polymerization, disproportionation. To 250°C, the dependence of S^0_{aq} on the sulphur vapour pressure, like the dependence of Hg^0_{aq}, is close to Henry's law. At higher temperatures, the concentration of Hg in solutions increases further as its vapour pressure rises, while sulphur exhibits an extreme: its concentration decreases due to competitive interaction mechanisms in solution.

We predict that other substances of high vapour pressure will behave similarly to Hg and S in hydrothermal processes. Figure 4.21 illustrates the vapour pressure of a number of such native elements and minerals that are stable under hydrothermal conditions. In the upper right corner of the diagram Figure. 4.21 is a region, where, according to calculations based on vapour pressures, the corresponding concentrations in saturated aqueous solution may exceed 10^{-5} molal. Among the substances whose vapour pressure lies in this field are mercury, sulphur, arsenic

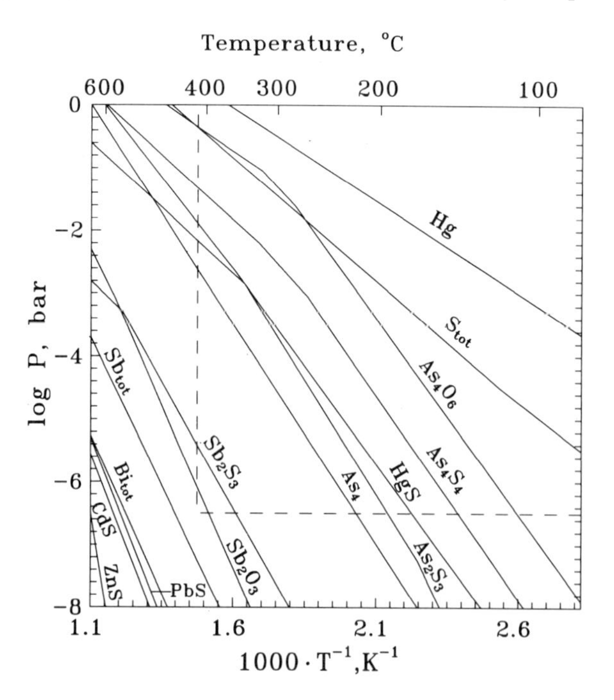

Figure 4.21. Vapour pressure of mineral forms, sulphides, and oxides that display stability in the hydrothermal process.

and some of their compounds. We suggest that the gas mechanism described here for dissolution of S and Hg also applies to the dissolution of As and Sb as As^0_{aq} and Sb^0_{aq}, and their oxides as $As_4O_6.nH_2O$, and $Sb_2O_3.nH_2O$. Similarly, we suggest that solvation of volatile compounds in the presence of hydrogen sulphide will result in the formation of compounds of the type: $As_4S_4.nH_2S$, $HgS.H_2S$, $As_2S_3.nH_2S$, $Sb_2S_3.nH_2S$. This is of particular relevance to low density aqueous fluids. Although there are no experimental data available on hydrated and solvated species in water vapour, field observations show that such transport is possible, e.g. Zotov *et al.* (1977) described high As contents of gases in hot springs from Uzon Caldera, Kamchatka. Existing experimental data (see, for example, Zotov *et al.*, Chapter 5, this volume) on the complexation of As, Hg, and Sb in high-density solutions seem to support this. Thus, the principal As- and Sb-bearing complexes recognized are $H_3AsO^0_3$, $Sb(OH)^0_3$ and their dissociates, $H_{3-n}AsO^{n-}_3$ and $Sb(OH)^{n-}_{3+n}$ (in other words, $As_4O_6.6H_2O$, $Sb_2O_3.3H_2O$, etc. as well as $H_2As_2S_4$, $H_2Sb_2S_4$ and their dissociates, i.e. $As_2S_3.H_2S$ and $Sb_2S_3.H_2S$, $As_2S_3.HS^-$ and $Sb_2S_3.HS^-$, etc.). The solubility values of these compounds correlate with the vapour pressures of the corresponding oxides and sulphides, as illustrated in Figure 4.21.

The other pair of components-analogues, SiO_2 and SnO_2, have similarities in chemical properties and geochemical history, but whereas Si is a typical non-metal and tetravalent only, Sn (and also Ge) shows some properties of both metals and non-metals and may be divalent. Thus, Sn may occur in solution as divalent hydroxy and chloride complexes, comparable to those of Pb (Seward, 1984) or as tetravalent $Sn(OH)^0_{4aq}$ complexes comparable to the dominant complex of silica, $H_4SiO^0_4$. As pressure rises, the solubility of cassiterite as Sn(IV) in hydrothermal solutions decreases, while that of SiO_2 increases. This opposing behaviour might be expected to act against the joint transport and deposition of SnO_2 and SiO_2. The widespread formation of quartz–cassiterite veins in nature demonstrates that the deposition of these minerals is not strongly affected by changes in pressure. Rather, the formation of such veins seems to be primarily the result of drop in temperature and changes in redox leading to the oxidation of Sn(II) to Sn(IV). It is likely that the behaviour of germanium in hydrothermal processes is intermediate between that of Si and Sn, i.e. divalent Ge is less important that divalent Sn.

ACKNOWLEDGEMENTS

The authors are grateful to G.A. Kashirtseva for her invaluable help during the preparation of the paper.

REFERENCES

Aidin'yan, N.Kh. (1969) *Handbook of Mercury Measurement in Natural Objects*, USSR's Academy of Sciences IGEM Press, Moscow (in Russian).

Alexander, G.B., Heston, W.M. and Iler, R.K. (1954) The solubility of amorphous silica in water. *J. Phys. Chem.*, **58**, (6) 453–5.

Alekseyev, V.A., Medvedeva, L.S. and Rafalskii,R.P. (1985) Kinetics of the sulphur–water interaction at elevated temperatures. *Geochem. Int.*, **22** (1).

Anderson, G.M. and Burnham, C.W. (1965) The solubility of quartz in supercritical water. *Amer. J. Sci.*, **263** (6), 494–511.

Anderson, G.M. and Burnham, C.W. (1967) Reactions of quartz and corundum with aqueous chloride and hydroxide solutions and high temperatures and pressures. *Amer. J. Sci.*, **265** (1), 12–27.

Barsukov, V.L., Volosov, A.G., Ryzhenko, B.N., *et al.* (1991) Calculated equilibria in the the Sn–Cl–F–O–H–Na system and the thermodynamic parameters of tin compounds. *Geochem. Int.*, **28** (7), 1–11.

Busey, K.H. and Mesmer, R.E. (1977) Ionisation equilibria of silicic acid and polysilicate formation in aqueous sodium chloride solutions to 300°C. *Inorg. Chem.*, **16** (10), 2444–50.

Choi, S.S. and Tuck, D.G. (1962) A neutron activation study of the solubility of mercury in water. *J. Chem. Soc.*, **797**, 4080–8.

Clever H.L. (ed.) (1987) *Mercury in Liquids, Compressed Gases, Molten Salts and Other Elements*, Pergamon Press, New York.

Dadze, T.P., Sorokin, V.I. and Nekrasov, I.Ya. (1981) Solubility of SnO_2 in water and in aqueous solutions of HCl, HCl + KCl, and HNO_3 at 200–400°C and 1013 kbar. *Geochem. Int.* **18** (5), 142–52.

Dadze, T.P. and Sorokin, V.I. (1986) Solubility of SnO_2 in water at 200–400°C and 16–1500 bar. *Doklady Akademii Nauk SSSR*, **286** (2), 426–8 (in Russian).

Dadze, T.P and Sorokin, V.I. (1993) Experimental determination of concentrations of H_2S, HSO_4^-, SO_{2aq}, $H_2S_2O_3$, S^0_{aq} and S_{tot} in the aqueous phase of the system S–H_2O at elevated temperatures. *Geokhimia*, **1**, 38–53 (in Russian).

Drummond S.E. (1981) Boiling and mixing of hydrothermal fluids: chemical effects on mineral precipitation. Ph. D. thesis, State Univer., Pennsylvania.

Egorov, V.M. and Ikornikova, N.Yu. (1973) Partial molar volumes of aqueous chloride solutions at high temperatures and high pressures. *Zapiski Vsesoyuznogo Mineralogicheskogo Obshchestva*. **102** (3), 272–81 (in Russian).

Ellis, A.J. and Giggenbach, W. (1971) Hydrogen sulphide ionization and sulphur hydrolysis in high temperature solution. *Geochim. Cosmochim. Acta*, **35** (3), 247–60.

Elmer, T.H. and Nordberg, M.E. (1958) Solubility of silica in nitric acid solutions. *J. Amer. Ceram. Soc.*, **41** (12), 517–20.

Fournier, R.O. and Rowe, I.J. (1977) The solubility of amorphous silica in water at high temperatures and high pressures. *Am. Miner.*, **62** (9–10), 1052–6.

Glew D.N. and Hames D.A. (1971) Aqueous nonelectrolyte solutions. Part X. Mercury solubility in water. *Canad. J. Chem.*, **49** (19), 3114–18.

Gushchina, L.V., Belevantsev, V.I. and Obolenskiy, A.A. (1989) A high-temperature spectrophotometric study of Hg_{liq} solubility in water. *Geokhimia*, **2**, 274–81 (in Russian).

Helgeson, H.C. and Kirkham, D.H. (1974) Theoretical prediction of the thermodynamic behavior of aqueous electrolytes at high pressures and temperatures II. Debye–Hückel parameters for activity coefficients and relative partial molar properties. *Amer. J. Sci.*, **274** (10), 1089–1261.

Helgeson, H.C. Kirkham, D.H. and Flowers G.C. (1981) Theoretical prediction of the thermodynamic behavior of aqueous electrolytes at high pressures and temperatures: IV Calculation of activity coefficients, osmotic coefficients and apparent molal and standard and relative partial molal properties to 600°C and 5 kb. *Amer. J. Sci.*, **281** (10), 1249–1493

IVTANTERMO (1984) *Tables of Thermodynamic Properties of Individual Substances.* USSR Academy of Sciences Institute for High Temperatures.

Jackson, K.J. and Helgeson, H.C. (1985) Chemical and thermodynamic constraints on the hydrothermal transport and deposition of tin: I. Calculation of the solubility of cassiterite at high pressures and temperatures. *Geochim. Cosmochim. Acta*, **49** (1), 1–22.

Johnson, J.W., Oelkers, E.H. and Helgeson, H.C. (1992) SUPCRT92: A Software Package for Calculating the Standard Molal Thermodynamic Properties of Minerals, Gases, Aqueous Species and Reactions from 1–5000 bar and 0–1000°C. Computers Geoscience, **18**, 889–947.

Jones, M.M. and Pytkowicz, R.M. (1977) Solubility of silica in seawater at high pressures. *Bull. Soc. Roy. Sci. Liege*, **42**, 118–20.

Kennedy, G.C. (1950) Pressure–volume–temperature relations in water at elevated temperatures and pressures. *Amer. J. Sci.*, **248** (3), 540–64.

Khodakovsky, I.L., Popova, M.Ya. and Ozerova, N.A. (1977) Transport forms of mercury in hydrothermal solutions, in *Geochemistry of Migration Processes of Ore Elements.* Nauka Press, Moscow, pp. 86–118 (in Russian).

Kishima, N. (1989) A thermodynamic study on the pyrite–pyrrhotite–magnetite–water system at 300–500°C with relevance to the fugacity: concentration quotient of aqueous H_2S. *Geochim. Cosmochim. Acta*, **53**, (9) 2143–55.

Kitahara, S. (1960) The solubility of quartz in the aqueous sodium chloride solution at high temperatures and high pressures. *Rev. Phys. Chem. Japan*, **30** (2), 115–21.

Klintsova, A.P and Barsukov, V.L. (1970) On cassiterite solubility in water and aqueous NaOH solutions at 25°C. *Geokhimia*, **10**, 1268–71 (in Russian).

Klintsova, A.P. and Barsukov, V.L., (1973) Solubility of cassiterite in water and in aqueous NaOH solutions at elevated temperatures. *Geochem. Int.*, **10** (5), 540–7.

Klintsova, A.P., Barsukov, V.L., Shemarykina, T.P., *et al.* (1975) Measurement of the stability constants for Sn(IV) hydroxofluoride complexes. *Geochem. Int.*, **12** (2) 207–15.

Kovalenko, N.I., Ryzhenko, B.N., Barsukov, V.L., *et al.* (1986) The solubility of cassiterite in in HCl and HCl + NaCl (KCl) solutions at 500°C and 1000 atm under fixed redox conditions. *Geochem. Int.*, **23** (7), 1–16.

Kovalenko, N.I., Ryzhenko, B.N., Dorofeyeva, V.A., *et al.* (1992) The stability of $Sn(OH)_4^{2-}$, $Sn(OH)_2F^-$ and $Sn(OH)_2Cl^-$ at 500°C and 1 kbar. *Geochem. Int.*, **29** (8), 84–94.

Kozintseva, T.N. (1964) Solubility of hydrogen sulfide in water at elevated temperatures. *Geochem. Int.*, **1** (4) 750–6.

Laptev, Yu.V., Sirkis, A.A. and Kolonin, G.R. (1987) *Sulphur and Sulphide Formation in Hydrometallurgic Process*, Nauka Press, Novosibirsk (in Russian).

Naumov, G.B., Ryzhenko, B.N. and Khodakovskiy, I.L. (1971) *Handbook of Thermodynamic Data*, U.S.Geol.Surv. Transl. USGS-WRD-74-001.

Ohmoto, H. and Lasaga, A.C. (1982) Kinetics of reactions between aqueous sulfates and sulfides in hydrothermal systems. *Geochim. Cosmochim. Acta*, **46**, 1727–45.

Plyasunov, A.V. (1989) Experimental and theoretical investigation of zinc oxide solubility in alkaline and chloride solutions. Ph.D. Thesis, Vernadsky Institute for Geochemistry and Analitical Chemistry, Moskow (in Russian).

Rafalskii, R.P., Medvedeva, L.S., Prisyagina, N.I., *et al.* (1983) Sulfur–water interaction at elevated temperatures. *Geokhimia*, **5**, 665–76 (in Russian).

Reichardt, H. and Bonhoefer, K.F. (1931) Uber das Absorptions spektrum von gelostem Quecksilber. Ztschr. Phys., **67** (11/12), 780–89.

Robinson, B.W. (1973) Sulphur isotope equilibrium during sulphur hydrolysis at high temperatures. *Earth Planet. Sci. Lett.*, **18**, 443–50.

Sadek, H. (1952) On compound formation between silicic and hydrochloric acid. *J. Ind. Chem. Soc.*, **29** (7), 507–10.

Sanemasa, I. (1975) The solubility of elemental mercury vapour in water. *Bull. Chem. Soc. Japan*, **48** (6), 1795–8.

Sorokin, V.I. (1973) Solubility of mercury in water over the temperature–pressure range 300–500°C and 500–1000 atm. *Doklady Akademii Nauk SSSR*, **213** (4), 852–5 (in Russian).

Sorokin, V.I. and Gruzdev, V.S. (1975) Mercury solubility in water over the temperature–pressure range 300–500°C and 500–1000 atm, and the problem of transport of some metals as vapours of elements under hydrothermal conditions, in *Proceedings of the 4th All-Union Meeting on Experimental and Technical Mineralogy and Petrography*, Nauka Press, Moscow, pp. 199–203. (in Russian).

Sorokin, V.I., Alekhin, Yu.V. and Dadze, T.P. (1978) Solubility of mercury in the systems Hg–H_2O, HgS–(Cl)–H_2O and forms of its existence in sulphide–forming thermal waters at Kamchatka and Kunashir Island, in *Contributions to Physico–chemical Petrology*, **VIII**, (ed. V.A. Zharikov), Nauka Press, Moscow, 133–49. (in Russian).

Sorokin, V.I. and Dadze, T.P. (1980) Solubility of amorphous SiO_2 in water and aqueous HCl and HNO_3 solutions at 100–400°C and 101.3 MPa. *Doklady Akademii Nauk SSSR*, **254** (3), 735–9 (in Russian).

Sorokin, V.I., Dadze, T.P. and Osadchii, Eu.G. (1986) Geochemistry of antimony–arsenic–mercury mineralization, in *Experiment in the Solution to the Burning Problems of Geology* (eds V.A. Zharikov and V.V. Fed'kin), Nauka Press, Moscow 368–86. (in Russian).

Sorokin, V.I., Pokrovskii, V.A. and Dadze, T.P. (1988) *The Physical-chemical Conditions of Formation of Antimony–Mercury Deposits*, (ed. V.I.Smirnov), Nauka Press, Moscow (in Russian).

Stock A., Cucuel F., Gerstner F., *et al.* (1934) Uber Verdampfung, Loslichkeit und Oxydation des metallischen Quecksilber. *Ztschr. Anorg. und Allg. Chem.* **217**, (3), 241.

Varekamp, I.C. and Buseck, P.R. (1984) The speciation of mercury in hydrothermal systems, with applications to ore deposits. *Geochim. Cosmochim. Acta*, **48** (1), 177–86.

Vukalovich, M.P., Ivanov, A.I., Fokin, P.R., *et al.* (1971) *Thermodynamic Properties of Mercury*, Izdatelstvo Standartov, Moscow (in Russian).

Walther, J.V. and Helgeson, H.C. (1977) Calculation of the thermodynamic properties of aqueous silica and the solubility of quartz and its polymorphs at high pressures and temperatures. *Amer. J. Sci.*, **277** (10), 1315–51.

West, J.R. (1950) Thermodynamic properties of sulphur. *Industrial and Engineering Chemistry*, **42** (4), 713–18.

Wildly, J.D. (1974) The effect of pressure on the solubility of amorphous silica in seawater at 0°C. *Marine Chem.*, **2**, 239–50.

Wilson, G.A. and Eugster, H.P. (1990) Cassiterite solubility and tin speciation in supercritical chloride solutions. *The Geochem. Society, Special Publication*, **2**, 179–95.

Zotov, A.V., Volchenkova, V.A., Kotova, Z.Yu., *et al.* (1977) Physico-chemical conditions of present-day formation of arsenic sulphides in Uzon Caldera, Kamchatka, in *Present-day Hot Springs and Mineral Formation*, Nauka Press, Moscow, pp. 77–103 (in Russian).

GLOSSARY OF SYMBOLS

A, B, å, b_γ	parameters in the Debye–Hückel equation
f	fugacity
f^0	fugacity on the saturation curve
$\Delta G^0_{f(T)}$	apparent molal Gibbs free energy of formation
ΔG^0_n	apparent molal Gibbs free energy of reaction n
$\Delta H^0_{f(T)}$	apparent molal enthalpy of formation
K_{diss}	dissociation constant
K^0_H	Henry's law constant
K_n	equilibrium constant for reaction n
m	molality
P	pressure in bars or kilobars
R	gas constant (1.98717 cal . $mol^{-1}.K^{-1}$; 82.0556 cm^3.atm. $mol^{-1}.K^{-1}$
sat	designation of liquid–vapour equilibrium for aqueous system
T	temperature in **K**
V^0	standard molal volume
ΔV^0	molal volume of reaction
α	degree of dissociation of acid
γ_{st}	stoichiometric activity coefficient of electrolyte
$\gamma_{\mp}$	mean ionic activity coefficient of electrolyte

CHAPTER FIVE

Experimental studies of the solubility and complexing of selected ore elements (Au, Ag, Cu, Mo, As, Sb, Hg) in aqueous solutions

Alexander V. Zotov, Aleksey V. Kudrin, Konstantin A. Levin, Nadezhda D. Shikina and Lidia N. Var'yash

5.1 INTRODUCTION

The importance of complexing in controlling the solubility, transport and precipitation of ore metals in aqueous solutions is a recurrent theme of this book. In this chapter experimental results obtained at the Institute of the Geology of Ore Deposits, Petrography, Mineralogy and Geochemistry of the Russian Academy of Sciences, Moscow during the past ten years are reviewed. The results presented here, on the solubility and complexing of a disparate range of metals in aqueous solutions, are linked together by a common methodological approach: the investigation, where possible, of simple systems to determine equilibrium constants for specific complex formation. The scope of this paper is unashamedly idiosyncratic. Some elements have received detailed and extensive study, others only limited investigation. However, the authors attempt to describe the results of many years work to characterize some

Fluids in the Crust: Equilibrium and transport properties.
Edited by K.I. Shmulovich, B.W.D. Yardley and G.G. Gonchar.
Published in 1994 by Chapman & Hall, London. ISBN 0 412 56320 7

of the major features of ore metal complexation under hydrothermal conditions.

5.1.1 Experimental procedures and data retrieval

Practically all the experiments have been based on solubility measurements. Below 100 °C, experiments were generally conducted in sealed glass capsules, whereas at higher temperature, Ti-alloy autoclaves were used. Below 300 °C, the autoclaves were fitted with Teflon inserts. Pressure was controlled by the degree of filling of the autoclave (Chapter 3). Solutions were sampled after quench, at which point solid–liquid separation occurred. For some elements (e.g. Au, Cu, Mo, Sb) redox conditions were controlled, and this was done in one of two ways: conventional solid-media buffers or by adding a weighed amount of metallic aluminium to evolve hydrogen into the system. Aluminium was loaded directly into the investigated solution. Non-ideality of the resulting H_2–H_2O mixture was allowed for using the procedure of Kishima and Sakai (1984).

Further details of the experimental techniques for specific systems are given in the original publications, to which references are given. Where primary experimental results are not included in this chapter, they can likewise be found there.

Calculation of equilibrium constants from the experimental results generally utilized an extended Debye–Hückel equation:

$$\log K = \log K^* + \frac{\Delta z^2 \cdot A \cdot I^{0.5}}{1 + \mathring{a} \cdot B \cdot I^{0.5}},$$

where K is the equilibrium constant, i.e. an activity product, and K* is the corresponding concentration product. I is the ionic strength; Δz^2 is the algebraic sum of ionic charges squared; A and B are characteristic parameters of the solvent, and are functions of both temperature and pressure; and $\mathring{a}$ is empirical parameter, normally set equal to 4.5 Å for all species at all temperatures. Activity coefficients for neutral species were set equal to unity; the confidence level was taken as 95% throughout this study.

5.2 GOLD

Not surprisingly, the solubility of gold in hydrothermal fluids has been studied more intensively than many other metals. Existing studies have included work on both chloride and bisulphide complexing (Seward, 1973; Nikolaeva *et al.*, 1972; Henley, 1973; Renders and Seward, 1989; Shenberger and Barnes, 1989; Hayashi and Ohmoto, 1991). For this study, the solubility of gold in both pure water and aqueous chloride solutions was measured at 300–500°C and 0.5–1.5 kbar. As a result, thermodynamic properties of the species $Au(OH)^0_{(aq)}$ and $AuCl_2^-$, have been retrieved over a wide range of temperatures and pressures. Original results and further details are given in Zotov *et al.* (1985, 1989, 1991) and Zotov and Baranova (1989).

5.2.1 The solubility of gold in water

At the start of this project, there were little reliable data on gold solubility in water at elevated temperatures under controlled redox conditions. Therefore, the solubility of gold in pure H_2O was first studied as a function of oxygen fugacity at 450°C and 500 bar (Baranova *et al.*, 1983).

The results (Figure 5.1) show a positive correlation between $\log m_{Au}$ and $\log f_{O_2}$ with a slope of 0.22 ± 0.04. This correlation indicates that gold goes into solution as the $Au(OH)^0_{(aq)}$ complex, formed according to the reaction:

$$Au_{(aq)} + 0.5\,H_2O + 0.25\,O_{2(g)} = Au(OH)^0_{(aq)}. \qquad (5.1)$$

Within the entire range of f_{O_2} under investigation ($\log f_{O_2}$ from -29 to -0.8), gold is chiefly monovalent, i.e. Au(I).

The equilibrium constant for reaction 5.1 at 300–500°C and 0.5–1.5 kbar (Zotov *et al.*, 1985) is independent of pressure within experimental uncertainty, and shows a dramatic increase with temperature (Table 5.1):

$$\log K_{(5.1)} = \frac{(3620 \pm 510)}{T} - (6.49 \pm 0.70),$$

where T is in Kelvins.

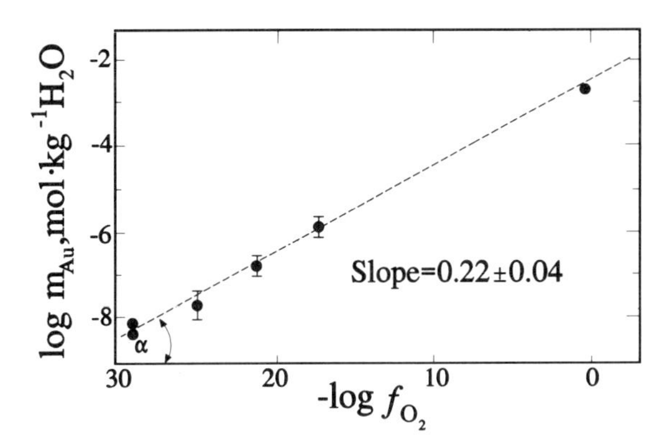

Figure 5.1. The solubility of $Au_{(c)}$ in water as a function of oxygen fugacity at 450°C and 500 bar.

Table 5.1. The solubility of $Au_{(c)}$ in water from 300 to 500 °C and 500 to 1500 bar

T°C	P (bar)	*Buffer*	$-\log f_{O_2}$	*Duration (days)*	*Runs*	$-\log m_{Au}$	$-\log K_{(5.1)}$
300	500	MH	30.6	45–60	4	7.42 ± 0.30	− 0.23 ± 0.30
		Cu–Cu_2O	23.5	60	3	6.75 ± 0.30	0.88 ± 0.30
							0.3 ± 0.5
400	500	NNO	27.7	27	3	7.91 ± 0.20	0.98 ± 0.20
		MH	23.9	27–32	3	7.32 ± 0.50	1.34 ± 0.50
							1.1 ± 0.2
450	500	NNO	25.1	8–45	7	7.71 ± 0.25	1.44 ± 0.25
		MH	21.3	7–45	9	6.79 ± 0.20	1.46 ± 0.20
		Cu–Cu_2O	17.3	6–33	3	5.89 ± 0.10	1.56 ± 0.10
		charges of Al and Fe	29.2	15	2	8.26 ± 0.04	0.96
							1.5 ± 0.2
	1000	NNO	25.1	21	3	8.13 ± 0.50	1.86 ± 0.50
		MH	21.3	22	3	6.46 ± 0.30	1.14 ± 0.30
							1.5 ± 0.4
	1500	NNO	25.2	21	3	7.65 ± 0.30	1.38 ± 0.30
		MH	21.3	22	3	7.36 ± 0.30	2.04 ± 0.30
							1.7 ± 0.3
500	500	NNO	22.9	13	3	7.39 ± 0.25	1.66 ± 0.25
		MH	19.0	13–14	5	6.72 ± 0.38	1.97 ± 0.38
							1.8 ± 0.2

5.2.2 The solubility of gold in chloride solutions

Existing experimental data on the behaviour of gold in a chloride medium at elevated temperatures (e.g. Henley, 1973; Wood *et al.*, 1987) cover only a limited range of conditions and are in poor agreement with the low temperature (25–80 °C) potentiometric measurements (Nikolaeva *et al.*, 1972).

In order to identify the dominant type of gold chloride complex, gold solubility was studied as a function of acidity, mCl^- and f_{O_2}, at 450 °C and 400 to 1000 bar (Zotov *et al.*, 1989; Zotov and Baranova, 1989).

The solubility of gold in the system H_2O–HCl (0.0006–0.06 m) was found to be independent of acidity and to be governed by the complex $Au(OH)^0$, despite the acidity of the fluid. The retrieved value of $\log K_{(5.1)} = -1.2 \pm 0.4$ fits the solubility data for water given in the previous section. However, the addition of up to 3.31 m KCl or 1.02 m NaCl into the system causes gold solubility to increase directly with the concentration of $HCl^0_{(aq)}$ (Figure 5.2) and Cl^- (Figure 5.3). The observed trends are consistent with the formation of the complex $AuCl_2^-$:

$$Au_{(c)} + HCl^0_{(aq)} + Cl^- + 0.25\,O_{2(g)} = AuCl_2^- + 0.5\,H_2O. \quad (5.2)$$

The calculated expression for $K_{(5.2)}$ (Table 5.2) used dissociation constants for HCl from Frantz and Marshall (1984) and for KCl and NaCl from the SUPCRT92 program (Johnson *et al.*, 1992).

Table 5.2. The solubility of gold in chloride solutions at 350–500°C and 500–1500 bar

m (mol. kg H_2O^{-1})			[*Duration* (days)]	[*Runs*]	$-\log m_{AuCl_2^-}$	$\log K_{(5.2)}$
HCl	KCl	NaCl				
			350°C, 1000 bar, $\log f_{O_2} = -35.3$			
0.1	0.1	–	20	5	7.50 ± 0.28	3.73 ± 0.28
0.1	0.2	–	20	3	7.30 ± 0.25	3.67 ± 0.25
0.2	0.1	–	20–21	4	7.29 ± 0.30	3.42 ± 0.30
0.2	0.3	–	21	2	6.97 ± 0.20	3.44 ± 0.20
						3.60 ± 0.15
			450°C, 500 bar, $\log f_{O_2} = -28.1$			
0.1	0.1	–	11	4	6.82 ± 0.20	2.80 ± 0.20
0.1	0.1	–	10–16	6	6.74 ± 0.10	2.87 ± 0.10
0.1	0.2	–	16	3	6.51 ± 0.08	2.84 ± 0.08
						2.84 ± 0.15
			450°C, 1000 bar, $\log f_{O_2} = -28.4$			
0.1	0.1	–	9–11	8	6.65 ± 0.10	2.60 ± 0.10
0.006	0.1	–	21	2	7.60 ± 0.40	2.94 ± 0.40
0.0064	2.13	–	11	2	6.98 ± 0.20	2.36 ± 0.20
0.1	–	0.1	6–11	6	6.75 ± 0.11	2.57 ± 0.11
0.1	–	0.51	11	2	6.05 ± 0.10	2.72 ± 0.10
0.1	–	1.02	11	4	5.74 ± 0.15	2.79 ± 0.15
						2.67 ± 0.15
			450°C, 1500 bar, $\log f_{O_2} = -28.6$			
0.1	0.1	–	9	4	6.78 ± 0.18	2.47 ± 0.18
			500°C, 1000 bar, $\log f_{O_2} = -25.7$			
0.1	0.1	–	8	3	6.40 ± 0.10	2.32 ± 0.10
0.1	0.2	–	8	3	6.05 ± 0.10	2.44 ± 0.10
						2.38 ± 0.09

Experimental values for $K_{(5.2)}$ at 300 to 500°C (Zotov and Baranova, 1989; Zotov *et al.*, 1991) and the low-temperature data of Nikolaeva *et al.* (1972) have been treated together using the Helgeson, Kirkham and Flowers (HKF) model (Johnson *et al.*, 1992) and a set of correlations between thermodynamic ionic properties (Shock and Helgeson, 1988). A comparison between the calculated values and the experimental data for $K_{(5.2)}$ is shown in Figure 5.4. The state variables for the aqueous complex $AuCl_2^-$, retrieved with the HKF model, are listed in Table 5.3.

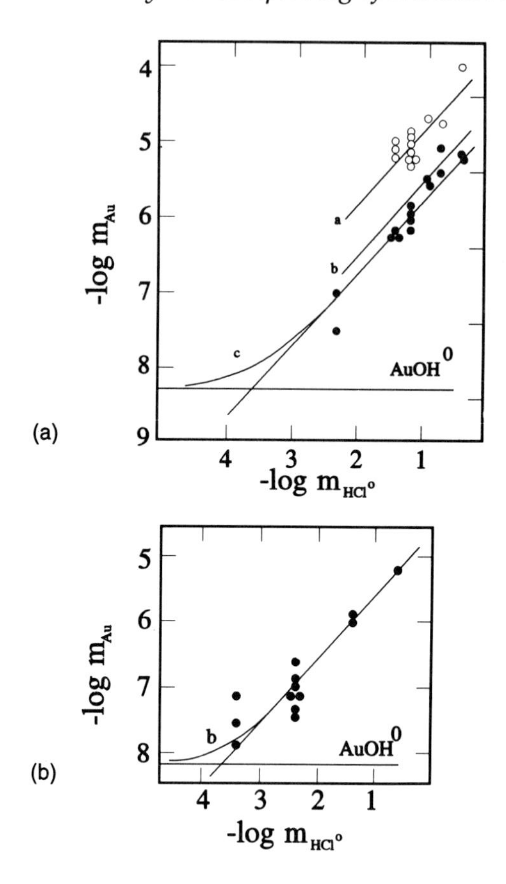

Figure 5.2. The solubility of gold in (0.00064–0.64)m HCl + 2.13m KCl solutions at 450°C as a function of HCl^0_{aq} concentration (Cl^- concentration is constant). Curve a: log f_{O_2} = – 26.1; curve b: log f_{O2} = –28.1; curve c: log f_{O2} = – 29.1. Part (**a**) shows experimental data points for curves a,c, part (**b**) shows data points for curve b. The filling factor of the autoclave is equal to 0.39; P is about 0.4 kbar.

The thermodynamic properties reported here for the complex $AuCl_2^-$ are significantly different from those available in the literature. At 250–350°C, the stability of $AuCl_2^-$ is 2–3 orders of magnitude lower than was proposed by Henley (1973) and Wood *et al.* (1987) (Figure 5.4). However compared to the data base of Turnbull and Wadsley (1988) (cited after Morrison *et al.*, 1991), its stability is 0.5–1.5 orders of magnitude higher.

In summary, these experiments demonstrate that in the absence of significant sulphide, the content of gold in hydrothermal solutions in the temperature range 350 to 450°C may nevertheless be as high as 10^{-5} to 10^{-7} molal, in the form of the $AuCl_2^-$ complex. At lower temperature, a gold concentration of about 10^{-8} molal is provided by the hydroxy complex $Au(OH)^0_{(aq)}$.

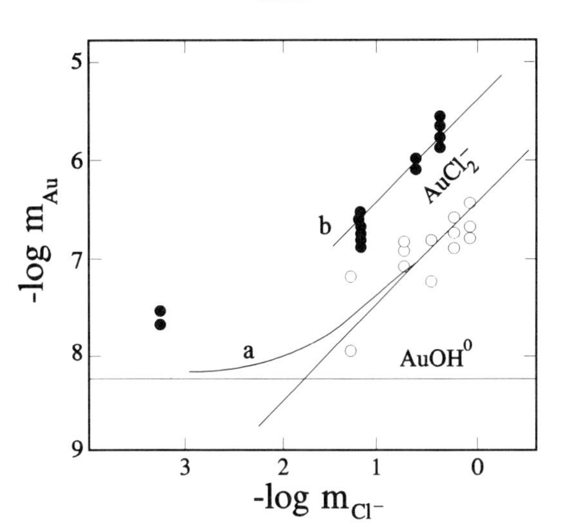

Figure 5.3. The solubility of gold as a function of Cl^- concentration at 450°C and constant $HCl^0_{(aq)}$ concentration. **a** In (0.006–0.0066)m HCl + (0–3.31)m KCl solutions, $\log f_{O_2} = -28$, the filling factor of the autoclave is equal to 0.61, P = 0.4–1 kbar. **b** In 0.1m HCl + (0.1–1.02)m NaCl, $\log f_{O_2} = -28.4$, P = 1 kbar.

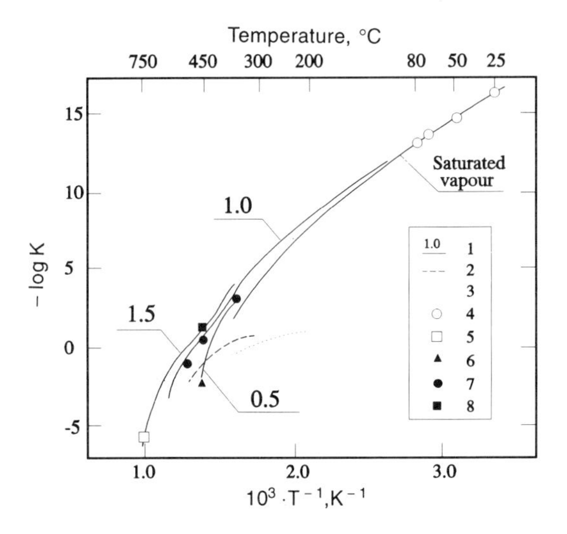

Figure 5.4. The variation of the equilibrium constant, log K, for reaction $Au_{(c)} + 2\,Cl^- + H^+ + 0.5\,NiO_{(c)} = AuCl_2^- + 0.5\,H_2O + 0.5\,Ni_{(c)}$ as a function of temperature and pressure . 1 – value calculated using a HKF model according to experimental data from this study and Nikolaeva *et al.* (1972) under the corresponding pressure in kbar. 2–8 – values calculated using experimental data: 2 Henley (1973), 1 kbar; 3 Wood *et al.* (1987), saturated vapour pressure; 4 Nikolaeva *et al.* (1972), saturated vapour pressure; 5 Ryabchikov and Orlova (1964), 1.5 kbar; 6 this study, 0.5 kbar; 7 *ibid*, 1.0 kbar; 8 *ibid*, 1.5 kbar.

5.3 SILVER

Although chloride complexes of silver are well known (e.g. Berne and Leden, 1953; Ivanenko and Pamfilova, 1975; Seward, 1976), their dissociation constants are not well known for many conditions of interest, and there are few data on the effect of pressure on their stability at supercritical conditions. A series of experiments on the solubility of $AgCl_{(c)}$ (Cerargyrite) in chloride solutions were conducted at temperatures in the range 150–450 °C, from the saturated vapour pressure to 1500 bar. In particular, a detailed study was carried out covering the compositional range from pure water to 7 m NaCl (Figures 5.5–5.7) at 300 °C and saturated vapour pressure (Zotov *et al.*, 1986a, 1986b; Levin, 1992).

The lack of data at low concentrations on the solubility curve (Figure 5.5) precludes distinction between the contributions from Ag^+ and $AgCl^0_{(aq)}$ to the total dissolved silver. The same constraint applies to data from the literature, in particular the 277 and 353 °C isotherms of Seward (1976). To overcome this difficulty, Zotov *et al.* (1986a) measured the solubility of $AgCl_{(c)}$ in a set of $NaClO_4$ solutions, ranging from 0.01 to 1 m, in order to make use of the fact that the equilibrium constants for the reactions

$$AgCl_{(c)} = Ag^+ + Cl^- \tag{5.3}$$

and

$$AgCl_{(c)} = AgCl^0_{(aq)} \tag{5.4}$$

show a different dependence on the ionic strength.

Using the Debye–Hückel equation (see Introduction) for description of the above dependence gives (Figure 5.6):

$$m_{Ag} = m_{Ag^+} + m_{AgCl^0} = (K_{(5.3)})^{0.5} \cdot 10^{\sqrt{\frac{\Delta z^2 \cdot A \cdot I^{0.5}}{1 + \mathring{a} \cdot B \cdot I^{0.5}}}} + m_{AgCl^0}.$$

The derived value $\log K_{(5.4)} = -3.02 \pm 0.20$ is close to Seward (1976), while the value of $\log K_{(5.3)} = -5.57 \pm 0.12$ is more than half an order of magnitude higher than Seward's (1976) estimate. Apart from the different way of separating fractions of species, this discrepancy could also be attributed to the shorter run times (several hours) in Seward's work because the longer-term experiments yielded a somewhat higher solubility.

At moderate NaCl concentrations, the solubility curve (Figure 5.5) reflects the stoichiometry of the complex $AgCl_2^-$:

$$AgCl_{(c)} + Cl^- = AgCl_2^-. \tag{5.5}$$

The value of $\log K_{(5.5)} = -0.75 \pm 0.03$ agrees with interpolation of Seward's (1976) data ($\log K_{(5.5)} \approx -1.1$ at 300 °C). At higher Cl^- concentrations, however, our interpretations of data differ. Seward inferred the existence of two complexes, $AgCl_3^{2-}$ and $AgCl_4^{3-}$, whereas we were able to

Table 5.3. Thermodynamic properties of some complexes of ore elements in aqueous solution

Species	$\Delta\overline{G}^0_{f,298}$ a)	$\overline{S}^0_{298}$ b)	$\overline{C}^0_{p,298}$ b)	$\overline{V}^0_{298}$ c)	$a_1 \cdot 10$ d)	$a_2 \cdot 10^{-2}$ a)	a_3 e)	$a_4 \cdot 10^{-4}$ f)	c_1 b)	$c_2 \cdot 10^{-4}$ f)	$\omega \cdot 10^5$ a)
$H_3AsO_3^0$	– 153.004	47.8	29.1	434.4	6.3337	7.6759	2.7294	– 3.0966	19.4362	2.8932	– 0.4111
$H_2AsO_3^-$	– 140.330	26.4	– 2.9	26.4	5.7935	6.3645	3.2486	– 3.0421	15.8033	– 3.6252	1.2304
$As_2S_3^0$	– 13.066	24.3	34.6	28.5	5.6549	6.0261	3.3810	– 3.0282	26.1711	4.0134	– 0.0301
$HAs_2S_4^-$	– 16.995	52.0	30.9	52.0	9.1539	14.6033	0.0115	– 3.3939	32.0507	3.2744	0.8413
$As_2S_4^{2-}$	– 5.712	53.0	– 28.1	66.4	11.6635	20.7266	– 2.3901	– 3.6329	11.8786	– 8.7476	2.4140
$Sb(OH)_3^0$	– 153.890	47.0	39.5	42.5	7.4932	10.5149	1.6163	– 3.2137	27.2207	5.0115	– 0.2300
$Sb_2(OH)_6^0$	– 307.885	104.1	79.0	84.7	14.9864	21.0298	3.2326	– 6.5462	54.4414	10.0230	– 0.4600
$Sb(OH)_4^-$	– 194.320	48.0	1.0	48.0	8.6736	13.3970	0.4833	– 3.3328	14.8793	– 2.8390	0.8946
$H_2Sb_2S_4^0$	– 33.800	82.7	– 11.6	74.2	11.5715	20.4729	– 2.2984	– 3.6254	– 8.9440	– 5.3975	– 0.9124
$HSb_2S_4^-$	– 27.350	63.3	– 82.7	63.3	10.7077	18.3638	– 1.4692	– 3.5382	– 35.6742	– 19.8806	0.7274
$Sb_2S_4^{2-}$	– 14.000	46.0	– 51.8	47.3	8.9334	14.0315	0.2339	– 3.3591	– 7.6536	– 13.5863	1.8153
$AuCl_2^-$	– 36.760	48.2	– 19.6	53.4	11.4088	9.9876	16.7245	– 8.4889	12.5752	– 14.3404	0.3364
$AgCl_2^-$	– 51.672	48.3	– 21.3	39.6	9.7925	6.0412	18.2756	– 8.3258	14.2458	– 12.8534	1.0384
$CuCl_2^-$	– 57.977	47.6	– 10.9	52.9	11.2770	9.6700	16.8490	– 8.4758	18.8345	– 13.8520	0.1703
Cu^+	11.950	20.0	3.0	9.9	3.2147	0.0680	5.7231	– 2.7818	10.1430	– 2.4235	0.2411

a) $kcal \cdot mol^{-1}$; b) $cal \cdot (mol \cdot K)^{-1}$; c) $cm^3 \cdot mol^{-1}$; d) $cal \cdot (mol \cdot bar)^{-1}$; e) $cal \cdot K \cdot (mol \cdot bar)^{-1}$; f) $cal \cdot K \cdot mol^{-1}$.

Table 5.3a. Thermodynamic properties of some solid phases from IVTANTERMO (1990)

Species	$\Delta G^0_{f,298}$ kcal · mol^{-1}	$\Delta H^0_{f,298}$ kcal · mol^{-1}	S^0_{298} cal · (mol.K)$^{-1}$	V^0_{298} cm^3 · mol^{-1}	$C^0_p = a + b \cdot T - c/T^2 + d \cdot T^2$			
					a	$b \cdot 10^3$	$c \cdot 10^{-5}$	$d \cdot 10^6$
As_2O_3 mon.	− 138.009	− 156.286	29.29	–	42.653	− 27.051	8.650	18.309
AS_2O_3 cub.	− 137.837	− 157.194	25.66	51.32	6.663	75.048	0.0	− 66.181
As_2S_3 orp.	− 21.866	− 22.156	39.05	70.52	42.761	− 52.625	− 4.310	60.286
As_4S_4 real.	− 31.721	− 33.533	58.68	119.20	51.766	− 0.053	5.581	0.0
Sb_2O_3 romb.	− 148.193	− 166.969	32.12	50.01	30.010	− 1.955	3.047	8.442
Sb_2O_3 cub.	− 149.967	− 169.347	30.09	52.21	28.261	− 1.955	3.047	8.442
Sb_2S_3 antim.	− 34.156	− 34.656	42.65	73.41	23.889	13.253	− 0.719	0.0

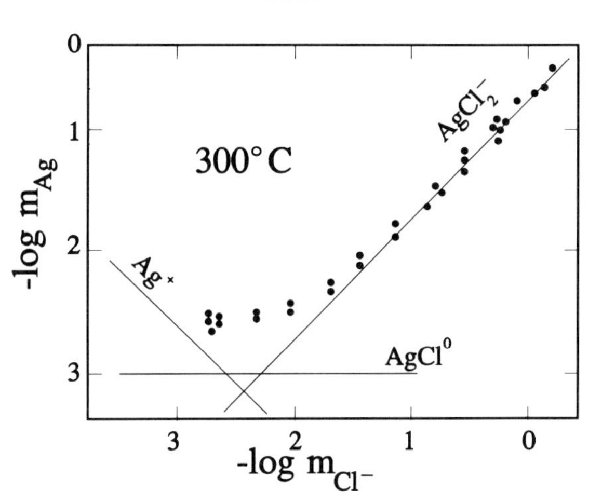

Figure 5.5. The solubility of $AgCl_{(c)}$ in water and NaCl solutions as a function of Cl^- concentration at 300°C and saturated vapour pressure.

show (Zotov *et al.*, 1986b; Levin, 1992) that the observed solubility of $AgCl_{(c)}$ can be best described (see Figure 5.7) as involving one complex only, $AgCl_3^{2-}$:

$$AgCl_{(c)} + 2\,Cl^- = AgCl_3^{2-}. \tag{5.6}$$

The 300°C runs showed a marked increase in solubility in KCl-bearing solutions (as opposed to those with NaCl) for salt concentrations higher

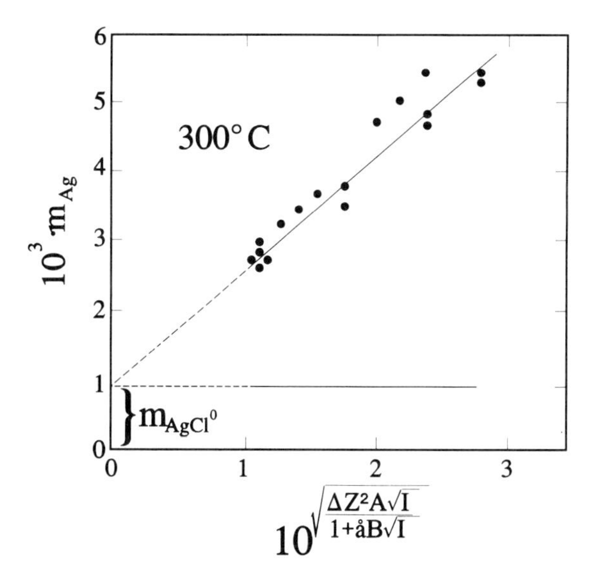

Figure 5.6. The solubility of $AgCl_{(c)}$ in water and $NaClO_4$ solutions at 300°C and saturated vapour pressure.

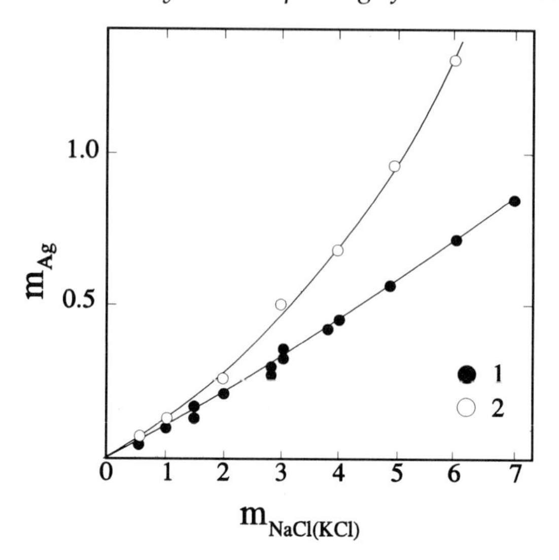

Figure 5.7. The solubility of $AgCl_{(c)}$ in chloride solutions at 300°C and saturated vapour pressure. 1 – NaCl, 2 – KCl.

than 2 m (Figure 5.7). It can be inferred that this results from the formation of the complex $KAgCl_3^-$:

$$AgCl_{(c)} + Cl^- + KCl^o_{(aq)} = KAgCl_3^- . \tag{5.7}$$

The value of the product of the concentrations of the aqueous species $K^*_{(5.7)}$ becomes a function of ionic strength in the pure experimental system, and it is found that $\log K^*_{(5.7)} = -1.15 + 0.14 \cdot I$ and $\log K_{(5.7)} = -1.15 \pm 0.11$ (Levin, 1992).

A separate series of experiments to investigate the solubility of $AgCl_{(c)}$ in water and in NaCl solutions up to 0.5 m, were conducted at 150, 200, and 250°C and at saturated vapour pressure (Zotov *et al.*, 1986a, 1986b). Bringing together all available information on silver chloride complexes (see Zotov *et al.*, 1986b; Levin, 1992), the temperature dependence of log K for reactions 5.3, 5.4, and 5.6 has been calculated over the temperature range from 20 to 350°C at saturated vapour pressure:

$$\log K_{(5.3)} = -431.83 + \frac{13984.6}{T} + 78.117 \cdot \ln T - 0.195 \cdot T + 8.692 \cdot 10^{-5} \cdot T^2,$$

$$\log K_{(5.4)} = -\frac{2826.9}{T} + 4.023 - 0.00342 \cdot T \ ,$$

$$\log K_{(5.6)} = -\frac{2230.4}{T} + 4.391 - 0.00528 \cdot T \ .$$

The solubility of $AgCl_{(c)}$ in water and 0.09 to 2.56 m NaCl solutions was measured at 450°C and pressures between 500 and 1500 bar (Table 5.4).

The solubility measurements indicate that the complex $AgCl_2^-$ is dominant (i.e. dissolution is according to Equation 5.5, above). The values of $\log K_{(5.5)}$ are 0.13, 0.18, and 0.07 at 500, 1000, and 1500 bar, respectively, i.e. the equilibrium constant is pressure independent within experimental uncertainty.

Table 5.4. The solubility of $AgCl_{(c)}$ in mol. kg H_2O^{-1} in water, 0.09–2.56 m NaCl and 0.09–0.46 m KCl solutions at 450 °C and different pressures

		Pressure (kbar)				
m_{NaCl}	m_{KCl}	0.5	1.0		1.5	1.75
0.00	0.00	0.0104	0.0084	–	0.0249	–
–	–	–	0.0084	(0.0078)	0.0513*	–
–	–	–	–	–	0.0253	–
–	–	–	–	–	0.0246	–
0.09	–	0.044	0.051	–	0.059	–
–	–	0.039	0.052	(0.055)	–	–
–	–	0.040	0.052	(0.050)	–	–
–	–	–	0.051	(0.051)	–	–
0.45	–	0.178	0.198	(0.242*)	0.185	0.192
–	–	0.247*	0.196	–	–	–
–	–	0.183	0.205	–	–	–
–	–	0.183	–	–	–	–
0.90	–	0.372	0.396	–	0.337	–
–	–	0.371	0.365	–	0.383	–
–	–	–	–	–	0.365	–
1.85	–	–	0.740	–	–	–
2.56	–	–	1.054	–	–	–
–	0.09	0.041	0.056	(0.054)	–	–
–	–	0.045	–	–	–	–
–	–	0.041	–	–	–	–
–	0.46	0.216	0.214	–	–	–

* Omitted from the calculation of equilibrium constants.

Additional experiments were conducted to determine the partial molar volume of $AgCl_2^-$ at 25 °C, by measuring the solubility of $AgCl_{(c)}$ in 0.5 m NaCl at pressures 1 and 1500 bar (Table 5.5). $\overline{V}^0_{(25°C)}$ $AgCl_2^-$ was calculated to be 38.8 ± 1.5 cm^3 mol^{-1}.

Table 5.5. The solubility of $AgCl_{(c)}$ in 0.5 m NaCl at 250 °C, 1 and 1500 bar, together with $\log K_{(5.5)}$ values

P (bar)	$m_{Ag} \cdot 10^5$			$-\log K_{(5.5)}$
	Run	*Blank*	*Average*	
1	1.78	0.008		
	1.65		1.75	4.43
	1.82			
1500	2.49			
	2.23		2.34	4.30
	2.30			

From the thermodynamic analysis of the experimental data for $AgCl_2^-$, parameters for the HKF model (Table 5.3) have been calculated, which provide a refinement of the parameters given in the SUPCRT92 data base (Johnson *et al.*, 1992), particularly in the supercritical region.

Over the temperature–pressure range that we considered experimentally, the retrieved thermodynamic properties of both silver ion and silver cloride complexes differ significantly from those recommended in the SUPCRT92 data base (Johnson *et al.*, 1992) at temperatures above 150–250°C. The main discrepancies are observed in those cases where SUPCRT92 data are obtained by extrapolation and are not supported experimentally. For Ag^+, agreement is to within 0.1 log units to 250°C, above which its stability is estimated as 0.15–0.3 log units higher. For the chloride complexes, it can be seen that experimental data are well described by assuming the existence of only three complexes ($AgCl^0_{(aq)}$, $AgCl_2^-$, $AgCl_3^{-2}$), without the need to invoke the presence of $AgCl_4^{-3}$, as considered earlier. According to the data, the stability of chloride complexes is 0.3–0.7 orders of magnitude lower than that estimated from the SUPCRT92 data. Compared to the data from Turnbull and Wadsley's (1988) data base (Morrison *et al.*, 1991), at 250–350°C, the stability of $AgCl_2^-$ at 250–350°C is 0.3–0.8 log units higher from our data.

It is clear from these results and earlier studies that the $AgCl_2^-$ complex is of major importance in the hydrothermal transport of silver. The possibility of the additional formation of mixed hydroxychloride complexes was investigated at 200–300°C by Zotov *et al.* (1982) who examined the equilibrium:

$$AgCl_{(c)} + OH^- = AgClOH^- . \qquad (5.8)$$

They obtained the expression:

$$\log K_{(5.8)} = \frac{-2966}{T} + 4.667,$$

valid in the temperature range 200–300°C. Extrapolation to 25°C yields $\log K_{(5.8)} = -5.28$. This value agrees well with $\log K_{(5.8)} = -5.05$, calculated from the data of Pouradier and Gadet (1973). Clearly, such mixed complexes predominate under strongly alkaline conditions only, and are therefore of little geochemical interest.

5.4 COPPER

The solubility method has been used to investigate: (a) Cu(II) and Cu(I) hydrolysis, and (b) Cu(I) chloride complexing at 150–450°C and at pressures from the saturated vapour pressure to 1500 bar. From the data in the literature (Nikolaeva, 1982; Crerar and Barnes, 1976) and the experiments below, it follows that Cu(I) complexes will generally predominate in hydrothermal solutions which are involved in copper ore formation (i.e.

f_{O_2} falls between the Ni–NiO and Fe_2O_3–Fe_3O_4 buffers). Aqueous complexes of Cu(II) are of interest in the study of the hypergenesis zone alone.

5.4.1 Cu(II) hydrolysis

The solubility of Cu(II) was investigated by studying the hydrolysis of tenorite (CuO) in the system $CuO_{(c)}$–HNO_3–NaOH–H_2O. Experiments were performed over the temperature range 200–350 °C at the saturated vapour pressure and at 450 °C and 0.5 kbar by Var'yash (1986).

The shape of the solubility curve (Figure 5.8) indicates that the dominant complexes are $Cu(OH)^+$ in acidic solutions (pH < 4), $Cu(OH)^0_{2\,(aq)}$ in near neutral to mildly alkaline solutions (solubility is here independent of pH), and $Cu(OH)_4^{2-}$ in more alkaline solutions (pH > 10). The presence of the complex $Cu(OH)_3^-$ noted in some studies under alkaline conditions (e.g. McDowell and Johnston, 1936; Buketov *et al.*, 1967; Solov'eva *et al.*, 1973) was not confirmed by the authors' investigations. This is in agreement with Baes and Mesmer's (1976) interpretation of the data of McDowell and Johnston (1936). The values of the equilibrium constants for the reactions:

$$CuO_{(c)} + H^+ = Cu(OH)^+, \tag{5.9}$$

$$CuO_{(c)} + H_2O = Cu(OH)^0_{2(aq)}, \tag{5.10}$$

$$CuO_{(c)} + H_2O + 2OH^- = Cu(OH)_4^{2-}, \tag{5.11}$$

are given in Table 5.6, while their temperature dependencies are in Figure 5.9.

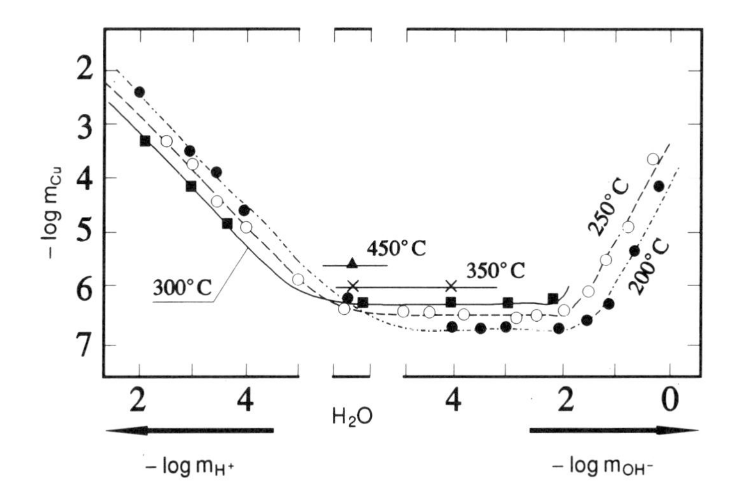

Figure 5.8. The solubility of $CuO_{(c)}$ in aqueous solutions at 200–450°C.

Table 5.6. Values of log K for reactions 5.9–5.16 in the systems $CuO_{(c)}$–H_2O and $Cu_2O_{(c)}$–H_2O at saturated vapour pressure and 0.5–1.0 kbar. The pressure magnitude is given in parentheses

Reaction	*Temperature* (°C)					
	25*	200	250	300	350	450
(5.9)	− 0.38 ± 0.04	− 0.50 ± 0.07	− 0.82 ± 0.18	− 1.17 ± 0.15	–	–
(5.10)	− 8.70 ± 0.15	− 6.80 ± 0.11	− 6.56 ± 0.13	− 6.26 ± 0.06	− 6.01 ± 0.11	–
(5.11)	− 3.99 ± 0.39	− 4.39 ± 0.34	− 3.74 ± 0.16	–	–	–
(5.12)	− 0.94	–	–	1.25 ± 0.17	1.32 ± 0.19	1.99 ± 0.10
(5.13)	–	–	–	7.19 ± 0.14	6.61 ± 0.17	6.31 ± 0.07
	–	–	–	(0.5)	(0.5)	(0.5)
	–	–	–	–	6.47 ± 0.16	–
	–	–	–	–	(1.0)	–
(5.14)	− 10.95	− 6.13 ± 0.15	− 5.57 ± 0.14	− 4.81 ± 0.18	− 3.61 ± 0.13	− 2.91 ± 0.26
	–	–	–	–	–	(0.5)
(5.15)	–	–	–	–	1.74 ± 0.37	0.89 ± 0.08
	–	–	–	–	(0.5)	(0.5)
	–	–	–	–	–	0.98 ± 0.32
	–	–	–	–	–	(1.0)
(5.16)	− 2.93	− 3.73 ± 0.14	− 3.26 ± 0.17	− 2.91 ± 0.17	–	–

* From literature data compiled by Var'yash (1985, 1989).

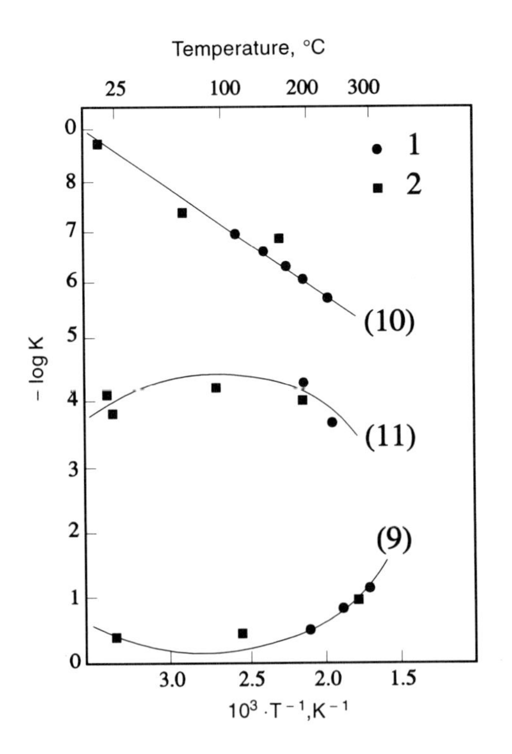

Figure 5.9. The temperature dependence of equilibrium constants for Cu(II) hydrolysis reactions (9)–(11). 1 – our data, 2 – data from the literature.

5.4.2 Cu(I) hydrolysis

The solubility of Cu(I) was investigated by Var'yash (1989), who utilized two systems. Hydrolysis of cuprite (Cu_2O) was investigated under neutral to alkaline conditions in the system $Cu_2O_{(c)}$–HCl–NaOH–H_2O, with f_{O_2} buffered by a $Cu_{(c)}$–$Cu_2O_{(c)}$ pair, while metallic Cu in the system $Cu_{(c)}$–HCl–H_2O was used for runs under acidic conditions, because of the high solubility of cuprite at low pH. For the runs in acid fluids, redox conditions were controlled by the addition of a weighed amount of aluminium, which evolved hydrogen into the autoclave. Experiments were carried out over the temperature range 150–350°C at saturated vapour pressure, and at 450°C and 500 bars.

There are two regions to the solubility curve for $Cu_2O_{(c)}$ (Figure 5.10), indicating that $Cu(OH)^0_{(aq)}$ predominates under neutral and slightly alkaline conditions, while $Cu(OH)_2^-$ dominates at higher pH. Under acidic conditions, the Cu^+ ion is dominant (Figure 5.11). The values of the equilibrium constants for the following reactions:

$$0.5\,Cu_2O_{(c)} + H^+ = Cu^+ + 0.5\,H_2O, \tag{5.12}$$

$$Cu_{(c)} + 0.25\,O_2 + H^+ = Cu^+ + 0.5\,H_2O, \tag{5.13}$$

$$0.5\,Cu_2O_{(c)} + 0.5\,H_2O = Cu(OH)^0_{(aq)}, \tag{5.14}$$

$$Cu_{(c)} + 0.25\,O_2 + 0.5\,H_2O = Cu(OH)^0_{(aq)}, \tag{5.15}$$

$$0.5\,Cu_2O_{(c)} + 0.5\,H_2O + OH^- = Cu(OH)_2^-, \tag{5.16}$$

are listed in Table 5.6, while their temperature dependencies are illustrated in Figure 5.12.

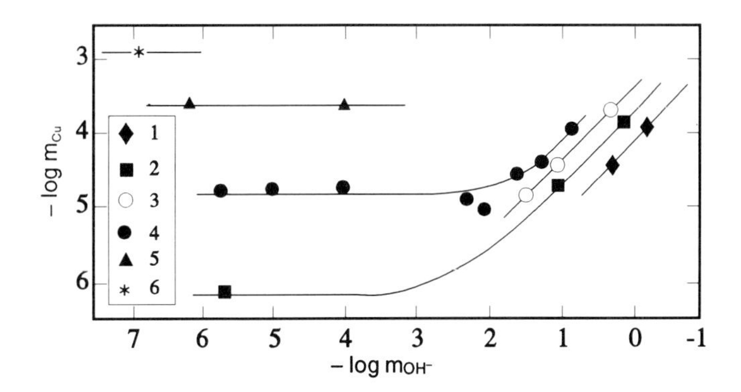

Figure 5.10. The solubility of $Cu_2O_{(c)}$ in water and aqueous solutions at 150 (1), 200 (2), 250 (3), 300 (4), 350 (5), and 450°C (6).

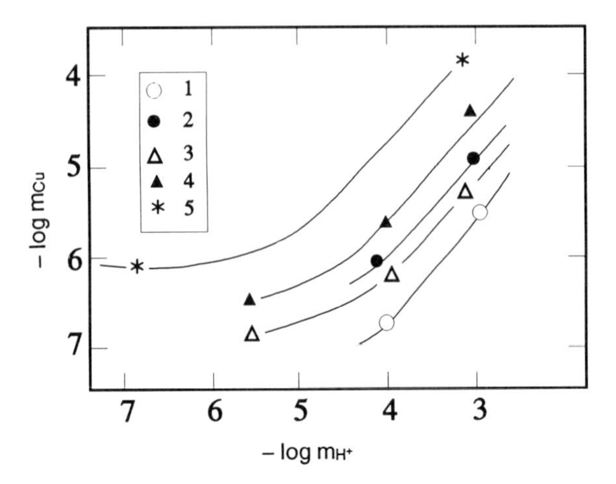

Figure 5.11. The solubility of $Cu_{(c)}$ in HCl solutions and water at 300 (1,2), 350 (3,4), and 450°C, 0.5 kbar (5). $-\log f_{O_2}$ are 39.5 (1), 37.5 (2), 35.1 (3), 32.6 (4), 28.2 (5).

5.4.3 Cu(I) complexing in chloride solutions

Var'yash (1991) has investigated the system $Cu_{(c)}$–HCl–NaCl–NaOH–H_2O at 250–450°C at saturated vapour pressure and 0.5–1.5 kbar. She showed from the dependence of solubility on the oxygen fugacity, concentration of NaCl and acidity, that $CuCl_2^-$ is the dominant complex at all temperatures. Complexes such as $CuCl^0_{(aq)}$ and $CuCl_3^{2-}$ (Utkina *et al.*, 1966; Crerar and Barnes, 1976) were present in subordinate amounts, if at all. Values of the equilibrium constant for the reaction:

$$Cu_{(c)} + H^+ + 0.25\,O_2 + 2\,Cl^- = CuCl_2^- + 0.5\,H_2O, \qquad (5.17)$$

are tabulated in Table 5.7.

Table 5.7. Values of log K for reaction 5.17 as a function of temperature and pressure

T(°C)	*Pressure*			
	sat. vapour	0.5 kbar	1.0 kbar	1.5 kbar
25	17.2 ± 0.29*	–	–	–
50	16.3 ± 0.14*	–	–	–
100	14.5 ± 0.22*	–	–	–
150	13.8*	–	–	–
200	12.8	–	–	–
250	12.4 ± 0.5	–	–	–
300	–	12.47 ± 0.5	–	–
350	–	12.70 ± 0.41	11.91 ± 0.40	11.4 ± 0.40
400	–	13.35 ± 0.42	12.3 ± 0.48	–
450	–	14.86 ± 0.52	13.36 ± 0.48	–

* From literature data compiled by Var'yash (1992).

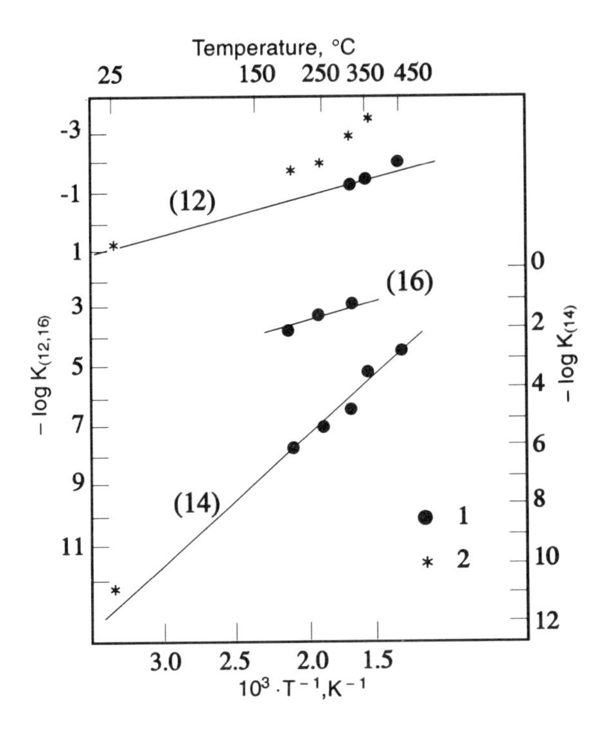

Figure 5.12. The temperature dependence of equilibrium constants for Cu(I) hydrolysis reactions (12), (14), (16). 1 – our data, 2 – data from the literature.

5.4.4 Thermodynamic properties of Cu^+ and $CuCl_2^-$

The HKF model (above) has been used to fit thermodynamic properties to the experimental data for Cu^+ and $CuCl_2^-$. For Cu^+, the results of our experiments at 300–450 °C were fitted with the value of $\Delta\overline{G}^0_{f,\,25\,°C}$ for $Cu^+ = 11.94\ \text{kcal} \cdot \text{mol}^{-1}$ (Naumov *et al.*, 1974). Parameters for $CuCl_2^-$ were calculated using the HKF model keeping in consideration numerous thermodynamic data at 15–150 °C (see Var'yash, 1992) and values of $\Delta\overline{G}^0_f$ for $CuCl_2^-$ at 250–450 °C calculated from our experiments (Figure 5.13, Table 5.3). The discrepancies thus obtained do not exceed 0.3–0.4 $\text{kcal} \cdot \text{mol}^{-1}$.

The results obtained by us for the chloride system are in conflict with the conclusions of Crerar and Barnes (1976), who argued that $CuCl^0_{(aq)}$ is the dominant complex in 0.1–6 m NaCl solutions at 200–350 °C. This issue is discussed at length in Var'yash and Rekharskiy (1981).

In conclusion, this experimental study, and the thermodynamic analysis of the results, confirm that the $CuCl_2^-$ complex is the dominant form of Cu in aqueous fluids under hydrothermal conditions, and show that

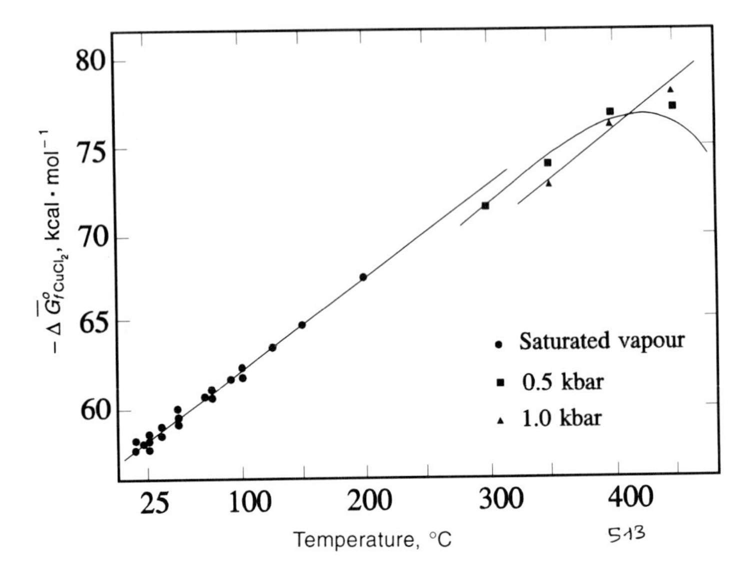

Figure 5.13. The temperature dependence of the apparent standard partial molal Gibbs free energy of formation of $CuCl_2^-$ at different pressure.

there is a major increase in Cu solubility at temperatures above 300–350 °C in aqueous solutions.

5.5 MOLYBDENUM

The solubility and complexing of molybdenum in hydrothermal solutions have been investigated by experimental studies of the solubility of the crystalline phases MoO_2 and MoS_2 (molybdenite) in water–salt solutions over the temperature–pressure range 300–450 °C and 88–1000 bar. The systems used were of four types:

1. dilute solutions,
2. NaCl–KCl solutions,
3. sulphide solutions,
4. Si-bearing solutions.

5.5.1 MoO_2 solubility in dilute solutions

The solubility of $MoO_{2(c)}$ was measured in water, in HCl solutions (10^{-3} to 1.0 m), and NaOH–KOH solutions (10^{-5} to 10^{-3}m). Log f_{O_2} was buffered in the range NNO − 4 to NNO + 4 (Kudrin, 1985). Calculation of the speciation in the experimental solutions used dissociation constants from Bryzgalin (1986) and Marshall and Franck (1981).

The dependence of solubility on the oxygen fugacity and pH is shown in Figures 5.14 and 5.15. Throughout the redox conditions investigated, molybdenum has a valency of six.

The change in solubility of MoO_2 with pH suggests a change in the dominant dissolution reaction from:

$$MoO_{2(c)} + H_2O + 0.5O_{2(g)} = HMoO_4^- + H^+ \qquad (5.18)$$

under alkaline to weakly acidic conditions, to

$$MoO_{2(c)} + H_2O + 0.5O_{2(g)} = H_2MoO^0_{4(aq)} \qquad (5.19)$$

under moderately acid conditions.

Equilibrium constants for reactions 5.18 and 5.19 are presented in Table 5.8. Increased solubility in more concentrated HCl solutions is probably due to the formation of chloride complexes of molybdenum.

5.5.2 NaCl–KCl solutions

The solubility of $MoO_{2(c)}$ was measured in NaCl–KCl solutions (0.05 to 1.0 m) whose acidity ranged from 10^{-2} m HCl to $5 \cdot 10^{-4}$ m NaOH or KOH. Temperatures were in the range 300–450°C and 500 bars pressure, and $\log f_{O_2}$ was buffered between NNO and NNO – 4 (Kudrin, 1989). Solubility increases with increasing pH, f_{O_2}, and with the total concentration of salts (Figure 5.16). The solubility in the KCl system is typically 1.5 to 2 times greater than that in the NaCl system under the same experimental conditions.

Table 5.8. Thermodynamic equilibrium constants (log K) for dissolution reaction of $MoO_{2(c)}$ in water and KCl–NaCl solutions and formation constants for alkali–molybdate and silicon–molybdate complexes at 350–450°C

Reaction	*P* (kbar)	*Temperature* (°C)			
		300	350	400	450
(5.18)	0.09	7.70 ± 0.5	–	–	–
	0.5	7.87 ± 0.3	5.68 ± 0.2	3.55 ± 0.3	0.48 ± 0.5
	1.0	–	6.06 ± 0.3	4.60 ± 0.5	2.96 ± 0.2
(5.19)	0.5	–	≤ 8.0	–	6.28 ± 0.2
(5.22)	0.5	9.84 ± 0.4	7.98 ± 0.3	–	4.55 ± 0.3
(5.23)	0.5	9.60 ± 0.3	7.63 ± 0.3	–	4.32 ± 0.4
(5.24)	0.5	1.97 ± 0.5	2.30 ± 0.4	–	4.07 ± 0.5
(5.25)	0.5	1.73 ± 0.4	1.95 ± 0.4	–	3.84 ± 0.6
(5.26)	0.5	1.5 ± 0.3	2.3 ± 0.3	2.2 ± 0.2	1.7 ± 0.2
	1.0	–	–	2.5 ± 0.3	2.5 ± 0.2
(5.27)	0.5	< 1.0	< 1.5	–	3.0 ± 0.3
	1.0	–	–	2.1 ± 0.3	2.8 ± 0.2

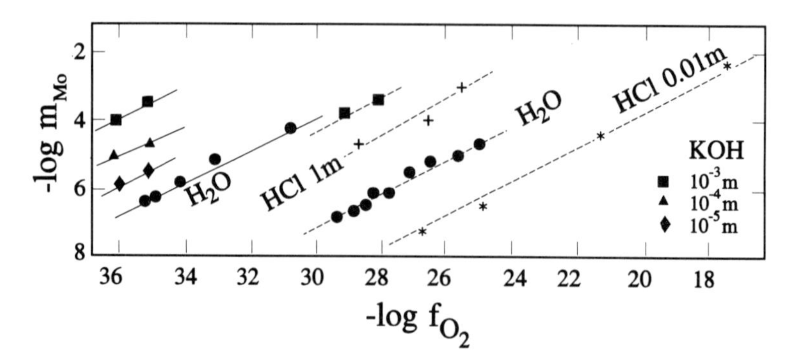

Figure 5.14. The solubility of $MoO_{2(c)}$ as a function of oxygen fugacity at 350 (solid lines) and 450°C (dashed lines); P = 0.5 kbar.

In the analysis of experimental data, the following reactions have been considered, together with reaction 5.18:

$$MoO_{2(c)} + H_2O + 0.5O_{2(g)} + Cl^- = HMoO_4Cl^{2-} + H^+, \quad (5.20)$$

$$MoO_{2(c)} + H_2O + 0.5O_{2(g)} + Na(K)Cl^0_{(aq)} = Na(K)ClHMoO_4^- + H^+, \quad (5.21)$$

$$MoO_{2(c)} + H_2O + 0.5O_{2(g)} + K^+ = KHMoO^0_{4(aq)} + H^+, \quad (5.22)$$

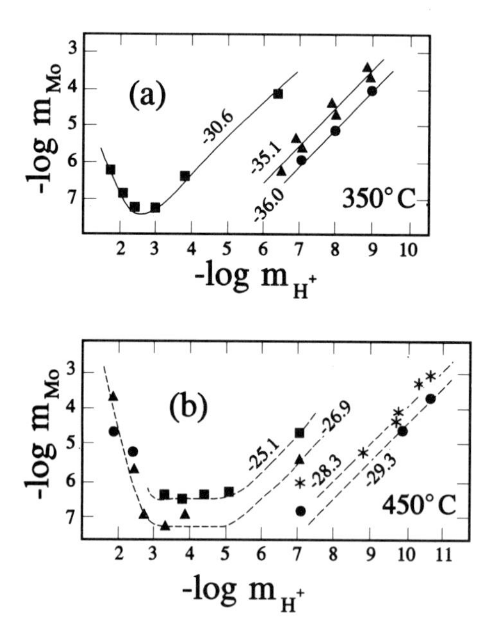

Figure 5.15. The solubility of $MoO_{2(c)}$ as a function of log m_{H^+} (a) at 350 (solid lines) and (b) at 450°C (dashed lines) and varying f_{O_2}; P = 0.5 kbar; the curves are labelled with log f_{O_2} values.

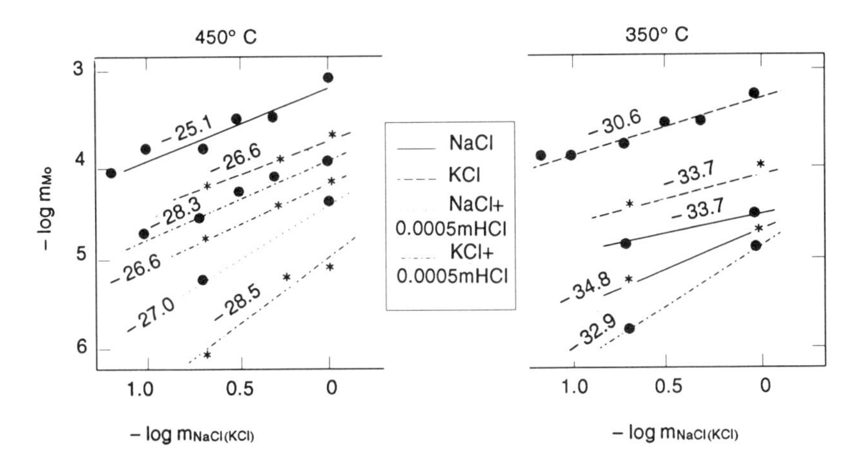

Figure 5.16. The effect of the concentration of alkali chlorides on $MoO_{2(c)}$ solubility at 350 and 450°C and varying f_{O_2}; numbers are $\log f_{O_2}$ values.

$$MoO_{2(c)} + H_2O + 0.5O_{2(g)} + Na^+ = NaHMoO^0_{4(aq)} + H^+. \qquad (5.23)$$

Reactions 5.22 and 5.23 best describe the relationship between Mo concentration in solution, pH and $\log f_{O_2}$. The values of equilibrium constants for these reactions are summarized in Table 5.8 together with formation constants of alkaline–molybdate complexes defined by:

$$K^+ + HMoO_4^- = KHMoO^0_{4(aq)}\,, \qquad (5.24)$$

$$Na^+ + HMoO_4^- = NaHMoO^0_{4(aq)}, \qquad (5.25)$$

which have been retrieved from reactions 5.18, 5.22, and 5.23. The stability of the molybdate complexes decreases rapidly with decreasing temperature.

5.5.3 Sulphide solutions

The solubility of molybdenite, MoS_2, was measured in water and in 1 m KCl solutions, with $m_{total\,H_2S} \approx m_{H_2S^0_{(aq)}}$, in the range 10^{-4} to $10^{-1.5}$ for temperatures of 350 and 450°C at a pressure of 0.5 kbar. The $\log f_{O_2}$ ranged from NNO to NNO − 3. At the conditions of the runs, Mo occurs as Mo(VI) in solution, and the solubility of molybdenite decreases with increasing H_2S concentration (Figure 5.17). Experimental solubility values for molybdenite are in reasonable agreement with values calculated from the experiments on $MoS_{2(c)}$ solubility (above), assuming the

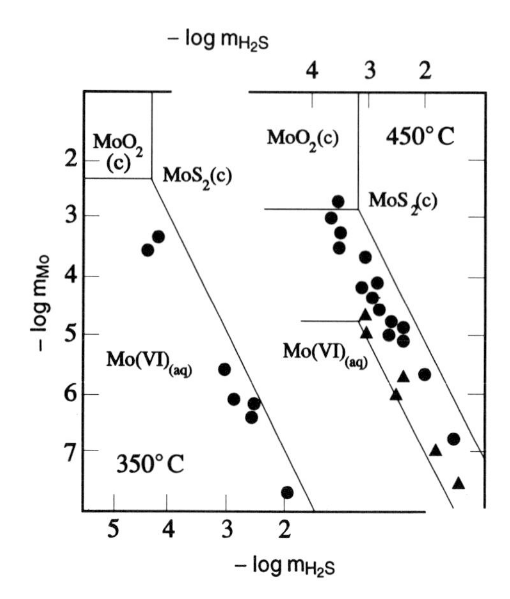

Figure 5.17. Comparison of the calculated (solid lines) and measured solubility of $MoS_{2(c)}$ in water (triangles) and 1m KCl (circles) at 350 and 450°C. 350°C, $\log m_{H}^{+} = -6$, $\log f_{O_2} = -32.0$; 450°C, $\log m_{H}^{+} = -6$ (1 m KCl), $\log f_{O_2} = -26.0$.

occurrence of Mo in solution as $HMoO_4^-$ and $KHMoO_{4(aq)}^0$ in pure water and in 1 m KCl solutions, respectively. Thus, under the conditions of the runs, sulphide complexes do not form in significant concentrations.

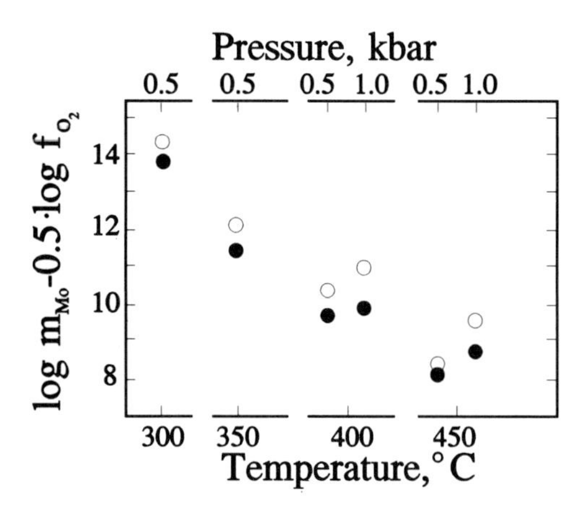

Figure 5.18. Comparison of the solubility of $MoO_{2(c)}$ in water without quartz (filled circles) and with excess quartz (open circles).

5.5.4 Si-bearing solutions

The effect of SiO_2 on MoO_2 solubility in water, in HCl solutions (10^{-3} and 10^{-4} m), and in 1 m NaCl solutions of varying pH, was investigated at 300–450 °C and 0.5–1 kbar. Throughout this range of temperatures, the solubility of MoO_2 in water and HCl solutions in the presence of excess quartz is higher than that measured in parallel runs without quartz (Figure 5.18). Note that at 400 and 450 °C, the magnitude of the difference is pressure dependent. In NaCl solutions, a similar effect becomes noticeable at temperatures above 350 °C. A series of runs with variable Si concentrations at 450 °C, 0.5–1 kbar, shown in Figure 5.19, indicate a linear increase in $MoO_{2(c)}$ solubility with increasing Si content with $\frac{\Delta \log_{Mo}}{\Delta \log m_{Si}} \approx 1$. The effect of SiO_2 on the solubility of Molybdenum is assumed to be associated with the formation of the following associates at high temperatures (Table 5.8):

$$HMoO_4^- + Si(OH)^0_{4(aq)} = HMoO_4 \cdot Si(OH)_4^-, \qquad (5.26)$$

$$NaHMoO^0_{4(aq)} + Si(OH)^0_{4(aq)} = NaHMoO_4 \cdot Si(OH)^0_{4(aq)}. \qquad (5.27)$$

On the basis of these experiments we can reach some general conclusions regarding the hydrothermal transport of molybdenum at 300–450 °C:

1. Hexavalent molybdenum complexes predominate for all common redox conditions.

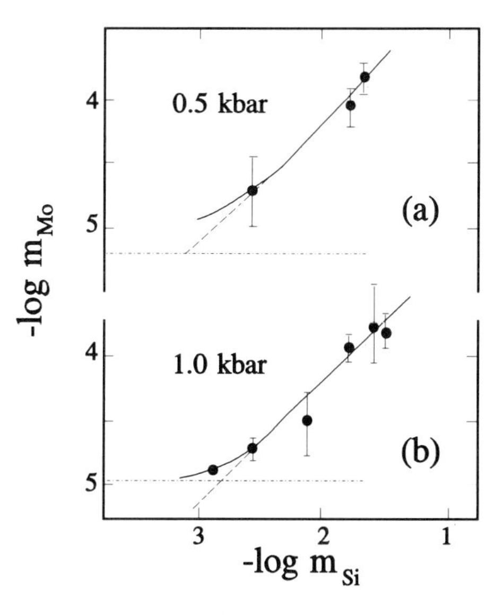

Figure 5.19. The effect of silica on $MoO_{2(c)}$ solubility in $Na_2SiO_3 + 10^{-3}$ m HCl solutions: (a) at 0.5 kbar, (b) at 1.0 kbar; NaCl = 0.05 m, T = 450°C, $\log f_{O_2} = -26.0$. The solubility without adding silica is shown as a dot-dash line.

2. At 300–350°C, molybdenum is predominantly transported as $NaHMoO^0_{4(aq)}$ or $KHMoO^0_{4(aq)}$ complexes.
3. At 400–500°C and probably also at higher temperatures, the $HMoO_4 \cdot Si(OH)^-_4$ and $NaHMoO_4 \cdot Si(OH)^0_{4(aq)}$ associates become more important.
4. Chloride and sulphide complexes of molybdenum are not significant; nor is the hydroxy complex, $HMoO^-_4$, except at temperatures below 300°C in solutions of low NaCl concentrations.

5.6 ARSENIC

The solubility of orpiment, As_2S_3, in water and in sulphide solutions was measured in the temperature–pressure range 25–150°C and 1–5 bar. The presence of the complexes $H_3AsO^0_{3(aq)}$, $As_2S^0_{3(aq)}$, $HAs_2S^-_4$, and $As_2S^{2-}_4$ has been confirmed, and their thermodynamic properties have been determined.

5.6.1 The solubility of As_2S_3 in sulphide solutions

In order to determine the composition of sulphide complexes, the solubility of native orpiment was investigated at 25 ± 5 and 90 ± 0.5°C, as a function of pH and total sulphur concentration (Figures 5.20–5.22) (Mironova and Zotov, 1980; Mironova *et al.*, 1990). At 150°C, experiments were performed with acidic fluids only (pH 1.1–2).

The curves of the pH dependence of orpiment solubility (Figure 5.20) contain three portions of different slope that correspond to the predominance of three different As species in solution.

At low pH (< 4), the solubility of $As_2S_{3\,(c)}$ is independent of pH (Figure 5.20) and also of sulphide concentration (Figure 5.22) at both 25 and 90°C. In this region, the dominant species was first identified as $H_2As_2S_3O^0_{(aq)}$ (Mironova *et al.*, 1990) and then, less specifically, as the hydrated species, $As_2S^0_{3(aq)}$ (Akinfiev *et al.*, 1992), which is formed according to the reaction:

$$As_2S_{3(c)} = As_2S^0_{3(aq)}. \quad (5.28)$$

At 150°C and high total sulphur (0.025–0.056 m), the solubility is also independent of pH and dissolved H_2S concentration; however, it increases dramatically with decreasing H_2S concentration below 10^{-2} m (Figure 5.23). This behaviour may be accounted for by the formation of the hydroxy complex $H_3AsO^0_{3(aq)}$ (Mironova *et al.*, 1984) according to the reaction:

$$As_2S_{3(c)} + 6H_2O = 2H_3AsO^0_{3(aq)} + 3H_2S^0_{(aq)}. \quad (5.29)$$

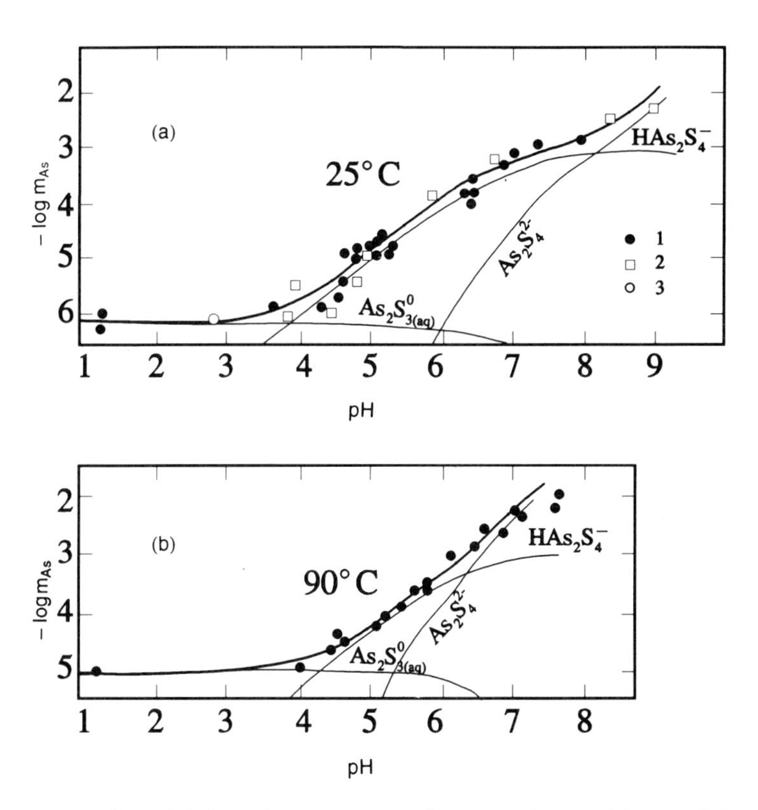

Figure 5.20. The solubility of orpiment as a function of pH at (a) 25 and (b) 90°C and constant sulphide concentration. 1 – original data, $m_{(H_2S+HS^-)} = 0.011 \pm 0.001$; 2 – various ΣS concentrations recalculated to $m_{(H_2S+HS^-)} = 0.011$; 3 – original data, $m_{(H_2S+HS^-)} = 0.035$ mol. kg H_2O^{-1}.

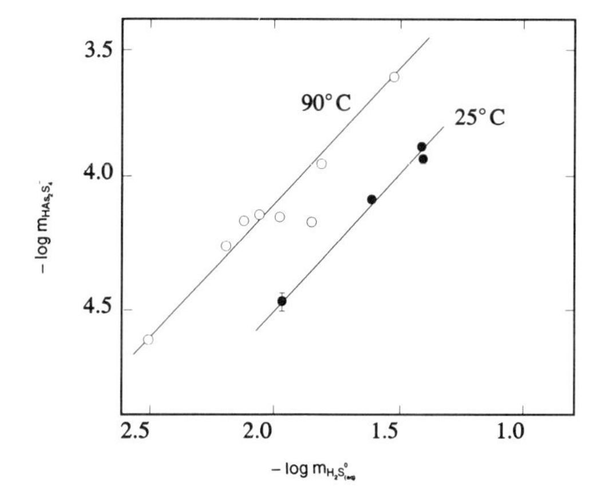

Figure 5.21. The concentration of $m_{HAs_2S_4^-}$ recalculated to pH = 5.5 from experimental data as a function of $H_2S^0_{(aq)}$ concentration at 25 and 90°C.

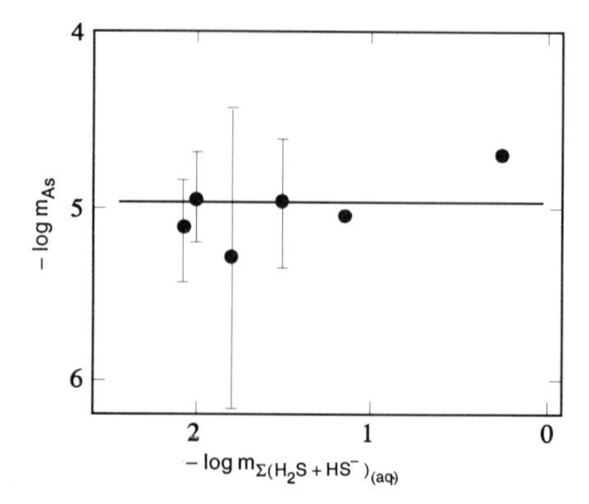

Figure 5.22. The solubility of orpiment as a function of sulphide content at 90°C and pH 1–4.

For the weakly acidic region, experimental results at pH 4.58–5.65 (90°C) and 5.03–6.90 (25°C) have been normalized to pH = 5.5 (Figure 5.21, Table 5.9), using the dependence on pH established in Figure 5.20, in order to examine the relationship between orpiment solubility and total sulphide concentration. The observed relationship between solubility, sulphide concentration, and pH may be accounted for by the formation of the dinuclear complex $HAs_2S_4^-$:

$$As_2S_{3(C)} + HS^- = HAs_2S_4^-. \tag{5.30}$$

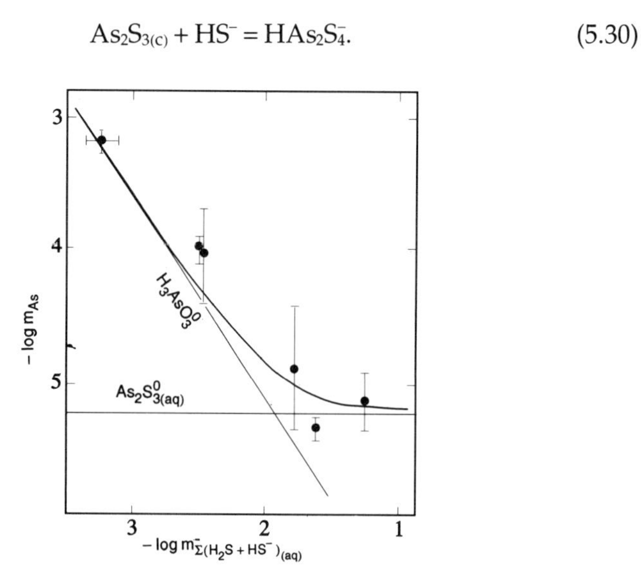

Figure 5.23. The solubility of orpiment as a function of sulphide content at 150°C and pH 1–2.

Table 5.9. The solubility of orpiment as a function of sulphide concentration recalculated to pH = 5.5

T (°C)	*Runs*	pH	m_{H_2S} (mol · kg H_2O^{-1})	$m \cdot 10^5$ (mol · kg H_2O^{-1})		
				$\Sigma As_2(aq)$	$HAs_2S_4^-$	$HAs_2S_4^-$ at pH = 5.5
25	1	5.03	0.040	4.68	4.64	13.14
	1	5.95	0.040	32.4	32.4	12.30
	1	6.90	0.025	104.7	104.7	8.32
	9	5–5.2	0.011	1.53 ± 0.32	1.49	3.39 ± 0.32
90	3	4.58	0.03	3.6	3.1	25.8
	3	5.00	0.016	4.5	3.8	12.0
	2	5.40	0.015	7.1	5.6	7.05
	2	5.40	0.0104	7.1	5.9	7.43
	3	5.65	0.0087	12.8	10.6	7.50
	4	5.39	0.0078	6.3	5.5	7.08
	3	5.50	0.0064	6.7	5.6	5.60
	4	5.82	0.0032	6.9	5.1	2.44

Previous studies have reached contradictory conclusions as to whether arsenic sulphide dimers were important in such solutions. Spycher and Reed (1989) concluded that the trimer $As_3S_6^{3-}$ is formed in sulphide solutions, although the more recent study of Webster (1990) did not attain sufficient accuracy to decide unambiguously between trimer and dimer. Eary (1992) did, however, observe a dependence of $As_2S_{3(c)}$ solubility on m_{H_2S} that was characteristic of trimer. The discrepancy between this conclusion and that of our work may be due to enhanced arsenic concentration resulting from amorphous phase dissolution during the course of Eary's experiments.

The relationship between orpiment solubility and total sulphur was not explored in detail in alkaline solutions, but over the range of S concentrations investigated (0.003–0.012 m) there is reasonable agreement between values calculated for the equilibrium constant for the hypothetical reaction:

$$As_2S_{3(c)} + OH^- + HS^- = As_2S_4^{2-} + H_2O, \qquad (5.31)$$

which is consistent with our hypothesis that arsenic sulphide dimers predominated.

The values of the equilibrium constants for reactions 5.28–5.31 are listed in Table 5.10.

5.6.2 The solubility of As_2S_3 in water and in acidic solutions

The solubility of orpiment was measured at 25–150 °C by Mironova *et al.* (1984), using a similar technique to Mironova and Zotov (1980). For pH values between 1.2 and 5.5, the solubility is independent of acidity, and is therefore interpreted as due to reaction 5.29. Equilibrium constants for this reaction are presented in Table 5.10.

Table 5.10. Thermodynamic equilibrium constants (log K) for reactions (5.28)–(5.31) at 25–150°C

Reaction	log K			
	25°C	90°C	150°C	*Temperature dependence*
(5.28)	-6.45 ± 0.5	-5.31 ± 0.12	-5.5 ± 0.2	$(-7618/T) + 34.84 - 0.0528T$
(5.29)	-25.8 ± 0.6	-20.2 ± 0.4	-16.2 ± 0.4	$(-12208/T) + 21.34 - 0.0208T$
(5.30)	-1.47 ± 0.08	-1.27 ± 0.07	–	$(-333/T) - 0.352$
(5.31)	4.05 ± 0.2	4.32 ± 0.12	–	$(-450/T) + 5.559$

5.6.3 Thermodynamic properties of arsenic complexes in solution

Thermodynamic properties of the aqueous species inferred to have been present in these experiments have been extracted from the experimental results of this study and from the literature, using the HKF model. Basic thermodynamic properties of $As_2S_{3(c)}$ (orpiment) and monoclinic $As_2O_{3(c)}$ (Table 5.3a) were taken from the IVTANTERMO data base (1990). The review of the literature and the logistics of the calculations are presented in Akinfiev *et al.* (1992). The results are given in Table 5.3.

It follows from these results that at temperatures below 150°C and at high sulphide content, the dominant arsenic species in solution are $As_2S^0_{3(aq)}$ at low pH, $HAs_2S_4^-$ at intermediate pH, and $As_2S_4^{2-}$ under alkaline conditions. The concentration of $As_4S_7^{2-}$, calculated using the data of Weissberg *et al.* (1966), is less than 10^{-5} m, and this species seems to be of no particular geochemical interest.

As the temperature increases, the complex $H_3AsO^0_{3(aq)}$ gradually gains importance. At the same time, the stability field of orpiment reduces greatly and moves to low pH. At temperatures 150–200°C and above, the major form of arsenic in hydrothermal solutions is $H_3AsO^0_{3(aq)}$. At lower temperature, there is competition between hydroxy and sulphide complexes, their relative importance being dictated by the actual sulphide concentrations in hydrothermal solutions.

5.7 ANTIMONY

Experimental studies were aimed at refining the thermodynamic properties of the species $Sb(OH)^0_{3(aq)}$, known to be one of the most important forms of antimony in hydrothermal solutions (Popova *et al.*, 1975; Sorokin *et al.*, 1988). The investigations therefore concerned:

1. the solubility of $Sb_2O_{3(c)}$ (senarmontite) in water in the temperature–pressure range 210–350°C and 20–1000 bar,
2. the solubility of Sb (metal) at 450°C and 500 bar,
3. the solubility of $Sb_2S_{3(c)}$ (antimonite) at 350°C and saturated vapour pressure.

The experimental procedure is described in detail in Shikina and Zotov (1991), and the results are presented in Tables 5.11 and 5.12 and Figure 5.24.

Table 5.11. The solubility of senarmontite, $Sb_2O_{3(c)}$, in water, with run duration 8–9 days

T (°C)	P (bar)	*Runs*	$m_{Sb}.10^3$ (mol · kg H_2O^{-1})
210	19.5	9	5.22 ± 0.39
	500	7	5.46 ± 0.51
	1000	8	5.80 ± 0.50
300	88	4	35.2 ± 2.8
	500	4	37.5 ± 1.4
	1000	4	38.4 ± 0.8
350	168	4	75.0 ± 2.5

The solubility of $Sb_2O_{3(c)}$ was found to be practically independent of pressure, but to increase sharply with temperature between 210–350 °C. The solubility of $Sb_{(c)}$ is closely related to the fugacity of hydrogen, and solubility contours have a slope of 1.50 ± 0.18 on a log m_{Sb} – log f_{H_2} diagram, which is indicative of the formation of Sb(III) complexes.

Table 5.12. The solubility of antimonite, $Sb_2S_{3(c)}$, at 350 °C and saturated vapour pressure, with run duration of 15 days

Solution	*Runs*	$m_{Sb} \cdot 10^3$ (mol · kg H_2O^{-1})
H_2O	2	7.6 ± 0.7
0.02 m H_2S + + 0.01 m HCl	3	6.7 ± 0.9

As a first approximation, $Sb_2O_{3(c)}$, $Sb_{(c)}$, and $Sb_2S_{3(c)}$ were assumed to dissolve according to the following reactions:

$$0.5Sb_2O_{3(c)} + 1.5H_2O = Sb(OH)^0_{3(aq)}, \quad (5.32)$$

$$Sb_{(c)} + 3H_2O = Sb(OH)^0_{3(aq)} + 1.5H_{2(aq)}, \quad (5.33)$$

$$0.5Sb_2S_{3(c)} + 3H_2O = Sb(OH)^0_{3(aq)} + 1.5H_2S^0_{(aq)}. \quad (5.34)$$

Simultaneous analysis of the solubility of these three phases (Akinfiev *et al.*, 1993) suggests that at high concentrations of antimony, the dinuclear complex $Sb_2(OH)^0_{6(aq)}$ is important. At 350 °C, the fraction of this complex in fluid saturated with $Sb_2O_{3(c)}$ may reach 30% of total dissolved Sb.

The HKF model has been used to extract thermodynamic data for Sb complexes from the results of our experiments, fitted together with solubility data for senarmontite in water at lower temperatures (Popova

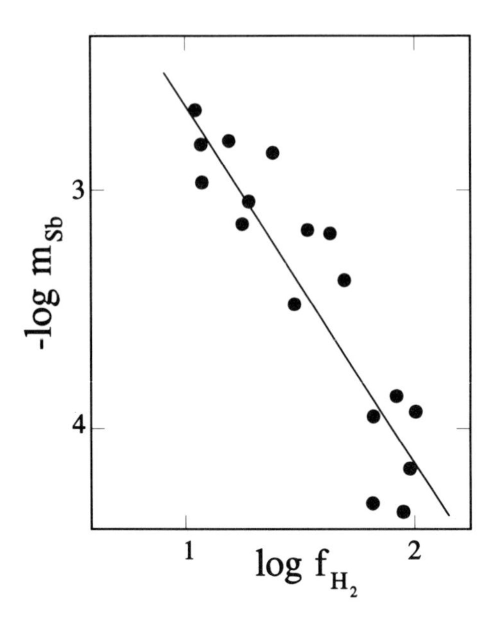

Figure 5.24. The solubility of $Sb_{(c)}$ in water as a function of hydrogen fugacity at 450°C and ≈ 500 bar.

et al., 1975), and with the data on antimonite solubility in water (Kozlov, 1982) and in sulphide solutions (Krupp, 1988; Learned *et al.*, 1974). The properties of solid phases (Table 5.3a) were taken from the IVTANTERMO data base (1990). The fitting process showed that our $Sb_{(c)}$ solubility experiments in water at 450 °C and 1 kbar pressure (Shikina and Zotov, 1991) are inconsistent with the rest of the data, probably due to loss of hydrogen during the course of the runs. These data were eliminated from the calculations. From the remaining results, the thermodynamic properties of 5 antimony complexes: $Sb(OH)^0_{3(aq)}$, $Sb_2(OH)^0_6$, $H_2Sb_2S^0_4$, $HSb_2S^-_4$ and $Sb_2S^{2-}_4$ were retrieved, over the temperature–pressure range 25–500 °C and 1–1500 bar (Akinfiev *et al.*, 1993). They are given in Table 5.3.

The thermodynamic properties of $Sb(OH)_{3(aq)}$ are considered reliable: they agree well with those from experiments by Wood *et al.* (1987) and calculations by Sorokin *et al.* (1988) and Spycher and Reed (1989) to 300–350 °C. In the sulphide system, however, experimental data of various workers are quite conflicting. Rather than attempt to bring them into agreement with each other, we have selected only those data which looked the most reliable.

Our results provide further evidence that the hydroxy complex $Sb(OH)^0_{3(aq)}$ is primarily responsible for the hydrothermal transport of antimony (Popova *et al.*, 1975; Sorokin *et al.*, 1988), especially at temperatures above 200 °C. Under these conditions, the complex seems to be predominant even at very high sulphide concentration (0.01 to 0.1 m total H_2S at 250 to 350 °C, respectively).

5.8 MERCURY

The behaviour of mercury under conditions of hydrothermal mineralization is relatively well understood. A detailed review of the data in the literature can be found in Sorokin *et al.* (1988). See also Chapter 4, this volume, for thermodynamic properties of the species $Hg^0_{(aq)}$.

Experiments were performed to address two particular problems which remain outstanding:

1. The system $HgS_{(c)}$–H_2S–H_2O was studied at 90 and 150 °C in order to better define the stoichiometry of the Hg sulphide complex formed under acidic conditions.
2. The system $HgO_{(c, red)}$–$NaHCO_3$–H_2O was investigated at 90 °C in order to assess the stability of the mixed complex $HgOHCO_3^-$.

5.8.1 The system HgS–H_2S–H_2O

The solubility of native cinnabar in acidic solutions (pH = 2–4) was investigated as a function of the concentration of H_2S; results are presented in Figure 5.25 and Table 5.13. There is a clear positive correlation

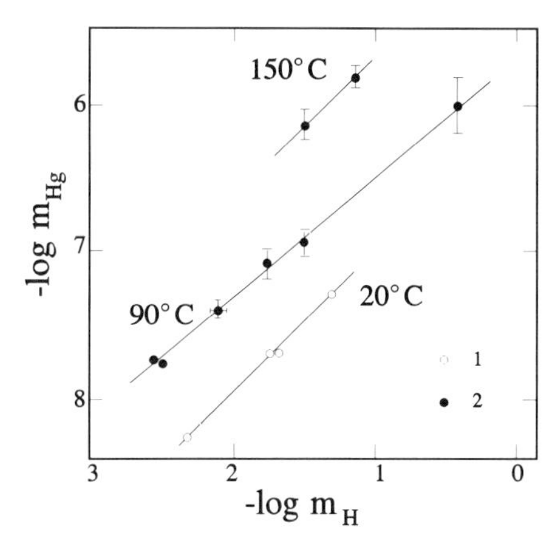

Figure 5.25. The solubility of cinnabar, $HgS_{(c)}$, as a function of H_2S concentration at 20, 90, 150°C and pH 2–4. 1 – Schwarzenbach and Widmer (1963); 2 – authors' data.

between log m_{Hg} and log m_{H_2S}, with slopes of 0.815 ± 0.08 and 0.97 ± 0.31 at 90 and 150°C, respectively. We interpret these results as indicating in both cases the formation of the complex $Hg(HS)^0_{2(aq)}$, according to:

$$\alpha\text{-}HgS_{(c)} + H_2S_{(aq)} = Hg(HS)^0_{2(aq)}. \qquad (5.35)$$

Our results are in agreement with the data of Schwarzenbach and Widmer (1963), although Barnes *et al.* (1967) have proposed a different stoichiometry of the complex $HgS(H_2S)^0_{2(aq)}$.

The values of $\log K_{(5.35)}$ are 5.31 ± 0.12 and 4.63 ± 0.05 at 90 and 150°C, respectively (Shikina *et al.*, 1981). Over the temperature range 25–150°C:

$$\log K_{(5.35)} = -\frac{(1474 \pm 130)}{T} - (1.16 \pm 0.33).$$

Table 5.13. The solubility of cinnabar at 150°C and saturated vapour pressure, with run duration of 12 days

$pH_{90°C}$	*Runs*	$-\log m_{H_2S}$	$-\log m_{Hg}$	$-\log K_{(5.35)}$
4	6	1.15	5.75 ± 0.05	4.60 ± 0.05
2	3	1.15	5.85 ± 0.01	4.70 ± 0.02
4	5	1.49	6.12 ± 0.04	4.62 ± 0.04
2	1	1.50	6.12	4.62
				4.63 ± 0.05

5.8.2 The system $HgO–NaHCO_3–H_2O$

The solubility of synthetic $HgO_{(c)}$ (montroydite) was investigated at 90°C, as a function of the concentration of HCO_3^- by Khodakovsky and Shikina (1981), and their results are given in Figure 5.26 and Table 5.14.

Table 5.14. The solubility of $HgO_{(c, red.)}$ in $NaHCO_3$ solutions at 90°C, with run duration of 20 days

$pH_{90°C}$		*Runs*	m ($mol \cdot kg\ H_2O^{-1}$)			$\log K_{(5.37)}$
			$NaHCO_{3,25°}$	$HCO^-_{3,90°}$	$m_{Hg,90°} \cdot 10^3$	
–		7	0	–	1.22 ± 0.06	–
8.37	0.01	3	0.00147	–	1.25 ± 0.03	–
8.48	0.02	3	0.0050	–	1.25 ± 0.02	–
8.57	0.02	3	0.0252	–	1.33 ± 0.01	–
8.65	0.02	3	0.050	–	1.40 ± 0.02	–
8.66	0.02	3	0.101	0.087	1.44 ± 0.01	– 2.60
8.61	0.02	3	0.246	0.190	1.77 ± 0.01	– 2.54
8.45	0.06	3	0.508	0.344	2.20 ± 0.05	– 2.54
8.28		1	1.030	0.595	2.63	– 2.63
						-2.56 ± 0.03

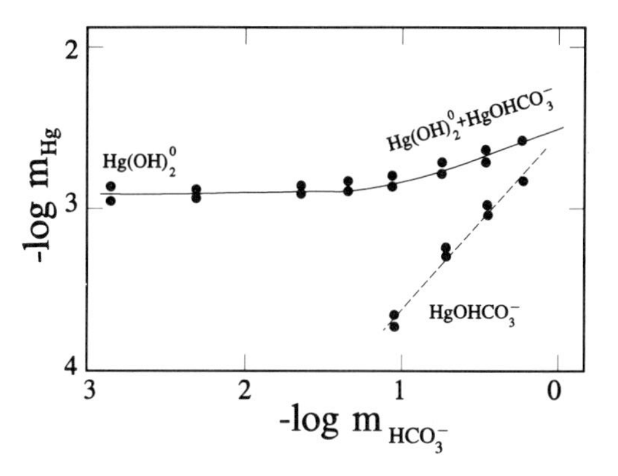

Figure 5.26. The solubility of montroydite, as a function of HCO_3^- concentration at 90°C.

At low HCO_3^- concentrations of up to 0.05 m, the solubility of $HgO_{(c)}$ remains practically constant. This may be accounted for by dissolution to $Hg(OH)^0_{2(aq)}$, according to the reaction:

$$HgO_{(c,\ red)} + H_2O = Hg(OH)^0_{2(aq)}, \tag{5.36}$$

for which $\log K_{90°C} = -2.91$. An increase in solubility with further increase in the concentration of HCO_3^- is probably due to the formation of the complex $HgOHCO_3^-$:

$$HgO_{(c,\ red)} + HCO_3^- = HgOHCO_3^- . \tag{5.37}$$

The value of $K^*_{(5.37)}$ is constant irrespective of ionic strength in the range $0.086 < I < 0.595$, and we therefore assume that the activity coefficients cancel and the equilibrium constant can be obtained directly from measured concentrations.

This assumption yields:

$$\log K_{(5.37)} = -2.56 \pm 0.03 \ \text{at} \ 90°C.$$

Over the temperature range 25 to 90°C:

$$\log K_{(5.37)} = -\frac{(318.79 \pm 235)}{T} - (1.68 \pm 0.65).$$

Thermodynamic calculations show that despite the high stability of carbonate complexes, including $HgOHCO_3^-$, they might be the dominant forms of mercury transport only at the very high activity of CO_2 in solution, since the stability of other aqueous Hg species also increases with temperature (Khodakovsky and Shikina, 1981).

5.9 DISCUSSION AND GEOLOGICAL IMPLICATIONS

The importance of complexing for the transport of ore metals in hydrothermal fluids and for enhancing the solubility of ore minerals is universally accepted, and the experimental results presented here are intended to complement existing data and improve models of ore transport and formation in supercritical aqueous fluids.

Precise knowledge of the form in which an ore constituent is present in solution is essential not only for characterization of its capacity for transport, but also to identify the conditions likely to lead to its precipitation as ore. For example, in the case of metals that are transported as chloride complexes, the precipitation of metal sulphides requires influx of H_2S. In contrast, H_2S loss is required to precipitate metals that are transported as hydrosulphide complexes.

Of the ore constituents studied, three elements, Hg, As, and Sb, are found in very low temperature ores. They are transported in aqueous solution without the presence of additional ligands for complexing. The high solubility of mercury reflects its high vapour pressure and the high solubility of its vapour in water as Hg^0 (Sorokin *et al.*, 1988). Under hydrothermal conditions, antimony and arsenic may be transported as very stable neutral hydroxy complexes $As(OH)^0_{3(aq)}$ and $Sb(OH)^0_{3(aq)}$. The solubility of the corresponding simple sulphides $As_2S_{3(c)}$ and $Sb_2S_{3(c)}$ in water at temperatures greater than 300 °C is so high (Figure 5.27) that they are unlikely to form at high temperatures, where complexes such as proustite, pyrargyrite, etc., commonly form instead. The strong temperature dependence of the solubility of $As_2S_{3(c)}$, $Sb_2S_{3(c)}$, and $HgS_{(c)}$ in water, accompanied with the formation of hydroxy complexes and the $Hg^0_{(aq)}$ species, means that temperature drop is the major mechanism for deposition of these ores. Other possible mechanisms include influx of H_2S into the system and lowering of pH.

Of the other elements investigated, Cu, Ag, and Au, which belong to group I of the periodic system, dissolve primarily as chloride and hydrosulphide complexes; hydroxy complexes are unimportant. The relative stability of their complexes, and its variation with temperature, are apparent from Figure 5.28. This illustrates that $MoCl_2^-$ is the most significant chloride complex, and that this type of complex becomes increasingly important relative to hydrosulphide complexes with increasing temperature and with decreasing mass from Au to Cu.

Hydrothermal ores of Cu, Ag and Au generally form in the range 200 to 500 °C, and chloride complexes are likely to predominate for transport of Ag and Cu throughout this range, especially where the ore fluids are brines. Au mineralization is often associated with dilute aqueous fluids and relatively low temperatures of mineralization, so that bisulphide complexes predominate (Seward, 1973, 1989). However, the effect of cooling in many types of ore systems means that sulphide complexes

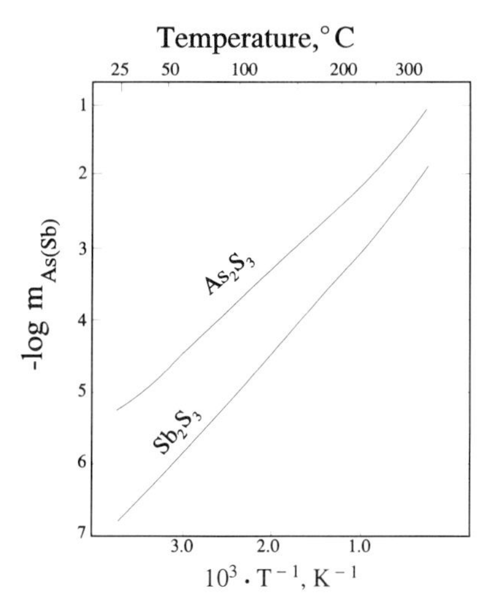

Figure 5.27. The solubility of the orpiment As_2S_3 and antimonyte Sb_2S_3 in water at saturated vapour pressure as a function of temperature.

of all three metals may play an important role in late stage redistribution of material in and around the deposit.

Of the elements studied here, Mo does not have close similarities to any of the others, but like arsenic and antimony, hydroxy complexes are

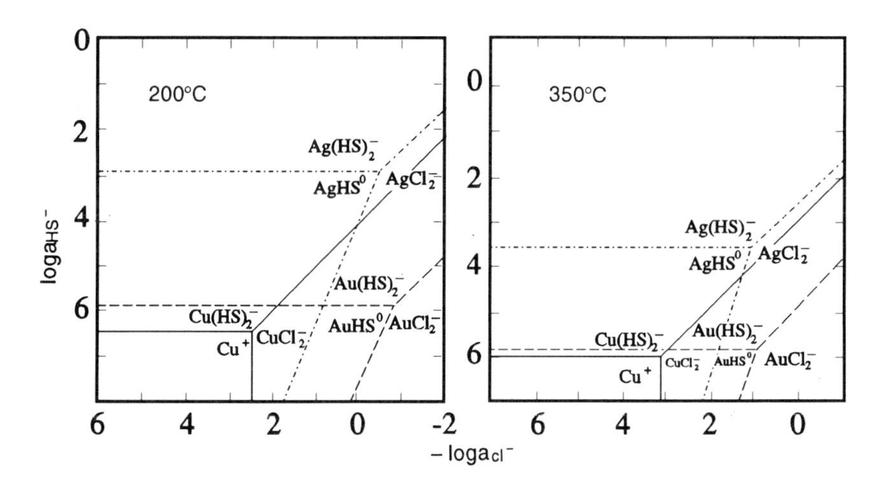

Figure 5.28. Fields of predominance of different species of gold, silver, and copper in aqueous solutions at 250 and 350°C and saturated vapour pressure.

important. Molybdenum occurs in hydrothermal solutions as Mo(VI) compounds, including both simple complexes, e.g. $HMoO_4^-$, and composite ones, such as $NaHMoO^0_{4(aq)}$ and $KHMoO^0_{4(aq)}$. Molybdenum sulphide (molybdenite) is however much less soluble than those of arsenic and antimony, and at temperatures below 300 °C, significant molybdenum concentrations may be transported in aqueous fluids only at very high pH values, which are unlikely to occur in nature. This may offer an explanation for the relatively high temperature of molybdenum precipitation in porphyry-type copper–molybdenum deposits (Roedder, 1971). Temperature decrease may be a major factor controlling the successive precipitation of molybdenum and copper from solution.

The evidence presented here and in previous studies (Khodakovsky, 1975; Barnes, 1979; Helgeson, 1970; Rafalsky, 1987) demonstrates that sulphide complexes do not play a leading role in the hydrothermal transport of most ore constituents. The corollary of this is that ore deposition should probably involve introduction of H_2S into the ore fluid, as well as a temperature decrease.

The thermodynamic information currently available allows complex computer models for hydrothermal systems to be developed, which potentially can allow for temperature and pressure gradients, a variety of hydrodynamic conditions, boiling and mixing of solutions, interaction with different country rocks, etc. The major difficulty that this creates is the need to obtain additional information on the parameters of natural ore-forming solutions and their evolution with space and time, in order to constrain models with large numbers of independent variables. To overcome this difficulty, new and elaborate geochemical investigations (e.g. paragenetic analysis, gas–liquid inclusions, permeability) of particular hydrothermal deposits will be needed.

ACKNOWLEDGEMENTS

The authors would like to thank N.N. Baranova and I.L. Khodakovsky of the Vernadsky Institute of Geochemistry and Analytical Chemistry, Russian Academy of Sciences, N.N. Akinfiev of the Moscow Geological and Prospecting Institute, and G.D. Mironova of the Institute of the Geology of Ore Deposits, Petrography, Mineralogy and Geochemistry who were involved in experimental investigations and treatment of results. The research described in this chapter was supported by the Russian Fund of Fundamental Investigations under Grant No. 93–058029.

REFERENCES

Akinfiev, N.N., Zotov, A.V. and Nikonorov, A.P. (1992) Thermodynamic analysis of equilibria in the As(III)–S(II)–O–H system. *Geochemistry International*, **29** (12), 109–21.

Akinfiev, N.N., Zotov, A.V. and Shikina, N.D. (1993) Experimental investigations and fitting of thermodynamic data in the system Sb(III)–S(II)–O–H. *Geokhimia*, (**12**) (in Russian).

Baes, C.F. and Mesmer, R.E. (1976) *The Hydrolysis of Cations*. Wiley Interscience, New York.

Baranova, N.N., Zotov, A.V., Bannykh, L.N., *et al.* (1983) The effect of redox conditions on the solubility of gold in water at 450°C and 500 atm. *Geochemistry International*, **20** (4), 117–22.

Barnes, H.L., Romberger, S.B. and Stemprok, K.M. (1967) Ore solution chemistry. II. Solubility of HgS in sulphide solutions. *Economic Geology*, **62** (7), 957–82.

Barnes, H.L. (1979) Solubility of ore minerals, in *Geochemistry of Hydrothermal Ore Deposits*, 2nd edn (ed. H.L. Barnes), Wiley–Interscience Publication, J. Wiley & Sons, N.Y–Chichester–Brisbane–Toronto.

Berne, E. and Leden, I. (1953). The solubility of silver chloride and the formation of complexes between silver and chloride ions. *J. Svensca. Kem. Tidskr.*, **65**, (5), 88–97.

Bryzgalin, O.V. (1986) Estimating dissociation constants in the supercritical region for some strong electrolytes from an electrostatic model. *Geochemistry International*, **23** (2), 84–95.

Buketov, E.A., Ugorets, M.S. and Ahmetov, K.M. (1967) The behaviour of cupric hydroxide and oxide in alkali solutions. *Trudy khimiko-metallurgicheskogo instituta AN Kaz. SSR*, **III**, 3–7 (in Russian).

Crerar, D.A. and Barnes, H.L. (1976) Ore solution chemistry. V. Solubilities of chalcopyrite and chalcocite assemblages in hydrothermal solutions at 200 to 350°C. *Economic Geology*, **71** (4), 772–94.

Eary, L.E. (1992) The solubility of amorphous As_2S_3 from 25 to 90°C. *Geochimica et Cosmochimica Acta*, **56**, 2267–80.

Frantz, J.D. and Marshall, W.L. (1984) Electrical conductances and ionization constants of salts, acids, and bases in supercritical aqueous fluids. I. Hydrochloric acid from 100 to 700°C and at pressures to 4000 bars. *American Journal of Science*, **284** (6), 651–67.

Hayashi, K. and Ohmoto, H. (1991) Solubility of gold in NaCl- and H_2S-bearing aqueous solutions at 250–350°C. *Geochimica et Cosmochimica Acta*, **55**, 2111–26.

Helgeson, H.C. (1970) A chemical and thermodynamic model of ore deposition in hydrothermal systems. *Special Paper of Mineralogical Society of America*, **3**, 155–86.

Henley, R.W. (1973) Solubility of gold in hydrothermal chloride solutions. *Chemical Geology*, **11** (2), 73–87.

Ivanenko, V.V. and Pamfilova, L.A. (1975) Determination of stability constants of chloride silver complexes by solubility method at 60–160°C. *Geokhimia*, **4**, 566–75 (in Russian).

IVTANTERMO data base. (1990) Institute of high temperatures, Russian Academy of Science.

Johnson, J.M., Oelkers, E.H. and Helgeson, H.C. (1991) SUPCRT 92: A Software Package for Calculating the Standard Molal Thermodynamic Properties of Minerals, Gases, Aqueous Species, and Reactions from 1 to 5000 bars and 0 to 1000°C. *Computers Geoscience*, **18**, 899–947.

Khodakovsky, I.L. (1975) Investigations of the thermodynamics of aqueous solutions at high temperatures and pressures. Ph.D. thesis, Moscow (in Russian).

Khodakovsky, I.L. and Shikina, N.D. (1981) The role of carbonate complexes in mercury transport in hydrothermal solutions (experimental studies and thermodynamic analysis). *Geochemistry International*, **18** (3), 32–43.

Kishima, N. and Sakai, H. (1984) Fugacity – concentration relationship of dilute hydrogen in water at elevated temperature and pressure. *Earth and Planetary Sciences Letters*, **67** (1), 79–86.

Kozlov, Ye.D. (1982) Migration of antimony and mercury in hydrothermal solutions based on experimental data. Ph. D. thesis, Moscow (in Russian).

Krupp, R.E. (1988) Solubility of stibnite in hydrogen sulphide solutions, speciation, and equilibrium constants, from 25 to 350°C. *Geochimica et Cosmochimica Acta*, **52** (12), 3005–15.

Kudrin, A.V. (1985) The solubility of tugarinovite MoO_2 in aqueous solutions at elevated temperatures. *Geochemistry International*, **22** (9), 126–37.

Kudrin, A.V. (1989) Behaviour of Mo in aqueous NaCl and KCl solutions at 300–450°C. *Geochemistry International*, **26** (8), 87–99.

Learned, R.E., Tunell, G. and Dickson, F.W. (1974) Equilibria of cinnabar, stibnite and saturated solutions in the system $Hg–Sb_2S_3–Na_2S–H_2O$ from 150 to 250°C at 100 bars with implications concerning ore genesis. *Journal of Research of the US Geological Survey*, **2** (4), 457–66.

Levin, K.A. (1992) Experimental and thermodynamic study of the stabilities of silver chloride complexes in KCl and NaCl solutions up to 7 m at 300°C. *Geochemistry International*, **29** (5), 103–8.

McDowell, L.A. and Johnston, H.L. (1936) The solubility of cupric oxide in alkali and the second dissociation constant of cupric acid. The analysis of very small amounts of copper. *Journal of American Chemical Society*, **53** (10), 2009–14.

Marshall, W.L. and Franck, E.U., 1981. An equation for the ion product of water, 0–1000°C; 1–1000 bars. *J. Phys. Chem. Ref. Data*, **10**, N 2, 295–304.

Mironova, G.D. and Zotov, A.V. (1980) Solubility studies of the stability of As(III) sulphide complexes at 90°C. *Geochemistry International*, **17** (2), 46–54.

Mironova, G.D., Zotov, A.V. and Gul'ko, N.I. (1984) Determination of the solubility of orpiment in acid solutions at 25–150°C. *Geochemistry International*, **21** (1), 53–9.

Mironova, G.D., Zotov, A.V. and Gul'ko, N.I. (1990) The solubility of orpiment in sulphide solutions at 25–150°C and the stability of arsenic sulphide complexes. *Geochemistry International*, **27** (12), 61–73.

Morrison, G.M., Rose, W.J. and Jaireth, S. (1991) Geological and geochemical controls on the silver content (fineness) of gold in gold–silver deposits. *Ore Geology Reviews*, **6** (4), 333–64.

Naumov G.B., Ryzhenko B.N. and Khodakovsky I.L. (1974) *Handbook of Thermodynamic Data*: U.S. Survey Rept. Inv. USGS–WRD–74–001.

Nikolaeva, N.M., Yerenburg, A.M. and Antipina, V.A. (1972) Temperature dependence of the standard potential of halide complexes of gold. *Izvestia Sibirskogo otdeleniya Akademii Nauk SSSR, ser. Khim*, **4**, 126–9 (in Russian).

Nikolaeva, N.M. (1982) *Chemical Equilibrium in Aqueous Solutions at High Temperature*. Nauka Press, Novosibirsk (in Russian).

Popova, M.Ya., Khodakovsky, I.L. and Ozerova, N.A. (1975) Measurement of the thermodynamic parameters of antimony hydroxo complexes and hydrofluoride complexes up to 200°C. *Geochemistry International*, **12** (3), Abstracts of translations not selected for publication.

Pouradier, I. and Gadet, M.C. (1973) *Complexe mixte chlorohydroxe de l'argent* (I). Comptes Rendus de l'Academie des Sciences, **277** C (16), 655–9.

Rafalsky, R.P. (1987) Ore elements in hydrothermal solutions. *Zapiski Vsesouznogo Mineralogicheskogo Obshchestva*, **116** (2), 178–91 (in Russian).

Renders, P.J. and Seward, T.M. (1989) The stability of hydrosulphido- and sulphido-complexes of Au(I) and Ag(I) at 25°C. *Geochimica et Cosmochimica Acta*, **53** (2), 245–53.

Roedder, E. (1971) Fluid inclusion studies on the porphyry-type ore deposits at

Bingham, Utah, Butte, Montans and Climax, Colorado. *Economic Geology*, **66** (1), 98–120.
Ryabchikov, I.D. and Orlova, G.P. (1984) Gold in magmatic fluids. In *Phiziko-khimicheskie Modeli Petrogeneza i Rudoobrazovaniya*. Nauka Press, Novosibirsk, 103–11 (in Russian).
Schwarzenbach, G. and Widmer, M. (1963) Die Zoslichkeit von Metallsulfieden. I. Schwarzes Quecksilbersulphide. *Helv. Chim. Acta*, **46** (7), 2613–28.
Seward, T.M. (1973) Thio complexes of gold and the transport of gold in hydrothermal ore solutions. *Geochimica et Cosmochimica Acta*, **37** (3), 379–99.
Seward, T.M. (1976) The solubility of chloride complexes of silver in hydrothermal solutions up to 350°C. *Geochimica et Cosmochimica Acta*, **40** (11), 1329–40.
Seward, T.M. (1989) The hydrothermal chemistry of gold and its implications for ore formation: Boiling and conductive cooling as examples. In *Economic Geology, Monograph 6, The geology of gold deposits: the perspectives in 1988*, 398–404.
Shenberger, D.M. and Barnes, H.L. (1989) Solubility of gold in aqueous sulfide solutions from 150 to 350°C. *Geochimica et Cosmochimica Acta*, **53** (2), 269–78.
Shikina, N.D., Zotov, A.V. and Khodakovsky, I.L. (1981) An experimental investigation of equilibria in the α–HgS–H_2S–H_2O system at 90 and 150°C. *Geochemistry International*, **18** (2), 109–17.
Shikina, N.D. and Zotov, A.V. (1991) Thermodynamic properties of $Sb(OH)^0_{3(sol)}$ up to 723.15 K and 1000 bar. *Geochemistry International*, **28** (7), 97–103.
Shock, E.L. and Helgeson, H.C. (1988) Calculation of the thermodynamic and transport properties of aqueous species at high pressures and temperatures: correlation algorithms for ionic species and equation of state predictions to 5 kb and 1000°C. *Geochimica et Cosmochimica Acta*, **52** (8), 2009–36.
Shock, E.L., Helgeson, H.C. and Sverjensky, D.A. (1989) Calculation of the thermodynamic and transport properties of aqueous species at high pressures and temperatures: Standard partial molal properties of inorganic neutral species. *Geochimica et Cosmochimica Acta*, **53** (9), 2157–83.
Solov'eva, V.D., Svirchevskaya, E.G., Bobrova, V.D. and El'ts, N.M. (1973) Solubility of cupric, cadmium and indium oxides in NaOH solutions. *Trudy instituta metallurgii i obogashcheniya AN Kaz. SSR*, **49**, 37–41 (in Russian).
Sorokin, V.I., Pokrovskii, V.A. and Dadze, T.P. (1988) *Physical–chemical Conditions of Antimony–Mercury Ore Formation*, Nauka Press, Moscow (in Russian).
Spycher, N.F. and Reed, M.N. (1989) As (III) and Sb (III) sulphide complexes: An evaluation of stoichiometry and stability from existing data. *Geochimica et Cosmochimica Acta*, **53** (9), 2185–94.
Turnbull, A.G. and Wadsley, M.W. (1988) *The CSIROSGTE Thermochemistry System*. CSIRO Div. Miner. Chem., Port Melbourne.
Utkina, I.N., Kunin, T.I., Shutov, A.A. (1966) Solubility of cuprous chloride in NaCl solutions. *Izvestia VUZov, Khimia i khimicheskaya tekhnologia*, **12** (6), 706–11 (in Russian).
Var'yash, L.N. and Rekharskiy V.I. (1981) On the behaviour of monovalent copper in chloride solutions. *Geochemistry International*, **18** (7), 1003–8.
Var'yash L.N. (1986) Hydrolysis of Cu (II) at 25–350°C. *Geochemistry International*, **23** (1), 82–91.
Var'yash, L.N. (1989) Equilibria in the Cu–Cu_2O–H_2O system at 150–450°C. *Geochemistry International*, **26** (10), 80–90.
Var'yash, L.N. (1991) An experimental study of Cu (I) complexing in NaCl solutions at 300 and 350°C. *Geokhimia*, (**8**), 1166–74 (in Russian).
Webster, J.G. (1990) The solubility of As_2S_3 and speciation of As in dilute and

sulphide-bearing fluids at 25 and 90 °C. *Geochimica et Cosmochimica Acta*, **54** (4), 1009–18.

Weissberg, B.G., Dickson, F.W. and Tunell, G.J. (1966) Solubility of orpiment (As_2S_3) in Na_2S–H_2O at 50–200 °C and 100–1500 bars, with geological applications. *Geochimica et Cosmochimica Acta*, **30** (8), 815–27.

Wood, F., Crerar, D.A. and Borcsic, M.P. (1987) Solubility of assemblage pyrite–pyrrhotite–magnetite–sphalerite–galena–gold–bismuthimite–argentite–molybdenite in H_2O–NaCl–CO_2 solution from 200 to 350 °C. *Economic Geology*, **82** (7), 1864–87.

Zotov, A.V., Levin, K.A., Kotova, Z.Yu. and Volchenkova, V.A. (1982) An experimental study of the stability of silver hydroxychloride complexes in hydrothermal solutions. *Geochemistry International*, **19** (4), 151–64.

Zotov, A.V., Baranova, N.N., Dar'yina, T.G., *et al.* (1985) The stability of $AuOH^0_{sol.}$ in water at 300–500 °C and 500–1500 atm. *Geochemistry International*, **22** (5), 156–61.

Zotov, A.V., Levin, K.A., Khodakovsky, I.L. and Kozlov, V.K. (1986a) Thermodynamic parameters of Ag^+ in aqueous solution at 273–573 K. *Geochemistry International*, **23** (3), 23–33.

Zotov, A.V., Levin, K.A., Khodakovsky, I.L. and Kozlov, V.K. (1986b) Thermodynamic parameters of Ag (I) chloride complexes in aqueous solution at 273–623 K. *Geochemistry International*, **23** (9), 103–16.

Zotov, A.V. and Baranova, N.N. (1989) Thermodynamic properties of the aurochloride solute complex $AuCl_2^-$ at temperatures of 350–500 °C and pressures of 500–1500 bars. *Sciences Geologiques, Bulletin*, **42** (4), 335–42.

Zotov, A.V., Baranova, N.N. Dar'yina, T.G. and Bannykh, L.N. (1989) Gold (I) complexing in the KCl–HCl–H_2O system at 450 °C and 500 atm. *Geochemistry International*, **26** (11), 66–75.

Zotov, A.V., Baranova, N.N., Dar'yina, T.G. and Bannykh, L.N. (1991) The solubility of gold in aqueous chloride fluids at 350–500 °C and 500–1500 atm: thermodynamic properties $AuCl^-_{2(sol)}$ up to 750 °C and 5000 atm. *Geochemistry International*, **28** (2), 63–71.

GLOSSARY OF SYMBOLS

T	temperature
P	pressure
K	equilibrium constant
K*	concentration constant
z	ionic charge
I	ionic strength
A, B, *å*	parameters in the Debye–Hückel equation
a_1–a_4, c_1, c_2, ω	parameters in the revised HKF equation of state
f	fugacity
a	activity
m	molal concentration (mol.kg H_2O^{-1})
(aq)	aqueous solution
(c)	crystal phase
(g)	gaseous phase
ΔG^0_f	standard molal Gibbs energy of formation

ΔH_f^0	standard molal enthalpy of formation
$\underline{S}^0$	standard molal entropy
$\overline{S}^0$	standard partial molal entropy
$\underline{V}^0$	standard molal volume
$\overline{V}^0$	standard partial molal volume
$\underline{C}_P^0$	standard molal isobaric heat capacity
$\overline{C}_P^0$	standard partial molal isobaric heat capacity

CHAPTER SIX

The influence of acidic fluoride and chloride solutions on the geochemical behaviour of Al, Si and W

Georgiy P. Zaraisky

6.1 INTRODUCTION

Hydrothermal ore deposits are often accompanied by extensive aureoles of metasomatic alteration in the surrounding country rocks and provide a clear demonstration of fluid activity in the crust. Among the most extreme examples of chemical interaction between infiltrating hydrothermal fluids and rocks are zones of acid metasomatism ('acidic leaching' of Korzhinskiy, 1953; or 'hydrogen metasomatism' of Hemley and Jones, 1964). The ultimate result of the most intense acidic metasomatism is the loss of most principal rock-forming components, leading to a final product made up of silica and aluminosilicate minerals or of quartz alone. Nevertheless, SiO_2 and at times Al_2O_3 are also involved in the hydrothermal transport during acid metasomatism.

The aim of this study is to consider the influence of chloride and fluoride in hydrothermal fluids on the transport of Si, Al, and W. As a suitable natural example to model, we have chosen the well-studied W greisen deposit at Akchatau, Central Kazakhstan (Zaraisky *et al.*, in press), for which fluid inclusion studies (Doroshenko *et al.*, 1981) have

Fluids in the Crust: Equilibrium and transport properties
Edited by K.I. Shmulovich, B.W.D. Yardley and G.G. Gonchar.
Published in 1994 by Chapman & Hall, London. ISBN 0 412 56320 7

established the involvement of chloride and fluoride solutions in the hydrothermal process. To explain the geochemical behaviour of Si, Al, and W during the formation of such deposits, we have invoked the results of our experiments on quartz, corundum and wolframite solubilities, on metasomatic zoning modelling, as well as on diffusion and infiltration mass transport in rocks during acidic metasomatism.

6.2 GAIN AND LOSS OF ROCK-FORMING COMPONENTS DURING GREISENIZATION: AN EXAMPLE FROM THE AKCHATAU W–Mo DEPOSIT

The metasomatic formation of greisens in granites and other rocks is generally accompanied by redistribution of large amounts of rock-forming components, notably SiO_2, Na_2O and K_2O, and in a few cases, Al_2O_3. Mass transfer of silica is common during many types of metasomatism and is inherent in processes of acidic leaching, typically leading to intense silicification of the parent rocks. In general, alumina is one of the most inert components, and its content remains practically unchanged during most metasomatic processes, providing a basis for the method by which gains and losses of other components may be evaluated quantitatively by comparing the proportion of Al_2O_3 in primary and altered rock (Kazitsin and Rudnik, 1968; Gresens, 1967). For metasomatic greisenization, however, such an approach is generally invalid due to the considerable mass transfer of alumina involved in the alteration of granitoid rocks to greisens.

The ores of the Akchatau W–Mo deposit of Central Kazakhstan occur in domes or cupolas of upper Permian leucocratic granites. They comprise subvertical vein-like greisen bodies with quartz veins in the axial part. The greisen bodies show a distinct horizontal and vertical zoning with the sequence: quartz greisen, quartz–topaz greisen, quartz–muscovite greisen, and greisenized granite (Figure 6.1).

The migration of rock-forming components during greisenization was evaluated quantitatively using the data of Gotman and Malakhova (1965) who carried out furrow sampling from greisen zones in one of the major ore bodies at Akchatau. Samples were taken from 5 levels on a continuous basis across the strike of the greisen body, including all the metasomatic zones, from the axial quartz vein to unaltered granite. Using their data, average values for the chemical composition of the greisen metasomatic zones have been calculated for the upper, barren portion of the deposit (at 0 m level), for the major production levels (20, 60, and 120 m), and for the underlying barren zones (at 240 m level) (Table 6.1).

In the main production zone (at 30–120 m level) the greisen body shows a single type of lateral zoning throughout its length and this

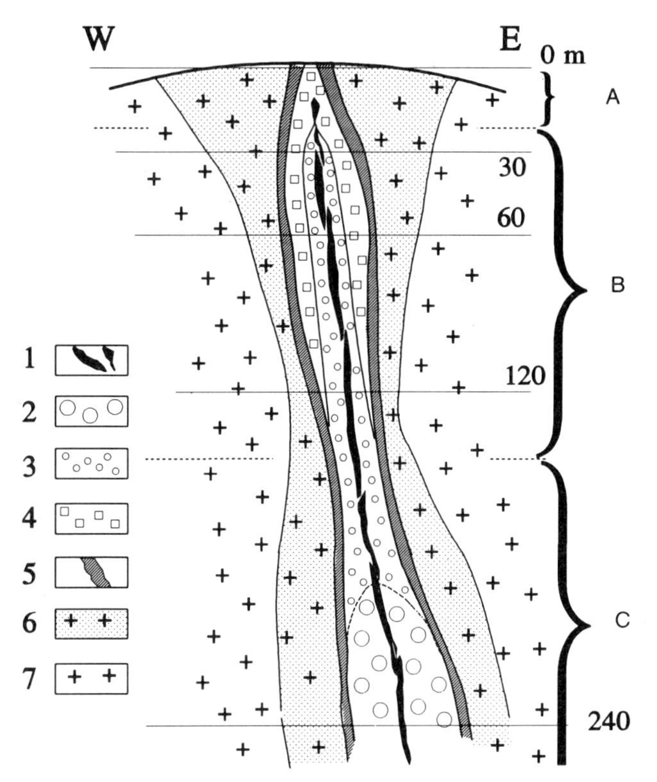

Figure 6.1. Schematic diagram showing zoning in the Akchatau greisen deposits: (1) quartz vein, (2) porous quartz greisen, (3) dense quartz greisen, (4) quartz–topaz greisen, (5) quartz–muscovite greisen, (6) greisenized granite, (7) leucogranite. A 'above-ore' level, B major ore-productive levels, C 'under-ore' level.

zoning is symmetric about the axial quartz vein (Figure 6.2). The metasomatic column depicted in the figure reflects the averaged greisen zoning corresponding to the thickness and average chemical compositions of the zones included in Table 6.1. The results of the calculations for gains and losses of rock-forming components are shown in Figure 6.3 for each of the metasomatic zones. Table 6.1 presents average values of density, specific gravity and porosity of the initial granite and of metasomatic rocks formed instead of this granite: greisenized granite, quartz–muscovite greisen, quartz–topaz greisen, dense and porous quartz greisens. The density of rocks was measured as the ratio of the sample weight in grams to its volume in cubic cm, i.e. including pores. Specific gravity was calculated from the specific gravity of the mineral matter with regard to the quantitative mineral composition of the rocks, i.e. excluding pores. The calculations for gains and losses of rock-forming components used the measured values of rock densities (including pores), assuming that their volume remains constant during the process of greisenization of granite.

Table 6.1. Some physical properties and averaged chemical composition (Wt %) of Greisen zones at the earth's surface, in the major greisen body and at the lower level (calculated after Gotman and Malakhova, 1965)

Rock	Thick (m)	Samp quan	Den-sity	Specf gravt	Poros (vol%)	SiO_2	TiO_2	Al_2O_3	Fe_2O_3	FeO	Fe	MgO	CaO	Na_2O	K_2O	P_2O_5	H_2O	S	F	CO_2	Total
						1) 0 m upper level (surface, trial trench)															
Quartz–topaz greisen	0.55	1	2.60	2.96	12.2	78.79	0.10	14.01	1.21	2.57	0.14	0.17	no	0.12	0.77	ck	1.51	0.16	0.76	no	100.32
Quartz–muscovite greisen	0.45	2	2.57	2.74	6.2	77.05	0.13	12.71	1.30	1.19	0.05	0.16	no	1.63	3.78	0.01	1.87	0.06	0.22	0.02	100.09
Greisenized granite	13.0	2	2.49	2.65	6.0	76.20	0.13	12.75	0.71	0.83	0.03	0.11	0.26	3.45	4.72	0.02	0.66	0.04	0.17	no	100.07
						2) 30, 60 and 120 m intermediate levels (level working)															
Dense quartz greisen	1.09	10	2.64	2.69	1.86	90.52	0.13	3.27	0.08	1.41	1.02	0.08	0.78	0.14	0.50	0.04	0.56	1.16	0.42	0.07	100.18
Quartz–topaz greisen	1.10	10	2.84	3.21	11.0	72.61	0.13	12.52	0.00	0.92	4.34	0.08	1.04	0.14	0.76	0.05	1.93	4.96	0.71	0.10	100.29
Quartz–muscovite greisen	0.22	17	2.70	2.87	5.92	75.91	0.12	11.73	1.15	1.47	1.03	0.10	0.92	0.44	3.72	0.03	1.72	1.18	0.77	0.05	100.34
Greisenized granite	3.41	19	2.50	2.70	7.40	75.64	0.13	12.48	0.34	0.97	0.24	0.11	0.65	3.11	5.19	0.02	0.51	0.27	0.31	0.10	100.13
Medium-grained granite	–	8	2.57	2.68	4.10	75.75	0.12	12.06	0.48	1.06	0.28	0.08	0.70	3.56	4.92	0.03	0.41	0.32	0.33	0.04	100.14
						3) 240 m lower level (borehole 280)															
Porous quartz greisen	9.55	2	2.61	2.74	4.74	90.64	0.14	3.59	0.49	1.37	0.36	0.30	0.36	0.13	1.43	0.03	0.41	0.41	0.25	0.19	100.10
Quartz–muscovite greisen	0.95	2	2.67	2.78	4.00	82.58	0.15	7.86	1.25	1.26	0.62	0.17	0.52	0.12	2.88	0.03	1.44	0.70	0.44	0.18	100.20
Greisenized granite	7.50	2	2.53	2.68	5.60	77.18	0.15	11.36	0.31	1.05	0.14	0.21	0.57	2.52	5.32	0.04	0.71	0.16	0.29	0.11	100.13

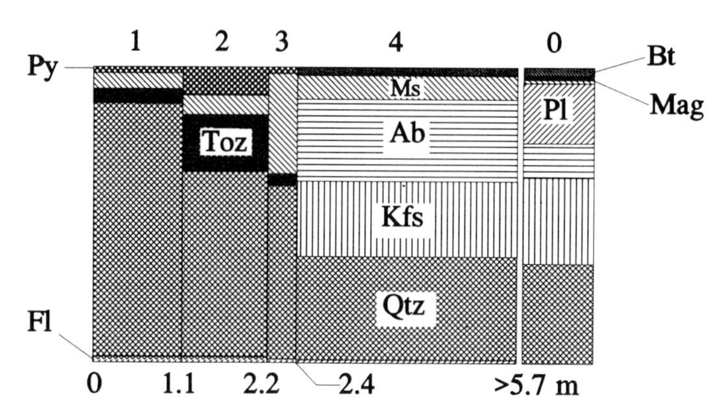

Figure 6.2. Metasomatic zoning and modal composition of the Akchatau greisen zones from the averaged data for furrow sampling (see Table 6.1; 30, 60, and 120 m levels).

As the degree of greisenization increases, the SiO_2 content of rock gradually increases, and the alkali metals K_2O and Na_2O are removed. Alumina is gained in significant amounts during the formation of the quartz–muscovite and quartz–topaz greisen in particular, but is strongly depleted in the inner zone of the quartz greisen. The greisen zones are all enriched in Fe and volatiles (H_2O, S, F, CO_2) relative to the initial granite. The total amount of CaO, MgO and TiO_2 in the Akchatau leucogranites and greisens does not exceed 1%, and so the role of these components in the overall mass transfer of a substance is negligible.

Overall, all the greisen zones at Akchatau show a net increase in material. Only in the outermost zone of the greisenized granite is there a small net loss, although, on the whole, the chemical composition here is little changed. Within a single greisen body, from top to bottom, marked changes in the content of rock-forming components (chemical zoning) may be observed, which correlate well with the vertical metasomatic zoning (mineralogical zoning) in greisens (see Figure 6.1). SiO_2 gain in different greisen zones at all depths varies from 33 to 440 $kg \cdot m^{-3}$ (see Table 6.1). Note that the degree of silicification and the total amount of added silica increase with depth. At intermediate and upper levels, SiO_2 is removed from the outer zone of the greisenized granite in quantities from 49 to 56 $kg \cdot m^{-3}$ which are sufficient to account for its accumulation in amounts up to 400 $kg \cdot m^{-3}$ in the narrower inner zones of the greisen body. However, at 240 m, there is a gain of silica in all the greisen zones, suggesting its introduction from a deeper-seated region of the granite massif.

Al_2O_3 exhibits opposite behaviour. At upper levels, it is added to different zones in amounts from 5 to 55 $kg \cdot m^{-3}$ during greisen formation. As the depth increases, its content in each zone decreases with large losses of 100 to 225 $kg \cdot m^{-3}$ at the barren 240 m level. Thus, the normally inert

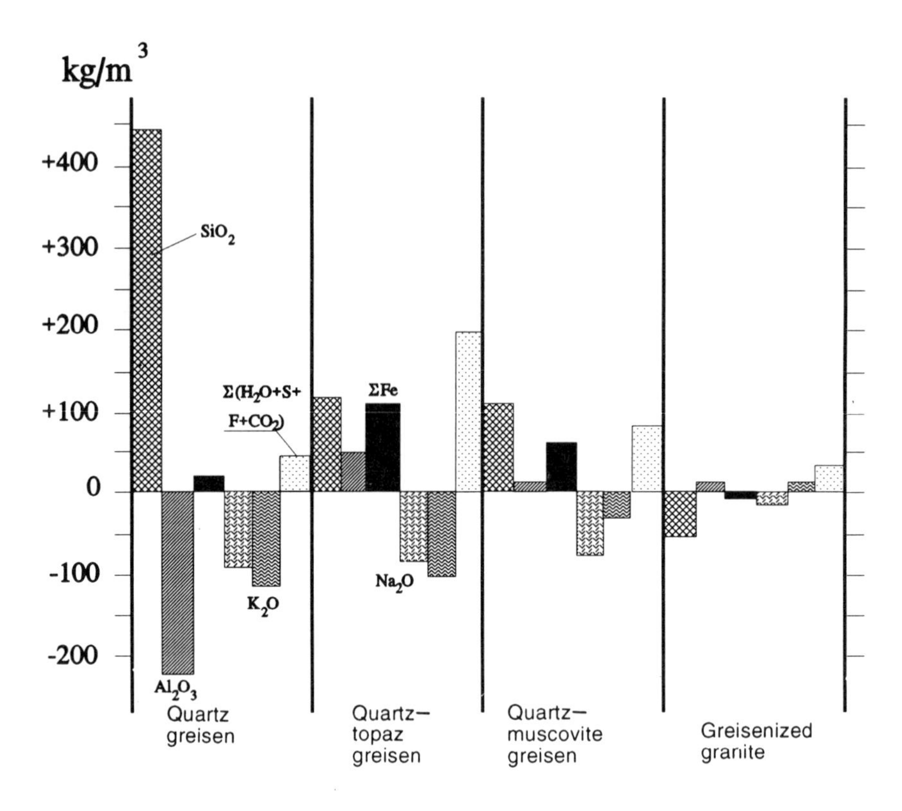

Figure 6.3. Gain and loss of principal rock-forming components in different greisen zones from the data in Table 6.1.

Al_2O_3 is rather mobile during the greisenization of the Akchatau leucogranites, being intensely removed from the deep-seated portions of the greisen bodies and redeposited at higher levels. One of the aims of this study was to evaluate whether this behaviour may result from the action of fluoride-rich solutions, uncommon in most metasomatic settings, but clearly important to greisenization because of the presence of topaz, fluorite and F-rich micas in greisens.

6.3 RESULTS OF MINERAL SOLUBILITY EXPERIMENTS

The mobility of chemical components during metasomatism is largely determined by the solubility of rock-forming minerals, which, in turn, is strongly dependent on the pH, redox state and ligand chemistry of the hydrothermal solution. The present study is primarily concerned with new results on the solubility of quartz, corundum and wolframite in acidic fluoride and chloride solutions.

6.3.1 The influence of acidic chloride and fluoride solutions on quartz solubility

The solubility of quartz in pure water at temperatures in the range 400–600 °C and 1 kbar pressure is $2.9–5.0 \times 10^{-2}$ molal (Walther and Helgeson, 1977). The chemical reaction for the dissolution of quartz proceeds with the formation of orthosilicic acid:

$$SiO_2 \text{ (quartz)} + 2H_2O = H_4SiO_4^0\text{(aq)}.$$

In more general terms, this reaction may be written as:

$$rSiO_2 \text{ (quartz)} + nH_2O = (SiO_2)_r \times nH_2O \text{ (aq)},$$

keeping in mind that aqueous silica may occur in both polymeric and monomeric forms and that the hydration number of water molecules may also be variable (Weill and Bottinga, 1970; see also Chapter 5, this volume). At room temperature and atmospheric pressure, silica in solution largely occurs as the monomeric neutral species $H_4SiO_4^0$ (Kopeikin and Mikhailov, 1970), which also predominates over a wide range of pH values from 8 to 1 (Krauskopf, 1956; Okamoto *et al.*, 1957). Data from hydrothermal experiments suggest that the same species predominates under supercritical conditions as well (Mosebach, 1957; Walther and Helgeson, 1977; Rimstidt and Barnes, 1980; see Chapter 5, this volume). Only at extreme temperatures and pressures, may other silica species become important (Weill and Fyfe, 1964; Anderson and Burnham, 1965).

The independence of quartz and amorphous silica solubilities from acidity has been shown experimentally for dilute HNO_3, HCl and other solutions (Elmer and Nordberg, 1958; Balitsky, 1978; Kamiya *et al.*, 1974; Sorokin and Dadze, 1980, Ch. 4, this volume). It results from the very weak dissociation of orthosilicic acid in neutral and acidic media and the lack of silica complexes with Cl, NO_3, SO_4 and many other common anions. At very low pH, the solubility of silica decreases compared to the solubility in pure water, possibly due to the decreased water activity, and it is clear that quartz solubility is decreased, not enhanced, in acid chloride solutions (Ostapenko and Arapova, 1971).

A very different situation arises with acidic fluoride solutions because F forms stable complexes with SiO_2. The increase in silica solubility in HF solutions formed the basis of classical silicate analysis and has been investigated for many years (Iler, 1979). However, for the high-temperature range characteristic of greisenization (300–600 °C), the chemistry of the system SiO_2–H_2O–HF remains poorly understood. In recent years new results have been obtained in our laboratory by Shapovalov and Balashov (1990). They measured the solubility of α-quartz over the HF concentration range 10^{-4} to 11.5 molal at 300, 400, 500 and 600 °C and 1 kbar. Experiments were carried out in sealed autoclaves (Chapter 3, 3.1) using gold- or platinum-lined, sealed containers. The solubility was determined from the weight loss of quartz single crystals after runs which lasted from 216 to 456 hours. For HF concentrations $>2\times10^{-2}$ molal, the solubility of quartz was found to increase sharply at all the temperatures investigated, but at lower HF concentrations, the solubility differed little from that in pure water (Figure 6.4). In 11.5 m HF solution at T = 600°C, the concentration of silica was very high approaching 5.6 m (or 25% by weight), approximately 100 times greater than the solubility in pure water under the same P–T conditions. Thermodynamic treatment of the experimental results, using an unpublished program 'x88' by Balashov showed that the data may be best described in the region of high HF concentrations if the dominant Si-bearing complex in solution is $Si(OH)_3F^0_{aq}$ (Shapovalov and Balashov, 1990). Nevertheless, these workers do not rule out the possibility that other complexes of the general formula $Si(OH)_m F^0_n$ (where m + n = 4) are present in solution as well. Haselton (1984), who investigated quartz solubility in dilute HF solutions at T = 600°C, P = 1 kbar, also found an increase in the SiO_2 solubility relative to its solubility in pure water. However, he inferred that the dominant Si complex was SiF_3OH^0. The solubility of quartz measured in the presence of KF and NaF at 100 to 350°C (or above) and at saturated vapour pressure remains almost the same as in pure water or only slightly higher (Balitsky *et al.*, 1971). This is attributed to weak NaF–KF hydrolysis under these conditions.

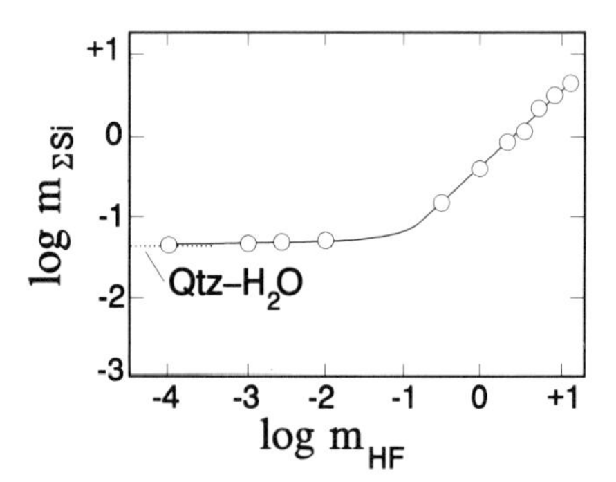

Figure 6.4. The dependence of quartz solubility on the concentration of HF at 600°C and 1 kbar, after Shapovalov and Balashov (1990).

6.3.2 The influence of acidic fluoride and chloride solutions on corundum solubility

The upper limit for aluminium concentration in natural waters is defined by corundum solubility at high temperatures and by diaspore solubility at lower temperatures. According to the data of Hemley *et al.* (1980), corundum and diaspore equilibrate at 394 ± 10°C at P = 1 kbar. The fact that the equilibrium concentration of aluminium in water in the presence of corundum or diaspore is about three orders of magnitude lower than the concentration of aqueous silica in equilibrium with quartz is of great geochemical significance. For example, at temperatures between 400 and 700°C and 1 kbar, $\Sigma Al_{(aq)} \approx 5 \times 10^{-5}$–$10^{-4}$ m (Ragnarsdottir and Walther, 1986). It is because of its low solubility that alumina is generally immobile during metasomatic events.

Unlike quartz solubility, the solubility of corundum increases with decreasing pH because of ionization of the uncharged complex $Al(OH)^0_{3(aq)}$, predominant in neutral fluids. This effect is most pronounced at low temperatures. Alumina solubility also increases in alkaline solutions, but the solubility minimum of crystalline alumina shifts to lower pHs as the temperature rises (Castet, 1991). Therefore, at high temperatures, increasing acidity from neutral first reduces Al solubility; only at very low pH values (< 2) does it increase. Overall, the influence of low pH on Al_2O_3 solubility is small and cannot change its geochemical inertness.

Another contrast between alumina and silica in solution is that Al can form complexes with Cl. However, Korzhinskiy (1987) showed that corundum solubility in HCl solutions at 450–700°C and 2 kbar was almost the same as in pure water ($\Sigma Al = n \times 10^{-5}$ m, where n = 1.2 to 8 at different temperatures), up to mHCl = 0.1. Only in very concentrated HCl solutions does the solubility increase dramatically, up to 10^{-2} m ΣAl in 1.9 m HCl solution. Based on the slope of solubility curves, Korzhinskiy suggested that the dominant Al-bearing species in solution is $Al(OH)_2Cl^0$ at mHCl = 0.1 to 0.5, and $Al(OH)Cl^0_2$ at mHCl > 0.5.

The presence of F has a much more pronounced effect on alumina solubility in aqueous solutions, although most experimental studies have been limited to low temperatures (e.g. Baumann 1961; Sazarin *et al.*, 1983, Couturier, 1986; Sanjuan and Michard, 1987).

Here, there is experimental evidence for the existence of neutral and charged aluminium complexes of the general formulas AlF_n^{3-n} (Baumann, 1961) and $Al(OH)_mF_n^{(3-m-n)}$ (Couturier, 1986; Sanjuan and Michard, 1987). Haselton *et al.* (1988) found high alumina solubility in experiments in the system K_2O–Al_2O_3–SiO_2–H_2O–HF at 400–700°C and suggested that a mixed hydroxyfluoride complex may exist in solution at high temperatures.

In this study, we have investigated the solubility of synthetic corundum in 10^{-4}– 0.5m aqueous HF solutions at 300, 400, 500 and 600°C and 1 kbar (Zaraisky, 1989; Soboleva and Zaraisky, 1990; Zaraisky and Soboleva, 1992). Corundum crystals of weight 0.1 to 1.7 g were loaded in the top of gold- or platinum-lined sealed containers, 7.6 or 23 cm^3 in volume, placed in autoclaves. The duration of runs was 1 to 3 weeks. The solubility was determined using a weight-loss method (the accuracy is c. 10^{-5}g) accompanied by a chemical analysis of the quench solution for Al and F.

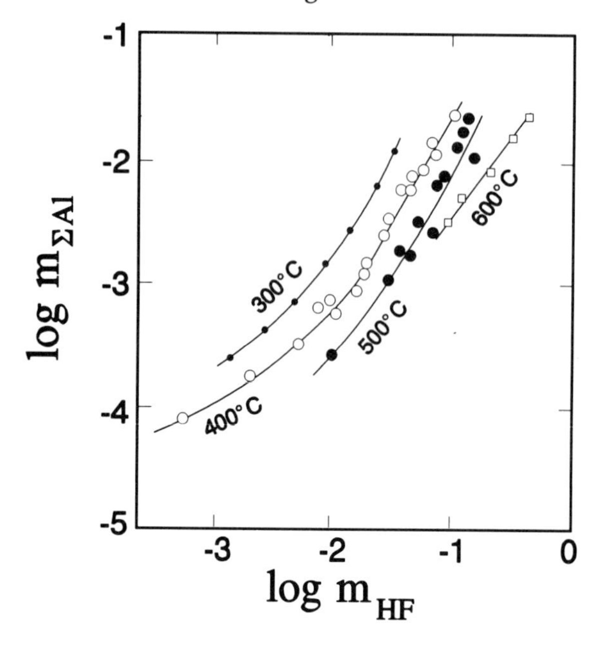

Figure 6.5. The dependence of corundum solubility on the concentration of HF and on temperature at 1 kbar.

Figure 6.5 shows the results for ΣAl concentration in solution as a function of HF concentration and temperature. Corundum solubility decreases with increasing temperature, but rises steeply with m HF. For example, at T = 400°C, P = 1 kbar, the solubility of corundum in 0.1 m HF solution reaches 2×10^{-2} m ΣAl, a value 400 times greater than the solubility in pure water under the same P–T conditions (Ragnarsdottir and Walther, 1986). As in the case of quartz, this may be due to the formation of both fluoride and hydroxyfluoride complexes of aluminium.

Treatment of our experimental data using Balashov's 'x88' program showed that at moderate HF concentrations from 10^{-4} to 10^{-2}m, the dominant Al-bearing complexes in solution may be $Al(OH)_2F^0$ and $Al(OH)F_2^0$. However, the steep slope of the solubility curves at higher HF concentrations (Figure 6.5) requires the presence of a F-rich complex, such as AlF_3^0, also. The thermodynamic calculations indicate that the fraction of charged species formed by ionization of these neutral complexes is small. This agrees with the general trend towards species association in aqueous electrolyte solutions at high temperatures.

6.3.3 The influence of chloride and fluoride solutions on wolframite solubility

W is one of the major ore metals in many greisen deposits, and so serves as a representative indicator of the physico-chemical conditions of ore metal deposition in greisens. This is particular true at Akchatau, where W is the main commercially useful metal, although Mo, Be, Bi, and some other metals are produced as by-products.

The solubility of W minerals, such as ferberite ($FeWO_4$), hubnerite ($MnWO_4$), scheelite ($CaWO_4$), tungsten trioxide (WO_3) and its hydrates ($WO_3 \cdot H_2O$ and $WO_3 \cdot nH_2O$) has been the subject of numerous experimental studies (Yastrebova *et al.*, 1963; Bryzgalin, 1976; Foster, 1977; Burkovskiy, 1981; Manning and Henderson, 1984; Wesolowski *et al.*, 1984; Galkin, 1985; Orlova *et al.*, 1987; Wood and Vlassopoulos, 1989; Khodarevskaya *et al.*, 1990; Zaraisky *et al.*, 1990; Wood, 1992). Nevertheless, many problems remain unresolved, in particular, the nature and extent of W complexation with Cl, F, K, Na and with other salt and acid components, of aqueous solutions at high temperatures. The role of Cl complexing for W is particularly controversial. In a recent study, Wood and Vlassopoulos (1989) concluded that the role of chloride complexes in the hydrothermal transport of tungsten is negligible. They found no increase in $WO_3(s)$ solubility with increasing HCl concentration from 0.49 to 5.37 m at 500 °C and 1 kbar, and concluded that the principal form in which W occurs in aqueous HCl solutions (as in pure water) is the tungstic acid neutral species that forms according to the reaction:

$$WO_3(s) + H_2O = H_2WO_4^0\ (aq).$$

This result accords with earlier work by Eugster and Wilson (1985), and Wood (1992) has supported it with a new set of experiments conducted at 1 kbar, 300 to 600 °C, and at HCl concentrations from 10^{-3} to 0.49 m, using an improved technique. In contrast, other workers have reported an appreciable effect of chloride concentration on the solubility of W minerals (e.g. Wesolowski *et al.*, 1984; Manning and Henderson, 1984; Galkin, 1985; Haselton and D'Angelo, 1986; Bryzgalin, 1986; Khodarevskaya *et al.*, 1990). In particular, Khodarevskaya *et al.* (1990) found a strong dependence of WO_3 solubility on the concentration of HCl in the range mHCl = 10^{-3} to 1, at 450 °C and pressure between 160 and 1000 bar.

Over a wide range of P–T conditions, a positive correlation between the solubility of W minerals and concentration of alkali chlorides, notably NaCl and KCl, has been widely reported (Yastrebova *et al.*, 1963; Bryzgalin, 1976; Foster, 1977; Burkovskiy, 1981; Wesolowski *et al.*, 1984; Galkin, 1985; Orlova *et al.*, 1987; Wood and Vlassopoulos, 1989). However, some workers believe that the reason for this lies in the formation of a cation-W ion pair of the type $NaHWO_4^0$, while others consider chloride complexes, such as WO_3Cl^-, $WOCl_4^0$, WCl_6^0, etc., to be responsible. Kolonin and Shironosova (1991), based on thermodynamic treatment of experimental data of different workers, concluded that mixed oxychloride complexes, such as HWO_3Cl^0, and $WO_2Cl_2^0$, rather than alkaline associates, such as $NaHWO_4^0$, are more likely to form in NaCl–KCl solutions.

The behaviour of W in fluoride solutions is less well understood. The experimental data available in this field are fragmentary, although complex formation between W and F has been reported for low temperatures (e.g. Jander and Fiedler, 1961; Yatsimirskiy and Prik, 1964; Glinkina, 1969).

In view of the uncertainties in the existing data, we have conducted a set of solubility experiments with natural wolframite (44.1 mol% Frb) in NaCl, HCl, and HF solutions at 1 kbar and 300 to 600 °C (Zaraisky *et al.*, 1990). The total solution concentrations ranged from 10^{-2} to 10 m. The experiments were carried out in gold-lined, sealed containers, 21 cm^3 in volume, placed in sealed autoclaves, in the presence of the NNO buffer contained in a squeezed Pt capsule (Chapter 3, 3.4.1). The duration of runs was 2 to 6 weeks. The solubility was determined from the weight loss of a wolframite single crystal (2–4 g in weight) accompanied by chemical analysis of the quench solution for W, Fe, and Mn. The crystal was placed such that the solution on quench was out of contact with it.

Wolframite solubility has been found to be almost congruent over the entire NaCl concentration range 3×10^{-2}– 8 m and in HF solutions with a concentration less than 0.5 m. For mHF = 1.0 to 10, the solution becomes enriched in Fe and Mn relative to W, while in 0.1–10 m HCl solutions, wolframite dissolves incongruently, with large quantities of Fe and Mn (up to 0.1–1.0 m) passing into solution.

The results of the experiments are shown in Figures 6.6 and 6.7. The temperature dependence of solubility in NaCl, HCl, and HF solutions is positive. For 1 m NaCl, m ΣW varies from 2×10^{-4} m at 300°C to 7×10^{-3} m at 600 °C. In 0.1 m HCl and HF solutions, the m ΣW values are 5×10^{-5}m and (1.5–2.0)10^{-3} m at the same temperatures. The concentration dependence of W passing into solution is also positive (Figure 6.7). In the region of concentrations from c. 10^{-2} to 1.0 m, the slope of the solubility curves is about the same for all the three solutions. The m ΣW values are here in the range from 10^{-4} to 10^{-3} m. At NaCl and HF concentrations greater than 4 and 0.5 m, respectively, the solubility of wolframite increases greatly, suggesting that new W-complexes might have formed in in this region. Our results do indicate a weak increase in solubility at high mHCl, although our data extend to more acid fluids than the work of Wood (1992).

Only two runs were conducted in alkaline medium, at 400 °C and 1 kbar, and these yielded a high wolframite solubility. For the run using 0.1 m NaOH solution, m ΣW is 4.8×10^{-2}m; for the run using 0.1 m Na_2SiO_3, m ΣW is 7.0×10^{-2} m.

The results of our experiments are difficult to compare directly with most other studies of W solubility, since they largely refer to the solubility of either WO_3, $CaWO_4$ or of the end members $FeWO_4$ and $MnWo_4$, rather than of intermediate wolframite. In general, however, our results are in good agreement with other data for wolframite solubility (Yastrebova *et al.*, 1963; Bryzgalin, 1976; Foster, 1977; Burkovskiy, 1981; Manning and

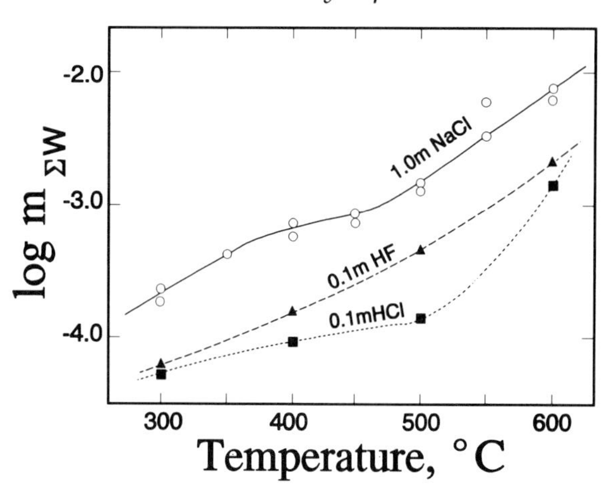

Figure 6.6. Temperature dependence of wolframite solubility in 1 m NaCl, 0.1 m HCl, and 0.1 m HF solutions at 1 kbar.

Henderson, 1984; Wesolowski *et al.*, 1984; Orlova *et al.*, 1987; Wood and Vlassopoulos, 1989). The exception is the work of Galkin (1985), who measured the solubility of $FeWO_4$ at 400 °C and 1 kbar in the NaCl concentration range 0.1–4.0 m and obtained values more than 10 times greater than the wolframite solubility measured by us under similar conditions.

The complexity of the investigated systems and the paucity of thermodynamic data make reliable theoretical analysis of the results difficult. Only generalizations can be made as to the dominant tungsten complexes in solution, but by analogy with Si and Al complexing in HF

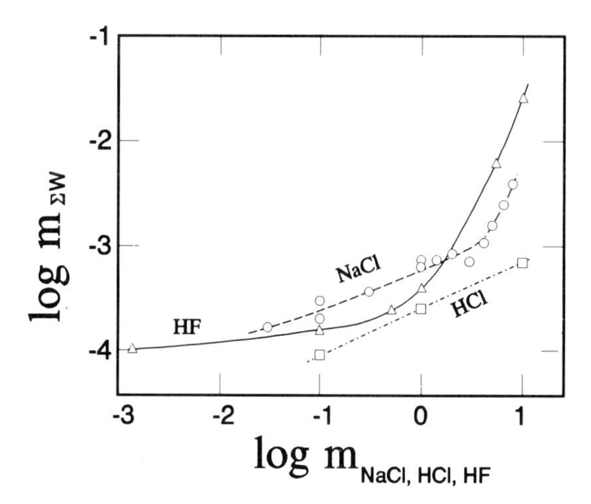

Figure 6.7. The dependence of wolframite solubility on the concentration of NaCl, HCl, and HF at 400°C, 1 kbar.

solutions, it is likely that similar tungsten hydroxyfluoride complexes of the type $W(OH)_nF^0_{6-n}$ may exist in these solutions.

The positive correlation of wolframite solubility with the concentration of HCl and NaCl may indicate the existence of W-chloride complexes. However, the fact that on dissolution of wolframite in HCl the concentration curve lacks a break of the type obtained in runs using NaCl does not allow us to state unequivocally that W-chloride complexes have become increasingly important in highly concentrated chloride solutions. The break of the concentration curve in runs using NaCl may be due to the formation of a Na-bearing complex of tungsten, such as $NaHWO_4^0$, as proposed by Wood and Vlassopoulos (1989).

An alternative interpretation, in terms of a dominant influence of chloride complexes, is that the incongruent dissolution of wolframite in the highly concentrated HCl solutions inhibits release of W to the fluid as chloride complexes because of the sharp increase in the Fe–Mn concentrations in equilibrium with wolframite. Because of this, the inherent break of the curve representing the dependence of WO_3 solubility on the concentration of HCl, as reported by Khodarevskaya *et al.* (1990), may be lacking in runs with wolframite. Thus, the question as to whether chloride or Na-bearing complexes of tungsten predominate in NaCl solutions is not finally resolved by our results.

6.4 SILICIFICATION OF ROCKS ASSOCIATED WITH ACID METASOMATISM

The redeposition of quartz in rocks is generally considered in the context of an equilibrium model and interpreted as being due changes in T, P, or neutralization of alkaline solutions. In many instances, such indeed may be the case. However, none of these reasons explains the common relation between the most intense silicification of rocks and acid metasomatism. As was shown above, quartz is highly soluble in alkaline solutions. Therefore, the deposition of large masses of quartz as a result of neutralization of alkaline solutions would be expected to occur specifically during alkaline metasomatism, for example, due to the formation of aegirite and riebeckite albitites in uranium deposits, of albitized and riebeckitized granites, as well as metasomatic carbonatites. In reality, alkaline metasomatites incorporate fewer quartz veins and zones of silicification than rocks that have experienced acidic metasomatism. Indeed, intense migration of SiO_2 and redeposition of giant quartz masses are restricted to acid metasomatism, as illustrated above for the greisens from Akchatau. A possible explanation could be the enhanced solubility of quartz in acidic fluoride solutions. However, from the results obtained by Shapovapov and Balashov (1990) it follows that a drastic increase of quartz solubility in acidic fluoride solutions relative to pure water is likely only when the HF concentration is greater than 0.1 m (see Figure 6.4), a value hardly likely in natural hydrothermal solutions.

Furthermore, the silicification of rocks during acid metasomatism is common not only in greisens, but in many other acidic metasomatites, including those which form without fluoride solutions being involved. Secondary quartzites, beresites, argillisites, for example, form primarily under the action of acidic chloride, carbonated, or sulphurous solutions; however, they all, as a rule, include quartz veins and zones of silicification.

To account for this phenomenon, we have proposed a hypothesis for the generation of solutions, which become supersaturated with respect to quartz during acid metasomatism (Zaraisky, 1989, 1992). The source of silica is aluminosilicate minerals, mostly feldspars, rather than quartz, which break down when acted upon by acidic solutions and are replaced by micas with the release of excess silica. There is reason to believe that excess SiO_2 is released as a metastable amorphous silica and not as quartz, the former being much more soluble than quartz. That is,

$$\underset{\text{Kfs}}{3KAlSi_3O_8} + 2HCl \Rightarrow \underset{\text{Ms}}{KAl_3Si_3O_{10}(OH)_2} + \underset{\text{amorphous silica}}{6SiO_2(am)} + 2KCl.$$

With time the metastable amorphous silica may recrystallize to a stable quartz; instead, reacting with water, it passes into solution far more rapidly than recrystallization occurs, supersaturating it with respect to quartz:

$$SiO_2(am) + 2H_2O = H_4SiO_4^0\,(aq).$$

As infiltration proceeds, these quartz-supersaturated solutions precipitate the excess silica as stable quartz:

$$H_4SiO_4^0\,(aq) = \underset{\text{Qtz}}{SiO_2} + 2H_2O.$$

This results in silicification, i.e. the formation of quartz pockets, veins and veinlets, and silica impregnations.

Such a sequence of events seem natural. After the crystal lattice of feldspar has been disrupted, the excess silica first remains solid and amorphous and then comes into equilibrium with solution, supersaturating it with respect to quartz. However, this is only an assumption, which requires further investigation.

During our extensive involvement in the experimental modelling of metasomatism we have repeatedly come up with evidence for such a process of supersaturating acidic chloride solutions with SiO_2 with respect to quartz after their interaction with granite and other aluminosilicate rocks at elevated P and T. A similar process has been inferred by other investigators also (e.g. Ostapenko, 1971; Balitsky, 1978; Rafalskiy *et al.*, 1987). Ostapenko (1971) and Balitsky (1978), for example, suggested that silica supersaturation of natural solutions may be of geological significance, particularly in forming rock crystal deposits.

6.4.1 Experimental studies of silica metasomatism

Experiments on metasomatic modelling provide further evidence for the proposed mechanism for equilibrium–non-equilibrium silicification of rocks during acid metasomatism (Zaraisky, 1989, 1992). Figure 6.8 shows the temperature dependence of silica concentration in the quench solution after interaction of 0.1 m HCl solution with plates of leucocratic granite, from Zaraisky (1989). The concentration is intermediate in value between quartz saturation and amorphous silica saturation, suggesting that it is not controlled by quartz solubility but has been enhanced by interaction with other minerals.

A further series of experiments has been devised to directly model silicification. As acidic chloride solutions attacked granite powder loaded into a small gold capsule left open at one end, at 400–500 °C, and 1 kbar, the granite was transformed into pyrophyllite rather than into silica-poorer andalusite stable under these P and T. According to the data of Hemley *et al.* (1980), pyrophyllite is stable under these conditions only when silica concentrations in solution are higher than quartz saturation. In these runs, an external (with respect to granite in the capsule) solution was saturated with SiO_2 by adding excess powdered quartz. Yet SiO_2 was removed from granite by diffusion and added to the external solution. According to chemical analysis data, the concentration of SiO_2 in the solution after the run was greater than quartz saturation, while the granite contained in the capsule was depleted in silica. All our attempts to prevent SiO_2 diffusion from the granite were unsuccessful, even when a 5 mm thick layer of powdered quartz was placed before the granite by the open capsule end. After the run, no evidence of the dissolution of quartz grains was found, whereas the ΣSiO_2 content of the granite gradually decreased.

Similar phenomena were observed in metasomatic column experiments that used microcline powder as an alternative to granite.

X-ray diffraction patterns of metasomatic zones in the experimental column for diffusion acid metasomatism show marked cristobalite reflections (Figure 6.9). Apparently, precipitation of silica from the supersaturated solution occurs with the formation of an intermediate metastable phase – cristobalite – which is more soluble than quartz, but is less soluble than amorphous silica. Note that even in the zone (zone 4) immediately next to unaltered microcline, cristobalite occurs in addition to the newly-formed muscovite. In zone 3, located by the capsule open

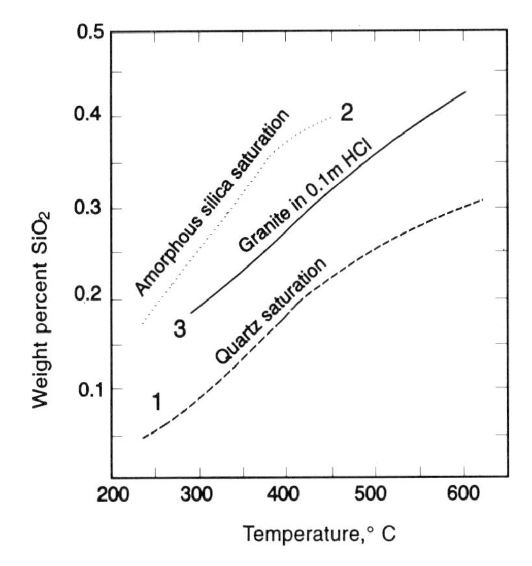

Figure 6.8. Comparison of quartz and amorphous silica solubilities in pure water (after Walther and Helgeson, 1977) with the solubility of leucogranite plates in 0.1 m HCl at 1 kbar (after Zaraisky, 1989): (1) temperature solubility curve of quartz, (2) solubility curve of amorphous silica, (3) ΣSi concentration in quench solution in runs on the interaction of leucogranite plates with 0.1 m HCl solution (the solution–rock ratio varies from 0.44 to 2.05; run times are from 336 hours at 600°C to 1296 hours at 300°C).

end, weak quartz reflections can be detected, along with strong cristobalite reflections.

The results of our infiltration experiments provide the clearest demonstration of silicification. It was possible to directly observe removal of silica by acidic chloride solutions from upflow rear zones in the metasomatic columns and its redeposition in downflow zones located farther forward. The experiments were carried out in a special device allowing slow flow of solution through the rock powder placed in a gold tube (Zaraisky, 1989). In order to presaturate the acidic chloride solution with silica, it was made to pass through a layer of powdered quartz prior to its percolation through granite or microcline. Temperature and pressure gradients were within the limits of measurement (± 1°C and 5 bar). Figure 6.10 illustrates the distribution of SiO_2 and Al_2O_3 in the zones of an experimental infiltration column for acidic metasomatism of granite. The figure is based on electron microprobe data obtained by scanning over profiles perpendicular to the column axis (Zharikov and Zaraisky, 1991). The silica content of the upflow Qtz + Ms zone is considerably lower than the initial silica content of the granite. The silica removed from this zone was evidently redeposited in the following downflow zones since their SiO_2 content is higher than in initial granite: they have effectively been silicified.

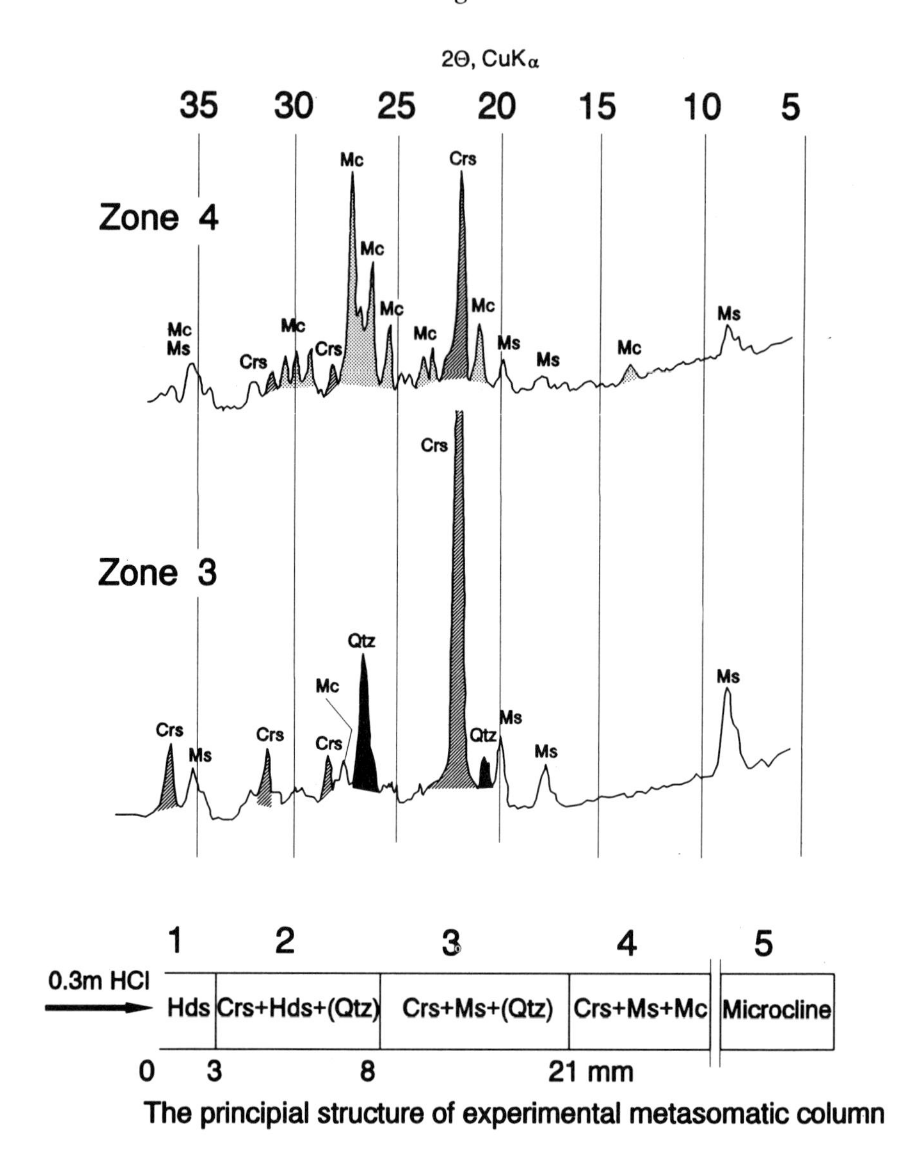

Figure 6.9. Formation of cristobalite and quartz in zones of the experimental column for the diffusion acidic metasomatism of microcline according to X-ray data (500°C, 1 kbar, 0.3 m HCl solution, the run time is 168 hours). Hds–hydzolsite $Al_2O_3 \cdot SiO_2 \cdot (0.5\text{–}1.0)H_2O$.

6.5 CONCLUSIONS

The experiments described here provide valuable information about the role of acid chloride and fluoride solutions in the mass transfer of Si, Al,

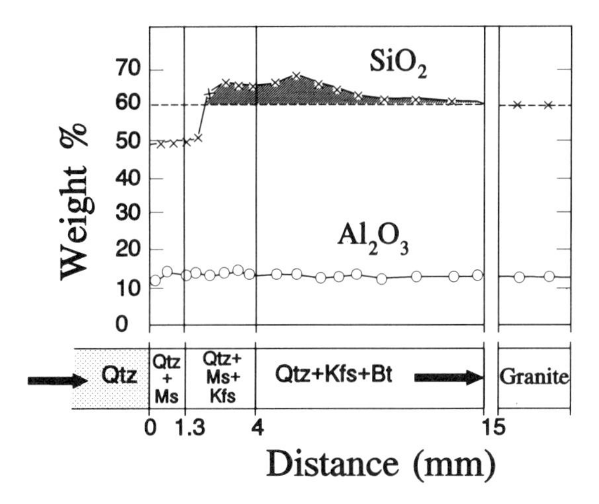

Figure 6.10. Distribution of SiO_2 and Al_2O_3 in zones of the infiltration experimental column for the acidic metasomatism of granite according to electron microprobe analysis (500°C, 0.8 kbar, 0.5 m KCl + 0.1 m HCl solution, the run time is 166 hours).

and W in greisen deposits. The data from quartz solubility experiments indicate that acidic chloride solutions should not behave significantly differently to pure water as agents for the mass transfer of SiO_2 and Al_2O_3. Because of the fairly high quartz solubility in water under hydrothermal solutions (c. $10^{-3}-10^{-2}$ m ΣSi), mass transfer of silica by fluids is not difficult to understand (Figure 6.11). At the same time, the low solubility of Al_2O_3 in both water and chloride solutions (c. 10^{-5} m ΣAl at 400–600 °C) prevents the migration of alumina during fluid–rock interaction except at geologically unreasonable HCl concentrations (> 0.1 m).

The situation is completely different when rocks interact with acidic fluoride solutions. Over the HF concentration range 10^{-4}–10^{-1} m, believed to be realistic for greisenization processes, the solubility of quartz increases only slightly when compared to its solubility in pure water, but corundum solubility increases by a factor of 400, from 5×10^{-5}m ΣAl in water to 2×10^{-2}m ΣAl in 0.1 m HF solution, at 400 °C and 1 kbar. From this it follows that acidic fluoride solutions whose effect on Si is negligible may dramatically alter the geochemical behaviour of the inert Al, transforming it into a mobile component like Si. Considerable mass transfer of Al_2O_3 at Akchatau and in other greisen deposits may be accounted for by this effect.

The metastable 'supersolubility' of silica demonstrated in the breakdown experiments of feldspars exposed to acidic chloride solutions may manifest itself in a similar way in nature, and lead to rock silicification. The breakdown of feldspars with the release of amorphous SiO_2 may

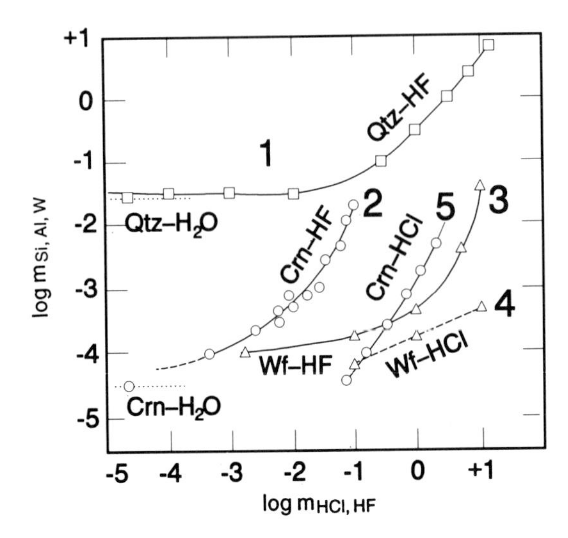

Figure 6.11. Comparison of quartz, corundum and wolframite solubilities in HF and in HCl solutions at 400°C and 1 kbar. Experimental data: (1) Shapovalov and Balashov (1990), (2–4) this study and Soboleva and Zaraisky (1990), (5) Korzhinskiy (1987) at 450°C.

occur throughout microcracks and pores in large rock volumes, whereas the deposition of stable quartz from supersaturated solutions should occur primarily in cracks and other permeable zones that focus fluid flow. Such a mechanism may be regarded as an alternative to the effect of temperature–pressure gradients in interpreting the origin of quartz veins.

The results of the wolframite solubility experiments show that factors conducive to the deposition of wolframite ores may include a temperature decrease, a decrease in the concentration of chloride or fluoride solutions, and neutralization of alkaline solutions. The experimental data permit the quantitative evaluation of the contribution from each of these factors.

ACKNOWLEDGEMENTS

I would like to thank my colleagues Yu.B. Shapovalov, Yu.B. Soboleva, and V.N. Balashov for assistance in performing this study and helpful discussions. The research described in this chapter was supported by the Russian Fund of Fundamental Investigations under Grant No. 93–05–9822 and International Soros Science Education Programme under Grant No. MUR000.

REFERENCES

Anderson, J.M. and Burnham, C.W. (1965) The solubility of quartz in supercritical water. *Amer. J. Sci.*, **263** (6), 494–511.

Balitsky, V.S. (1978) *An Experimental Study of the Rock Crystal Formation Processes.* Nedra Press, Moscow (in Russian).

Balitsky, V.S., Orlova, V.P., Ostapenko, G.T. and Khetchikov, L.N. (1971). Quartz solubility in hydrothermal solutions of sulphurous sodium and of potassium, sodium and ammonium fluorides. *Proceedings of the 8th Meeting on Experimental and Technical Mineralogy and Petrography*, pp. 220–4. Nauka Press, Moscow (in Russian)

Baumann, E.W. (1961) Determination of stability constants of hydrogen and aluminium fluorides with a fluoride selective electrode. *J. Inorg. Nucl. Chem*, **31**, 3155–62.

Bryzgalyn, O.V. (1976) On the solubility of tungstic acid in aqueous salt solutions at high temperatures. *Geochem. Int.*, **13** (3), 155–9.

Bryzgalyn, O.V. (1986) Tungsten hydroxy and chloride complexes in aqueous solutions at elevated temperatures. In *Minerals and Geochemistry of Tungsten Deposits.* Leningrad University Press, Leningrad (in Russian), pp. 209–14.

Burkovskiy, S.I. (1981) An experimental investigation of equilibria in the system $MnWO_4(FeWO_4)$–NaCl–H_2O at 75 at 300 °C. *Izvestia Akademii Nauk Kazakhskoi SSR, ser. Geol.*, **N 3**, 42–5 (in Russian).

Castet, S. (1991) Solubilite de la boehmite et speciation de l'aluminium dissous dans les solutions aqueous a haute temperature (90–300 °C). Determination experimentale et modelisation. Thèse présentée du grade de doctor de l'universite specialiste geochimie. Univ. Paul Sabatier de Toulouse.

Couturier, Y. (1986) Contribution a l'etude des complexes mixtes de l'aluminium (III) avec les ions fluorures et hydroxyde. *Bull. Soc. Chimique de France*, **2**, 171–7.

Doroshenko, Yu.P., Pavlun', N.N. and Simkiv., Zh.A. (1981) Chemical composition of mineral-forming solutions from the Akchatau deposit. *Mineralogicheskii Sbornik*, **35** (1), 77–80.

Elmer, T.H. and Nordberg, M.E. (1958) Solubility of silica in nitric acid solutions. *Amer. Ceramic Soc. J.*, **41** (12), 517–20.

Eugster, H.P. and Wilson, G.A. (1985) Transport and deposition of ore-forming elements in hydrothermal system associated with granites, in *High Heat Production Granites, Hydrothermal Circulation and Ore Genesis* (ed. C. Halls). Institute of Mining and Metallurgy, London, pp. 87–98.

Foster, R.P. (1977) Solubility of scheelite in hydrothermal chloride solutions. *Chemical Geology*, **20** (1), 27–43.

Galkin, A.V. (1985) Solubility of tungsten–iron–manganese minerals in high-temperature chloride solutions. Ph.D. thesis, Novosibirsk (in Russian).

Glinkina, M.I. (1969) Behaviour of tungsten and calcium in acid fluoride solutions at temperatures from 18 and 350 °C. *Geochem. Int.*, **6** (4), 641–8.

Gotman, Ya.D. and Malakhova, V.I. (1965) *Near-vein Alteration of Granitic Rock at the Tungsten Deposit in Kazakhstan.* Nedra Press, Moscow (in Russian).

Gresens, R.L. (1967) Composition–volume relationships of metasomatism. *Chemical Geology*, **2** (1), 47–65.

Haselton, H.T.Jr. (1984) The solubility of quartz in dilute HF solutions at 600 °C and 1 kbar (abs.). *Amer. Geoph. Union Trans.*, **65**, pp. 308.

Haselton, H.T.Jr. and D'Angelo, W.M. (1986) Tin and tungsten solubilities (500–700 °C, 1 kbar) in the presence of a synthetic quartz monzonite (abs.). *Amer. Geoph. Union Trans.*, **67**, pp. 386.

Haselton, H.T.Jr., Cygan, G.L. and D'Angelo, W.M. (1988) Chemistry of aqueous solutions coexisting with fluoride buffers in the system K_2O–Al_2O_3–SiO_2–H_2O–F_2O_{-1} (1 kbar, 400–700°C). *Econ. Geol.*, **83** (1), 163–73.

Hemley, J.J. and Jones, W.R. (1964) Chemical aspects of hydrothermal alteration with emphasis on hydrogen ion metasomatism. *Econ. Geol.*, **59**, 538–69.

Hemley, J.J., Montoya, J.W., Marinenko, J.W. and Luce, R.W. (1980) General equilibria in the system Al_2O_3–SiO_2–H_2O and some implications for alteration/mineralization processes. *Econ. Geol.*, **75** (2), 210–28.

Iler, R.K. (1979) *The Chemistry of Silica*. John Wiley & Sons Inc., New York.

Jander, G. and Fiedler, B. (1961) Uber der Molekulazustand von Fluorowolframaten in Wabrigen Losungen. *Z. Anorg. Und Allg. Chem.*, **B 308**, H. 1–6.

Kamiya, H., Ozaki, A. and Imanachi, M. (1974) Dissolution rate of powdered quartz in acid solution. *Geochem. J.*, **8**(1), 21–6.

Kazitsin, Yu.V. and Rudnik, V.A. (1968) *Manual for Calculating a Mass Balance and Internal Energy in Forming Metasomatic Rocks*. Nedra Press, Moscow (in Russian).

Khodarevskaya, L.I., Tikhomirova, V.I. and Postnova, L.E. (1990) Study of WO_3 solubility in HCl solutions at 450°C. *Doklady Akkademii Nauk SSSR*, **113** (3), 720–2 (in Russian).

Kolonin, G.R. and Shironosova, G.P. (1991) *The state of the art in the field of experimental studies on tungsten forms in hydrothermal solutions*. 12th All-Union Meeting on Experimental Mineralogy, Miass, USSR, 24–26 September 1991, Abstracts, pp. 58.

Kopeikin, V.A. and Mikhailov, A.S. (1970) Solubility and silica forms in dilute solutions under normal conditions. *Doklady Akademii Nauk SSSR*, **191** (4), 917–20 (in Russian).

Korzhinskiy, D.S. (1953) An essay of metasomatic processes. In *Key Problems in Magmatogene Ore Deposit Studies* (ed. Betekhtin, A.G.). Academy of Sciences Press, Moscow (in Russian), pp. 332–452.

Korzhinskiy, M.A. (1987) The solubility of corundum in an HCl fluid and forms taken by Al. *Geochem. Int.*, **24** (11), 105–9.

Krauskopf, K.B. (1956) Dissolution and precipitation of silica at low temperatures. *Geochim. Cosmochim. Acta*, **10** (1–2), 1–26.

Manning, D.A.C. and Henderson, P. (1984). The behaviour of tungsten in granitic melt–vapour systems. *Contr. Mineral. Petrol.*, **86** (3), 286–93.

Mosebach, R. (1957) Thermodynamic behavior of quartz and other forms of silica in pure water at elevated temperatures and pressures with conclusions of their mechanism of solution. *J. Geol.*, **65** (4), 347–63.

Okamoto, G., Okura, T. and Goto, K. (1957) Properties of silica in water. *Geochim. Cosmochim. Acta*, **12**, (1/2), 123–32.

Orlova, G.P., Lapin, A.A. and Ryabchikov, I.D. (1987). An experimental study of equilibria in the system scheelite–granite–fluid under conditions of hypabyssal magmatism. *Geologia Rudnykh Mestorizhdenii*, **4**, 107–10 (in Russian).

Ostapenko, G.T. (1971). Retarded dissolution and possible growth quartz crystals in the presence of kyanite in strong hydrochloric acid at elevated temperatures and pressures. *Geochem. Int.*, **8** (12), 919–21.

Ostapenko, G.T. and Arapova, M.A. (1971) Solubility of kyanite, corundum, quartz, and amorphous silica in aqueous hydrochloric acid solutions at 285°C and 450 bars. *Geochem. Int.*, **8** (7), 482–8.

Rafalskiy, R.P., Prisyagina, H.I., Alekseyev, V.A., *et al.* (1987) Reactions of amphibolite with aqueous solutions at 250°C. *Geochem. Int.*, **24** (6), 46–61.

Ragnarsdottir, K.V. and Walther, J.V. (1986) Experimental determination of corundum solubilities in pure water between 400–700°C and 1–3 kbar. *Geochim. Cosmochim. Acta*, **49** (10), 2109–16.

Rimstidt, J.D. and Barnes, H.L. (1980) The kinetics of silica–water reactions. *Geochim. Cosmochim. Acta*, **44** (11), 1683–99.

Sanjuan, B. and Michard, G. (1987) Aluminium hydroxide solubility in aqueous solutions containing fluoride ions at 50°C. *Geochim. Cosmochim. Acta*, **51** (7), 1823–31.

Sazarin, G., Couturier, Y. and Michard, G. (1983) The role of fluoride in the dissolved aluminium species in thermal waters. *WRI-4, Misasa, Japan, Abstracts*, pp. 576–9.

Shapovalov, Yu.B and Balashov, V.N. (1990) Quartz solubility in hydrofluoric acid solutions at temperatures between 300 and 600°C and 1000 bar pressure. In *Experiment-89. Informative Volume* (ed. Zharikov, V.A.). Nauka Press, Moscow, pp. 72–74.

Soboleva, Yu.B. and Zaraisky, G.P. (1990) The solubility of corundum in HF solution at elevated temperatures. In *Experiment-89. Informative Volume* (ed. Zharikov, V.A). Nauka Press, Moscow, pp. 77–9.

Sorokin, V.S. and Dadze, T.P. (1980) Solubility of amorphous SiO_2 in water and in aqueous HCl and HNO_3 solutions at 100–400°C and 101.3 MPa. *Doklady Akademii Nauk SSSR*, **254** (3), 735–9 (in Russian).

Walther, J.V. and Helgeson, H.C. (1977) Calculation of the thermodynamic properties of aqueous silica and the solubility of quartz and its polymorphs at high pressures and temperatures. *Amer. J. Sci.*, **277** (10), 1315–51.

Weill, D.F. and Fyfe, W.S. (1964) The solubility of quartz in H_2O in the range 1000–4000 bars and 400–550°C. *Geochim. Cosmochim. Acta*, **28** (8), 1243–55.

Weill, D.F. and Bottinga, G. (1970) Thermodynamic analysis of quartz and cristobalite solubilities in water at saturation vapor pressure. *Contr. Mineral. Petrol.*, **25** (2), 125–32.

Wesolowski, D., Drummond, S.E., Mesmer, R.E. and Ohmoto, H. (1984) Hydrolysis equilibria of tungsten (IV) in aqueous sodium chloride solutions to 300°C. *Inorganic Chemistry*, **23**, 1120–32.

Wood, S.A. and Vlassopoulos, D. (1989) Experimental determination of the hydrothermal solubility and speciation of tungsten at 500°C and 1 kbar. *Geochim. Cosmochim. Acta*, **53** (2), 303–12.

Wood, S.A. (1992) Experimental determination of the solubility of WO_3(s) and thermodynamic properties of H_2WO_4 (aq) in the range 300–600°C at 1 kbar: Calculation of scheelite solubility. *Geochim. Cosmochim. Acta*, **56**, 1827–36.

Yastrebova, L.F., Borina, A.F. and Ravich, M.I. (1963). Solubilities of calcium molybdate–calcium wolframate in aqueous potassium and sodium chloride solutions at high temperature. *Zhournal Neorganicheskoi Khimii*, **8** (1), 208–17 (in Russian).

Yatsimirskiy, K.B. and Prik, K.Ye. (1964) W^{6+} complexation with some inorganic ligands in dilute solutions. *Zhournal Neorganicheskoi Khimii*, **9** (1), 178–80 (in Russian).

Zaraisky, G.P. (1989) *Zoning and Conditions of Formation of Metasomatic Rocks*. Nauka Press, Moscow (in Russian).

Zaraisky, G.P., Tikhomirova, V.I. and Postnova, L.E. (1990) Wolframite solubility in chloride and fluoride solutions. In *Experiment-89. Informative Volume* (ed. Zharikov, V.A.). Nauka Press, Moscow, pp. 74–7.

Zaraisky, G.P., Shapovalov, Yu.B., Soboleva, Yu.B., *et al.* (in press) Physicochemical conditions of greisenization at the Akchatau deposit, based on geologic and experimental data. In *Experimental Problems of Geology*. Nauka Press, Moscow (in Russian).

Zaraisky, G.P. (1992) *Equilibrium – non-equilibrium model of silicification of rocks during acidic metasomatism*. Second International Symposium on Thermodynamics of Natural Processes, Novosibirsk, Russia, 13–20 September 1992, Abstracts, pp. 68.

Zaraisky, G.P. and Soboleva, Yu.B. (1992) *Solubility of corundum in HF aqueous solutions at temperatures from 300 to 600°C and 1 kbar pressure.* Fifth International Symposium on Solubility Phenomena, Moscow, 8–10 July 1992, Abstracts, pp. 217.

Zharikov, V.A. and Zaraisky, G.P. (1991) Experimental modelling of wall-rock metasomatism, in *Progress in Metamorphic and Magmatic Petrology* (ed. Perchuk, L.L.). Cambridge University Press, pp. 197–245.

CHAPTER SEVEN

The behaviour of components in complex fluid mixtures under high T–P conditions

Michael A. Korzhinskiy

7.1 INTRODUCTION

The chemical behaviour of salts in complex hydrothermal fluids is not well understood over most of the P–T conditions of geochemical interest. This study of the behaviour of electrolyte components in complex fluid mixtures over a range of pressures and temperatures is designed to help to decipher the physico-chemical conditions of natural hydrothermal processes. The information is also necessary to test and develop new equations of state for hydrothermal fluids.

Specifically, we have explored the variation in the concentration of dissolved components with the composition of the host solution, at fixed component activity. In this context, highly soluble substances such as chloride salts present experimental problems. Investigations of the relative behaviour of pairs of components whose activity ratio can be fixed, prove the simplest experimentally. An example of this is exchange equilibria of the type:

$$\underset{\text{wollastonite}}{2CaSiO_3} + MgCl_2 = \underset{\text{diopside}}{CaMgSi_2O_6} + CaCl_2, \tag{7.1}$$

for which

$$K_1 = \alpha_{CaCl_2} \cdot \alpha_{MgCl_2}^{-1}. \tag{7.2}$$

Fluids in the Crust: Equlibrium and transport properties.
Edited by K.I. Shmulovich, B.W.D. Yardley and G.G. Gonchar.
Published in 1994 by Chapman & Hall, London. ISBN 0 412 56320 7

The absolute component activity is much more difficult to fix. It is well known that each chloride salt may be represented as a product of interaction between a cation in oxide form and an acid component in accordance with the following stoichiometric reaction:

$$MeO_X + 2XHCl^0 = MeCl^0_{2X} + XH_2O. \qquad (7.3)$$

If the activities of both the oxide and the acid component can be fixed, we obtain a constant value for the salt activity, which depends solely on the water activity in the system at chosen P and T.

The activity of the acid component may be controlled using the buffer technique of Frantz and Eugster (1973). This utilizes an inner Ag–AgCl buffer in conjuction with an outer oxygen–hydrogen buffer (Ni–NiO–H_2O). These are separated from one another by a platinum membrane. In the outer system, for example a Ni–NiO pair, the following equilibria are realized:

$$0.5Ni + 0.25O_2 = 0.5NiO, \qquad (7.4)$$

$$0.25O_2 + 0.5H_2 = 0.5H_2O. \qquad (7.5)$$

For a platinum membrane permeable to hydrogen,

$$f^{(I)}_{H_2} = f^{(II)}_{H_2}, \qquad (7.6)$$

where I and II denote the inner and outer parts of the buffer system, respectively. For the inner system containing the Ag–AgCl pair, we have:

$$Ag + 0.5Cl_2 = AgCl, \qquad (7.7)$$

$$0.5H_2 + 0.5Cl_2 = HCl^0. \qquad (7.8)$$

Combining Equations 7.4–7.8 results in the generalized equilibrium:

$$0.5Ni + 0.5H_2O + AgCl = Ag + HCl^0 + 0.5NiO, \qquad (7.9)$$

for which the equilibrium constant at the chosen T and P has the form:

$$K_{(9)} = f_{HCl^0}, \qquad (7.10)$$

provided the pressure in the outer system is set by pure water. Under these conditions, the equilibrium constant for reaction 7.3 is defined as:

$$K_{(3)} = \frac{\alpha_{MeCl^0_{2X}} \cdot \alpha^X_{H_2O}}{f^{2X}_{HCl^0}}, \qquad (7.11)$$

where f_{HCl^0} is defined by Equation 7.10 and the activity of the cation in oxide form may be fixed at unity by either its oxide (or mixes of oxides, e.g. FeO or $Fe_3O_4 + Fe_2O_3$) or by an invariant assemblage which buffers oxide activity. For example:

$$NaAlSi_3O_8 = 0.5\ Al_2SiO_5 + 2.5\ SiO_2 + 0.5\ Na_2O. \qquad (7.12)$$

albite andalusite quartz

The results of hydrothermal-equilibria studies using this type of approach can be found in Frantz and Eugster (1973), Frantz and Popp (1979), Korzhinskiy (1981, 1986, 1987, 1992), Popp and Frantz (1979).

The utility of the approach is that the absolute activities of the cation in oxide form and associated HCl^0 remain constant in the system at the chosen T and P. However, the activity of salt components depends in this instance upon the activity of water. The equilibrium

$$3Fe_3O_4 + 2AgCl = FeCl_2^0 + 2Ag + 4Fe_2O_3, \qquad (7.13)$$

with the constant defined as

$$K_{13} = \alpha_{FeCl_2^0}, \qquad (7.14)$$

may serve as an example of buffering the absolute activity of the salt component in the system. Whatever the way of buffering the activity of the acid or salt components, their concentrations may vary when additional NaCl or CO_2 are introduced into the system. This is due to:

- variation in the ratio of associated to dissociated forms;
- the common ion effect that prevails in the system;
- intermolecular association processes;
- the solvation state of the salt substance dissolved in the hydrothermal phase.

7.2 EXPERIMENTAL PROCEDURES AND TECHNIQUES

Experiments to investigate the relative behaviour of salt components proved the simplest and most accessible. They used the capsule technique described by Hemley (1959) and Korzhinskiy (1986, 1987). All these experimental runs were carried out in gold or platinum capsules, $5 \times 0.2 \times 40$ (50) and $7 \times 0.2 \times 60$ (70) mm in size, respectively.

A schematic representation of the experimental procedure with the activity of associated HCl^0 and salt components buffered, is shown in Figure 7.1. A set of welded platinum capsules ranging in size from $3 \times 0.1 \times 40$ to $5 \times 0.2 \times 50$ mm were placed in a large open stainless steel container. If the behaviour of HCl was to be examined, the capsules contained only the Ag–AgCl buffer and H_2O–HCl solutions of different initial concentrations. If the behaviour of salt components was investigated, the capsules contained the Ag–AgCl buffer, an invariant assemblage of solid phases, and a chloride solution of the appropriate component. The initial solution volume ranged from 0.05 to 0.14 cm^3 depending on the T–P parameters and the mole fraction of CO_2. The

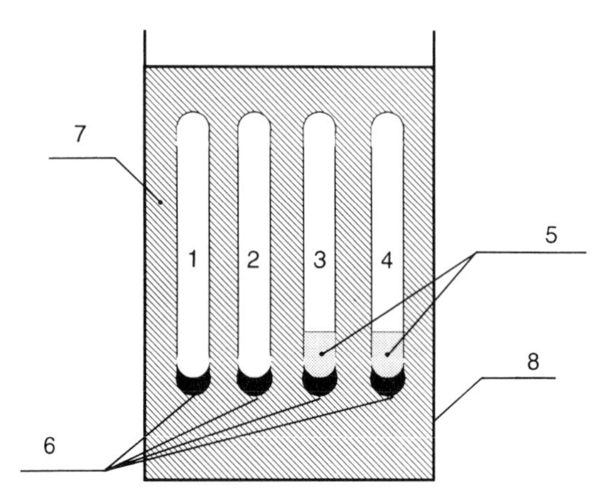

Figure 7.1. Schematic representation of the experimental runs, with activity of associated HCl^0 fixed. 1–4 are Pt capsules filled as follows: 1 – H_2O–HCl, 2 – H_2O–HCl–CO_2(NaCl), 3 – H_2O–$MeCl_x$, 4 – H_2O–$MeCl_x$–CO_2(NaCl). Each capsule also contains an Ag–AgCl buffer (6), numbers 3 and 4 also contain an invariant assemblage (5). They are inmersed in an oxygen buffer assemblage (7) contained within a stainless steel container (8).

concentration of the background electrolyte NaCl was set by a weight change method, whereas the mole fraction of CO_2 was controlled by a mix of oxalic acid plus Ag_2CO_3 in the mole ratio of 1:1. Such a way of introducing CO_2 into the fluid provided a neutral reaction with respect to hydrogen, preventing variation in its partial pressure during the run. All platinum capsules were also loaded with an f_{O_2} buffer (normally, Fe_3O_4–Fe_2O_3). The stainless steel container was then placed in an autoclave, where it was kept at the T and P of interest until equilibrium was attained. Accumulated experience in the use of the buffer technique (with platinum capsules) shows that the system equilibrates within 6–7 days at 700 °C and within 26–30 days at 400 °C (Frantz and Eugster, 1973; Frantz and Popp, 1979; Korzhinskiy, 1981, 1992; Popp and Frantz, 1979). The use of parallel capsules containing compositionally different initial solutions served as a criterion for attaining equilibrium. Because the capsules all contained the Ag–AgCl buffer and were run under the same redox conditions, the fugacity of associated HCl^0, by virtue of hydrogen diffusion through platinum, was also the same within all the capsules.

After the run, the capsules were opened and the solution extracted and diluted, if required. The concentration of the acid component was determined by pH measurement, while cation concentrations were analysed by atomic absorption spectrophotometry.

All runs were made in sealed autoclaves, 50 and 75 cm^3 in volume, which, in turn, were placed in two-section autoclave furnaces. Temperature was measured with a platinum–platinum-rhodium thermocouple with an uncertainty of ± 5°C. Pressure within the system was generated by loading the autoclave with measured amounts of water to give the desired P at the T of interest after calculating the volumes of the container, buffer mix and solution contained in the capsules. The uncertainties in the pressure measurements were within ± 5–10% of the desired value. At the end of the run the autoclave was quenched with cold running water within 3 to 5 minutes. Note that despite the uncertainties associated with total pressure, we have chosen the apparatus of this sort because of its large working volume. This permitted the entire run series (up to 12–14 capsules) to be carried out under the same T–P conditions, and, most importantly, at uniform hydrogen fugacity. Details of the results of individual runs are tabulated in Korzhinskiy (1981).

7.3 BEHAVIOUR OF THE ACID COMPONENT HCl

The behaviour of the acid component HCl in $NaCl–CO_2–H_2O$ fluids has been studied at temperatures between 400 and 700°C at 1 and 2 kbar. X_{CO_2} ranged from 0 to 0.2, and the NaCl contents were up to 6.16 mol · kg H_2O^{-1}. The experiments were done using a magnetite–hematite buffer.

7.3.1 The H_2O–HCl–CO_2 system

First consider the behaviour of HCl in a pure aqueous system. Figure 7.2 illustrates the behaviour of HCl in equilibrium with the buffer assemblage Ag–AgCl + Fe_3O_4–Fe_2O_3. As the temperature decreases, the equilibrium concentration of HCl first decreases, then rises. Such behaviour is the result of the greater degree of dissociation of HCl at low temperatures (Frantz and Eugster, 1973; Frantz and Popp, 1979). The available data on the dissociation constant of HCl permit calculation of the ratio of associated to dissociated forms using the relationships:

$$HCl^0 = H^+ + Cl^-, \tag{7.15}$$

$$K_{HCl} = \frac{\alpha_{H^+}.\alpha_{Cl^-}}{\alpha_{HCl^0}} = \frac{(m_{H^+})^2 \cdot (\gamma\pm)^2}{m_{HCl^0}}, \tag{7.16}$$

$$m_{HCl^\Sigma} = m_{HCl^0} + m_{H^+}, \tag{7.17}$$

where α is activity, $\gamma\pm$ the mean ionic activity coefficient, and m the molality. Values for the K_{HCl} were taken from Frantz and Marshall (1984):

$$K_{D_{HCl}} = -5.405 + \frac{3874.9}{T,K} + 13.93 \cdot \log \rho_{H_2O}. \tag{7.18}$$

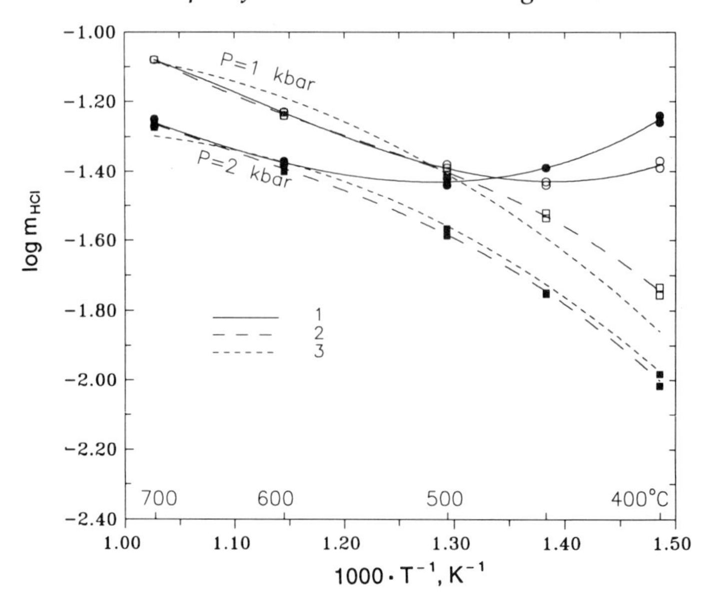

Figure 7.2. Molal concentrations of total HCl^{Σ} and associated HCl^0 in the system H_2O–HCl, which are in equilibrium with the Ag–AgCl + $Fe_3O_4Fe_2O_3$ buffer. 1 – experimental values of $m_{HCl^{\Sigma}}$; 2 – values of m_{HCl^0} calculated using $K_{D_{HCl}}$ 3 – values of m_{HCl^0} calculated from thermodynamic data.

Activity coefficients were calculated with the Debye–Hückel equation:

$$\log \gamma \pm = \frac{-A \cdot I^{1/2}}{1 + \mathring{a} \cdot B \cdot I^{1/2}}. \tag{7.19}$$

Values for the Debye–Hückel coefficients A and B and those for the dielectric constant of a medium were calculated with the equations from Helgeson *et al.* (1981) and Pitzer (1983), respectively. Parameter $\mathring{a}$, which characterizes the ionic 'distance of closest approach' in solution, was taken to be 4.89 for HCl following Helgeson *et al.* (1981). The calculated values for m_{HCl^0} are shown in Figure 7.2 as dashed lines.

The available thermodynamic data, in turn, provide a means of independently calculating equilibrium concentrations of associated HCl^0. To this end, by using Equations 7.4–7.10, as well as data on the Gibbs free energy (Robie *et al.*, 1978) and fugacity coefficient for water (Mel'nik, 1978), the equilibrium values for HCl^0 fugacity have been calculated. Then, by utilizing the relationship:

$$m_{HCl^0} = \frac{f_{HCl^0} \cdot 55.51}{P \cdot \gamma^*_{HCL^0} \cdot \varphi_{HCl^0} \cdot (1 - X_{HCl^0})}, \tag{7.20}$$

the molal HCl^0 concentration at the experimental T and P has been computed. Values of fugacity coefficients ($\gamma^*_{HCl^0}$) for pure HCl were taken from Mel'nik (1978), and HCl was assumed to mix ideally with H_2O. The calculated values for molal concentration of HCl^0 are shown in Figure 7.2 as dot-dash lines.

Analysis of the data plotted in Figure 7.2 shows that in the high-temperature region, where HCl^0 predominates, the calculated and experimental values of m_{HCl^0} are close. This points to the absence of strong hydration interactions of molecular HCl^0 with water. The difference between two HCl^0 data sets becomes larger with decreasing temperature and is most likely a consequence of non-ideal mixing of HCl and H_2O at low temperatures.

Figure 7.3 illustrates the behaviour of HCl in the system H_2O–HCl–CO_2 at 1 and 2 kbar, its activity being set by the buffer assemblage Ag–AgCl + Fe_3O_4–Fe_2O_3. There is a significant linear increase in HCl at high temperatures when CO_2 is added into the system. At low temperatures, however, the reverse is the case.

Using the approach of Marshall, which assumes that elecrolytic dissociation constants are functions of the volume concentration of water in the system (Marshall, 1970), and extending it to other parameters such as dielectric constant and Debye–Hückel parameters, the values of HCl^0 association in the presence of CO_2 have been calculated from values for total HCl concentration. By so doing, the value of the density of water in the mix defined as:

$$\tilde{\rho}_{H_2O} = \frac{I}{V^{(II)}}, \tag{7.21}$$

(where V^{II} is the total volume of the CO_2-bearing system with a constant mass of water of 1 g) was substituted into equations describing the functional dependence of these parameters on the density of pure water. These data are shown on Figure 7.3(a), (b) as dashed lines.

Assuming that there are no strong hydration interactions between molecular HCl^0 and water, the volume molar concentration of HCl^0 should remain constant when CO_2 is added to the system at f_{HCl^0} = const. For concentrations in molality at constant mass water in the system (this condition stems from the definition of concentration in molality), the following relationship should be valid:

$$m_{HCl^{0I}} \cdot V^{I^{-1}} = m_{HCl^{II}} V^{II^{-1}}, \tag{7.22}$$

where the superscript I refers to the molality of HCl^0 and volume for the pure aqueous system, and the superscript II indicates the same values for the system with CO_2. The volume of the pure aqueous system corresponds to its specific volume (in cm^3) if the mass of water is 1 g. The volume of the CO_2-bearing system at the Xth value of CO_2 and at the same mass of H_2O is defined as:

$$V^{II} = V^{T,P}_{H_2O} + V^{T,P,X}_{CO_2} + V^E, \tag{7.23}$$

where $V_{H_2O}^{T,P}$ is the volume of pure water at $\overline{m}_{H_2O} = 1$ g, $V_{CO_2}^{T,P,X}$ the volume of pure CO_2 at its Xth mole fraction, V^E is the excess volume of this system. Under these conditions,

$$V_{CO_2}^{T,P,X} = \overline{m}_{CO_2} \cdot v_{CO_2}^{T,P} = \frac{X_{CO_2} \cdot M_{CO_2} \cdot v_{CO_2}^{T,P}}{X_{H_2O} \cdot M_{H_2O}}, \tag{7.24}$$

where $\overline{m}_{CO_2}$ is the amount of CO_2 in the system, M_{CO_2} and M_{H_2O} the molecular masses of CO_2 and H_2O, respectively, and $v_{CO_2}^{T,P}$ is the specific volume of CO_2. The excess volume of the system is given by:

$$V^E = V^{E,\%} \cdot V^{II} \cdot 100^{-1}, \tag{7.25}$$

where $V^{E,\%}$ is the excess volume expressed as a percentage. The values for specific volumes of water and CO_2 necessary for such calculations were taken from Sato *et al.* (1984) and Shmulovich and Shmonov (1978), respectively. The values of $V^{E,\%}$ were calculated using the data of Shmulovich (1988).

By using relationship 7.22 and volume data for pure water and CO_2-bearing systems, variation in the molal concentration of HCl^0 in the presence of various amounts of CO_2 has been calculated. The calculated values are shown on Figure 7.3(a),(b) as dashed lines. Analysis of the obtained data shows that Equation 7.22 is valid for a fairly wide T–P range. The difference between the experimental and calculated values is not greater than 0.05 log units for the high-temperature region. However, at low temperatures and thus high densities of water (400 °C, 1 kbar and 450–400 °C, 2 kbar), the values of HCl^0 calculated from its total concentration and those calculated using relation 7.22 differ from one another. The reduction in the equilibrium volume concentration of HCl^0 and the inadequacy of Equation 7.22 in the high-density region is a consequence of 'salting out' HCl^0 under these conditions. The close agreement between the calculated and experimental data in the high-temperature region supports the validity of the assumption above that there are no strong hydration interactions between molecular HCl^0 and water.

In an earlier paper (Korzhinskiy, 1992), which covers all the experimental data for this system, we proposed that Equation 7.22 holds for the entire T–P range studied and that parameters of the system are determined by the relative partial density of water. However, a more sophisticated analysis of the data shows that, unlike the high-temperature region, a pressure increase in the system in the low-temperature region reduces the equilibrium volume concentration of HCl^0. Addition of CO_2 in the system, along with pressure decrease, however, reduces the density of water in the system. From this, it follows that Equation 7.22 holds for that T–P range in which the density of water is not high, i.e. its value does not exceed ~ 0.65–0.70 g · cm^{-3}. The approach developed in Korzhinskiy (1992) seems to be in error.

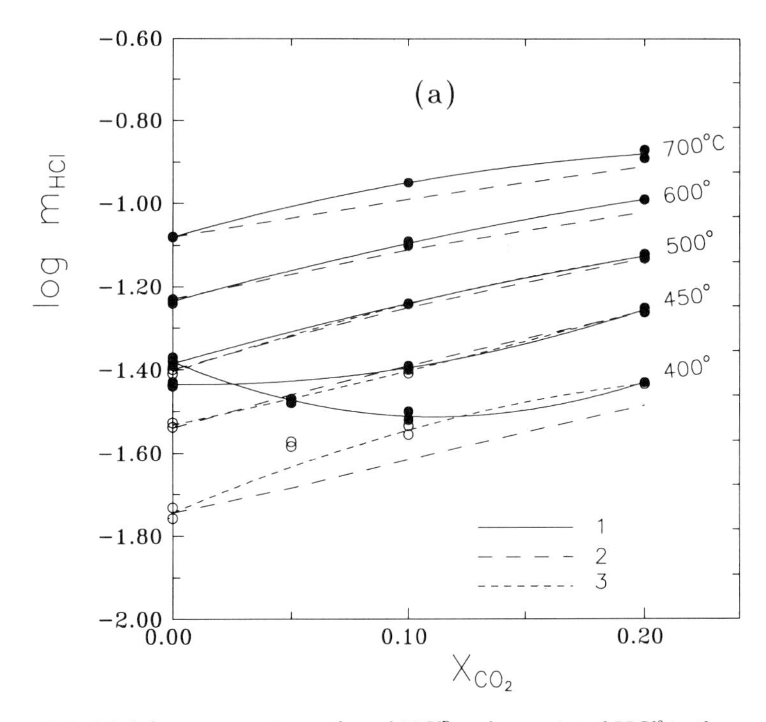

Figure 7.3. Molal concentrations of total HCl^{Σ} and associated HCl^0 in the system H_2O–HCl–CO_2, which are in equilibrium with the Ag–AgCl + Fe_3O_4–Fe_2O_3 buffer. (a) 1 kbar, (b) 2 kbar; 1 – experimental values of $m_{HCl^{\Sigma}}$, 2 – values of m_{HCl^0}, calculated from bulk concentrations of HCl^{Σ} using equations and 7.14–7.18, 3 – values of m_{HCl^0} calculated from relationship 7.21.

7.3.2 The H_2O–HCl–NaCl system

Figure 7.4 illustrates the behaviour of HCl in the system H_2O–HCl–NaCl at 1 and 2 kbar, its activity being controlled by the buffer assemblage Ag–AgCl + Fe_3O_4–Fe_2O_3. The results of the investigations are summarized in Table 7.1. The most striking feature of the figure is a steady decrease in mHCl with increasing NaCl content at high temperatures. At low temperatures, the effect becomes more pronounced.

Table 7.1. Conditions and results of experimental investigation of HCl behaviour in the system H_2O–HCl–NaCl in equilibrium with the Ag–AgCl + F_3O_4–Fe_2O_3 buffering assemblage

	Initial conditions		*Results*			*Calculations*	
Run	m_{NaCl}	log m_{HCl}	V extract (cm^3)	pH measured	log m_{HCl} (total)	log m_{HCl^0}	log Z
		700 °C, 1 kbar, 7 days					
155	–	−0.3	0.065	2.32	−1.08	−1.08	–
156	–	−1.3	0.069	2.30	−1.08	−1.08	–

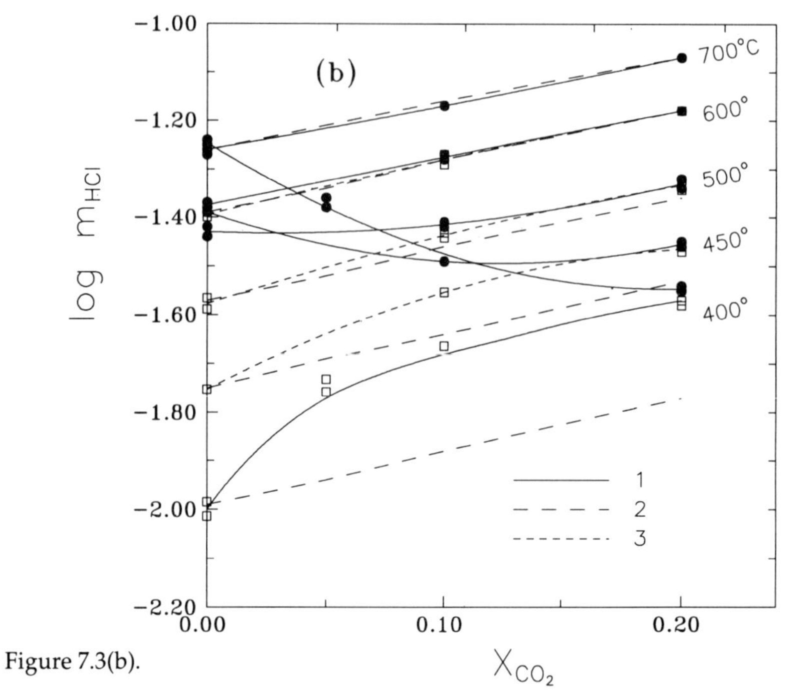

Figure 7.3(b).

Table 7.1 (Contd)

Run	*Initial conditions* m_{NaCl}	log m_{HCl}	*Results* V extract (cm^3)	pH measured	log m_{HCl} (total)	*Calculations* log m_{HCl°	log Z
161	2.93	−0.3	0.060	2.58	−1.31	−1.31	0.0675
162	2.93	−1.3	0.062	2.58	−1.32	−1.32	0.0675
163	6.16	−0.3	0.049	2.86	−1.48	−1.48	0.117
154	6.16	−1.3	0.051	2.86	−1.49	−1.49	0.117
		600°C, 1 kbar, 8 days					
81	–	−1.0	0.092	2.37	−1.23	−1.23	–
82	–	−2.0	0.081	2.40	−1.24	−1.24	–
86	2.93	−1.0	0.080	2.59	−1.47	−1.47	0.0792
87	2.93	−2.0	0.079	2.61	−1.48	−1.48	0.0792
88	6.16	−1.0	0.071	2.85	−1.67	−1.67	0.115
89	6.16	−2.0	0.066	2.86	−1.65	−1.65	0.115
		500°C, 1 kbar, 15 days					
41	–	−2.0	0.110	2.42	−1.39	−1.41	–
42	–	−1.0	0.104	2.44	−1.38	−1.40	–
47	2.93	−2.0	0.118	2.59	−1.59	−1.59	0.0646
48	2.93	−1.0	0.118	2.59	−1.59	−1.59	0.0646
49	6.16	−2.0	0.092	2.91	−1.78	−1.78	0.0994
50	6.16	−1.0	0.090	2.91	−1.80	−1.80	0.0994
		450°C, 1 kbar, 20 days					
122	–	−1.0	0.103	2.50	−1.44	−1.54	–

	Initial conditions		*Results*			*Calculations*	
Run	m_{NaCl}	log m_{HCl}	V extract (cm^3)	pH measured	log m_{HCl} (total)	log m_{HCl°	log Z
123	–	−2.0	0.092	2.54	−1.43	−1.52	–
128	2.93	−1.0	0.095	2.80	−1.72	−1.73	0.0459
129	2.93	−2.0	0.103	2.79	−1.74	−1.75	0.0459
130	6.16	−1.0	0.080	3.06	−1.88	−1.89	0.0690
131	6.16	−2.0	0.100	2.97	−1.88	−1.89	0.0690
	400 °C, 1 kbar, 28 days						
165	–	−1.0	0.090	2.50	−1.39	−1.74	–
166	–	−2.0	0.098	2.45	−1.37	−1.72	–
31	0.2	−1.0	0.097	2.65	−1.60	−1.74	0.0045
32	0.6	−2.0	0.108	2.70	−1.69	−1.79	0.0128
33	0.6	−1.0	0.091	2.76	−1.68	−1.78	0.0128
34	1.4	−2.0	0.076	2.92	−1.77	−1.85	0.0268
35	2.93	−1.0	0.098	2.93	−1.86	−1.92	0.0450
36	2.93	−2.0	0.093	2.93	−1.84	−1.90	0.0450
37	6.16	−1.0	0.081	3.14	−1.97	−2.02	0.0579
38	6.16	−2.0	0.084	3.11	−1.95	−2.00	0.0579
	700 °C, 2 kbar, 7 days						
51	–	−2.0	0.078	2.41	−1.25	−1.25	–
52	–	−1.0	0.073	2.46	−1.27	−1.27	–
57	2.93	−2.0	0.089	2.60	−1.49	−1.49	0.0513
58	2.93	−1.0	0.087	2.62	−1.50	−1.50	0.0513
59	6.16	−2.0	0.074	2.91	−1.70	−1.70	0.0828
60	6.16	−1.0	0.075	2.94	−1.73	−1.73	0.0828
	600 °C, 2 kbar, 10 days						
187	–	−1.0	0.095	2.45	−1.37	−1.39	–
188	–	−2.0	0.095	2.46	−1.38	−1.40	–
195	1.00	−1.0	0.100	2.54	−1.51	−1.51	0.0211
196	1.00	−2.0	0.100	2.57	−1.53	−1.53	0.0211
197	2.93	−1.0	0.090	2.78	−1.68	−1.68	0.0490
198	2.93	−2.0	0.090	2.76	−1.66	−1.66	0.0490
199	6.16	−1.0	0.080	2.98	−1.80	−1.80	0.0730
200	6.16	−2.0	0.080	3.02	−1.84	−1.84	0.0730
	500 °C, 2 kbar, 13 days						
11	–	−2.0	0.112	2.45	−1.42	−1.56	–
12	–	−1.0	0.110	2.47	−1.44	−1.58	–
17	2.93	−2.0	0.108	2.77	−1.74	−1.76	0.0362
18	2.93	−1.0	0.103	2.80	−1.75	−1.77	0.0362
19	6.16	−2.0	0.098	3.04	−1.94	−1.95	0.0517
20	6.16	−1.0	0.099	3.00	−1.91	−1.92	0.0517
	450 °C, 2 kbar, 20 days						
132	–	−1.0	0.090	2.49	−1.41	−1.76	–
133	–	−2.0	0.100	2.46	−1.42	−1.77	–
138	2.93	−1.0	0.088	2.94	−1.84	−1.89	0.0290
139	2.93	−2.0	0.105	2.87	−1.85	−1.90	0.0290
140	6.16	−1.0	0.088	3.07	−1.94	−1.98	0.0397
141	6.16	−2.0	0.087	3.10	−1.97	−2.01	0.0397

Table 7.1 (Contd)

	Initial conditions		*Results*			*Calculations*	
Run	m_{NaCl}	log m_{HCl}	V extract (cm^3)	pH measured	log m_{HCl} (total)	log m_{HCl°	log Z
		400 °C, 2 kbar, 28 days					
178	–	−1.0	0.078	2.43	−1.26	−1.98	–
179	–	−2.0	0.078	2.45	−1.24	−1.95	–
28	0.2	−1.0	0.087	2.64	−1.54	−1.96	0.0024
29	0.6	−2.0	0.097	2.76	−1.72	−2.02	0.0069
30	0.6	−1.0	0.080	2.84	−1.72	−2.02	0.0069
182	1.4	−2.0	0.098	2.86	−1.81	−2.04	0.0142
183	2.93	−1.0	0.048	3.23	−1.89	−2.07	0.0234
184	2.93	−2.0	0.068	3.10	−1.90	−2.08	0.0234
185	6.16	−1.0	0.068	3.23	−2.03	−2.17	0.0314
186	6.16	−2.0	0.068	3.20	−2.00	−2.14	0.0314

Before proceeding to a description and discussion of the results obtained, consider the P–V–T properties of the system H_2O–NaCl. Figure 7.5 shows how the volume of the system varies according to the NaCl content for a constant specified amount of water of 1 g. The lines representing variation in the system volume at the T and P considered in the present study are the result of averaging both the data available in the literature to 600 °C (Gehrig *et al.*, 1983; Hilbert, 1984) and experimental values retrieved at 700 °C and 1 kbar by a gold bag procedure (Shmonov and Shmulovich, 1978). The latter data are tabulated in Table 7.2. Also included in Figure 7.5 for T = 600 and 700 °C, P = 1 kbar, are data for the theoretical calculation of the variation in the volume properties of the system from Lvov and Wood (1990) using a molecular dynamics technique. It is evident that the latter data differ drastically from the experimental ones. The difference may be due to unmixing in the high-temperature region (Sourirajan and Kennedy, 1962), disregarded in Lvov and Wood (1990). According to the presented evidence, adding NaCl into the system at high temperature causes a progressive reduction in its volume. As the temperature is lowered (< 500°C) and NaCl added, the volume of the system passes through a minimum; beyond this the volume increases in proportion to a value equal to the volume of the crystalline NaCl added. As an illustration, for T = 150°C and P = 2 kbar, experimental data and data for the additive sum of the volumes of pure water and crystalline NaCl are presented which practically coincide. The change in the system volume due to interior interaction between species may be characterized in terms of the relative compressibility factor:

$$Z = V^{I} \cdot (V^{II} - V_{NaCl^{cr}})^{-1}. \qquad (7.26)$$

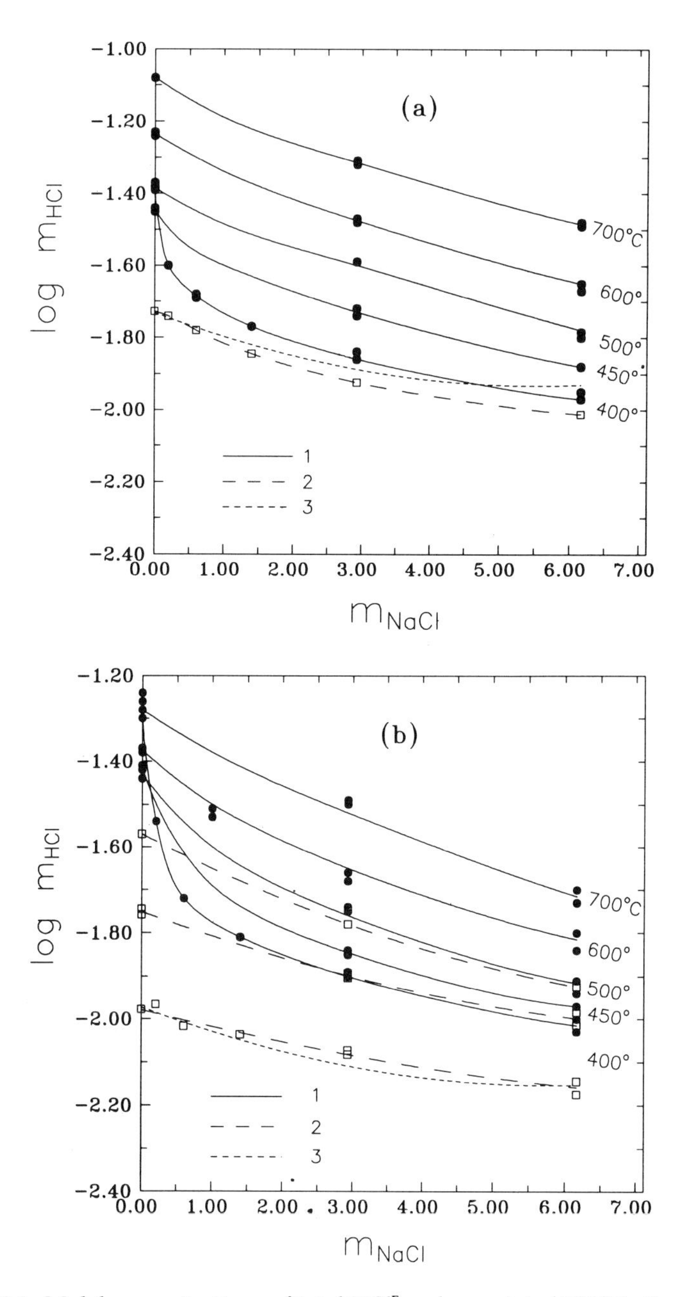

Figure 7.4. Molal concentrations of total HCl^{Σ} and associated HCl° in the system H_2O–HCl–NaCl, which are in equilibrium with the Ag–AgCl + Fe_3O_4–Fe_2O_3 buffer. (a) 1 kbar, (b) 2 kbar; 1 – experimental values of $m_{HCl^{\Sigma}}$, 2 – values of $m_{HCl^{0}}$, calculated from equations and relationships 7.14–7.18 and 7.29, 3 – values of $m_{HCl^{0}}$, calculated from relationship 7.30.

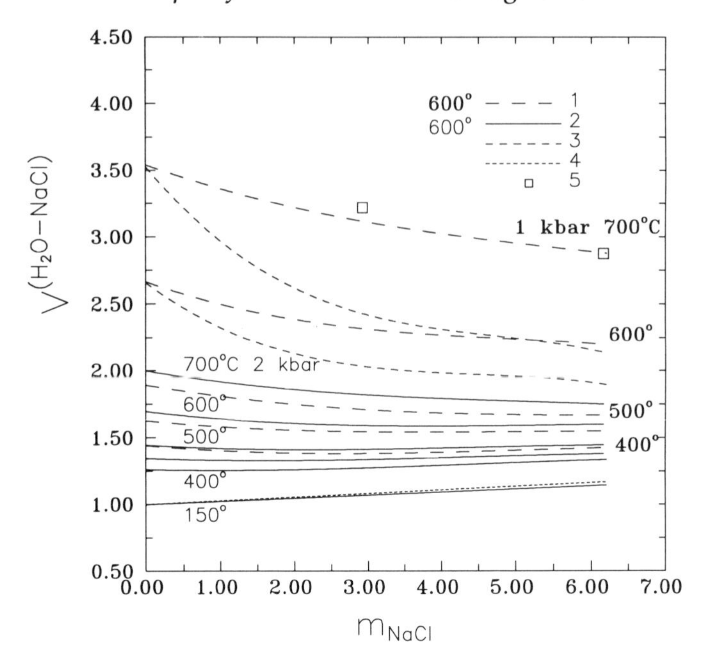

Figure 7.5. Variation in volume of the system H_2O–NaCl as a function of the NaCl content at constant $\overline{m}_{H_2O} = 1$ gram. 1 – P = 1 kbar, 2 – P = 2kbar, 3 – calculated data of Lvov and Wood (1990), 4 – line representing the additive sum of volumes of H_2O and NaCl for 150°C, P = 2 kbar, 5 – author's experimental data.

Table 7.2. Experimental values of volumes in the system H_2O–NaCl at T = 700 °C and P = 1 kbar at $m_{H_2O} = 1$ gram

m $mol \cdot kg\ H_2O^{-1}$	2.92	6.16	25	50
V cm^3	3.218	2.876	2.712	3.128

The concentration of water (its relative density, $g \cdot cm^{-3}$) in the system may be defined by:

$$\tilde{\rho}_{H_2O} = V^{I} \cdot (V^{II} - V_{NaCl^{cr}})^{-1}, \tag{7.27}$$

where V^I is the volume of 1 g pure water, V^{II} the volume of the system H_2O–NaCl at chosen NaCl concentration and $\overline{m}_{H_2O} = 1\ g \cdot V_{NaCl^{cr}}$ is the volume of crystalline NaCl at its mth concentration, given by:

$$V_{NaCl^{cr}} = m_{NaCl} \cdot 0.05854 \cdot \rho_{NaCl}^{-1}, \tag{7.28}$$

where m is molality, and ρ_{NaCl} the density of crystalline NaCl. In this instance, it is logical to assume that there should be simple linear relationships between parameters of state for the system (or the solute) and values of Z and $\tilde{\rho}_{H_2O}$.

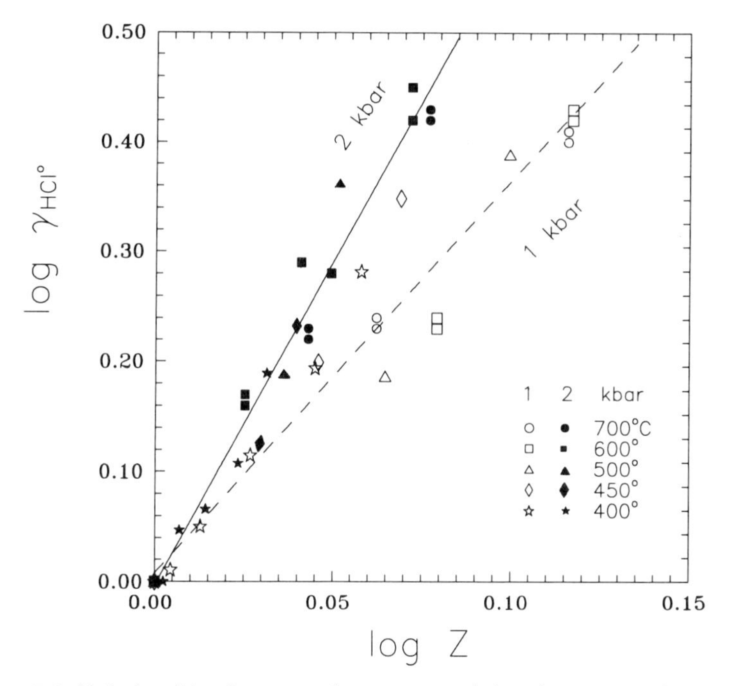

Figure 7.6. Relationships between the compressibility factor Z and γ_{NaCl^0} in the system H_2O–NaCl–HCl. Dashed line and open symbols refer to 1 kbar, solid line and filled symbols refer to 2 kbar.

Figure 7.6 shows the relationship between the relative degree of decreasing the equilibrium $mHCl^0$

$$\frac{m_{HCl^0}^{I}}{m_{HCl^0}^{II}} = \gamma_{HCl^0} \tag{7.29}$$

and compressibility factor, Z, both expressed in log units. As is evident from the figure, there is a linear relationship between the values considered at the chosen P:

$$\log \gamma_{NaCl^0} = S \cdot \log Z, \tag{7.30}$$

the relationship is weakly temperature dependent. For P = 1 kbar, S = – 3.1; for P = 2 kbar, S = – 4.5.

In order to calculate the distribution of species in the investigated system at low temperature, it is necessary to know the activity coefficients of a neutral form of $NaCl^0$ as well. They may be estimated from data on equilibria involving albite–andalusite–quartz (Hiroshi and Koichiro, 1989) and tantalum-bearing ($Na_2Ta_4O_{11}$–Ta_2O_5) assemblages (unpublished data) in H_2O–NaCl–HCl solutions at T = 600–700°C and P = 1–2 kbar. Note that the values of γ_{NaCl^0} may be defined by an equation such as 7.30. For P = 1 kbar, S = 2.8; for P = 2 kbar, S = 4.

Knowledge of the activity coefficients of uncharged species permits correct calculation of the species distribution in the investigated system. The calculation was carried out using mass balance equations and those for charges, as well as dissociation constants for HCl from Frantz and Marshall (1984) and for NaCl from Plyasunov (1989):

$$\log K_{D_{NaCl}} = -2.851 + 2.368 \cdot T,K \cdot 10^{-3}$$

$$+ \frac{503.861}{T,K} - 8.979 \cdot \log(\rho_{H_2O}). \qquad (7.31)$$

Activity coefficients for ions were calculated in much the same way as those of HCl (see previous section) using the Debye–Hückel equation (7.19), with the value of parameter $\mathring{a}$ characterizing the ionic 'distance of closest approach' in solution being equal to 3.72 for NaCl (Helgeson *et al.*, 1981). Figure 7.4 (a),(b) (T = 400°C) presents HCl^0 data calculated by relations 7.29 and 7.30 on the one hand and directly from experimental values of equilibrium total HCl concentration using dissociation constants and activity coefficients on the other. Analysis of the data shows that these values agree well. The discrepancy is not greater than 0.08 log units. In this case, there seems to be no need to replace the density of pure water by the relative density of water in the system defined by Equation 7.27 in equations reflecting a functional relationship between ionic activity coefficients and dissociation constants for HCl and NaCl, because activity coefficients for neutral species are taken into consideration.

7.4 BEHAVIOUR OF THE SALT COMPONENTS $MgCl_2$, $CaCl_2$ AND $FeCl_2$

The behaviour of divalent metal chlorides in the presence of NaCl, HCl, and CO_2 at fixed activities of the cation in oxide form has been studied in the temperature range 400–700°C at 2 kbar. The CO_2 contents of the fluid ranged from 0 to 0.2 mole fraction, the NaCl contents from 0 to 4 $mol \cdot kg\ H_2O^{-1}$. The studies were done in accordance with the following univariant equilibria:

$$\underset{\text{albite}}{NaAlSi_3O_8} + HCl = \frac{1}{2}\underset{\text{kyanite}}{Al_2SiO_5} + \frac{5}{2}\underset{\text{quartz}}{SiO_2} + NaCl + \frac{1}{2}H_2O, \qquad (7.32)$$

$$\frac{1}{3}\underset{\text{talc}}{Mg_3Si_4O_{10}(OH)_2} + 2HCl = +\frac{4}{3}\underset{\text{quartz}}{SiO_2} + MgCl_2 + \frac{4}{3}H_2O, \qquad (7.33)$$

$$MgWO_4 + 2HCl = WO_3 + MgCl_2 + H_2O, \qquad (7.34)$$

$$CaTiSiO_5 + 2HCl = TiO_2 + SiO_2 + CaCl_2 + H_2O, \quad (7.35)$$

sphene rutile quartz

$$CaWO_4 + 2HCl = WO_3 + CaCl_2 + H_2O, \quad (7.36)$$

scheelite

$$Fe_3O_4 + 2HCl = Fe_2O_3 + FeCl_2 + H_2O, \quad (7.37)$$

magnetite hematite

$$FeWO_4 + 2HCl = WO_3 + FeCl_2 + H_2O. \quad (7.38)$$

ferberite

α_{HCl^0} was fixed using the Ag–AgCl buffer in combination with Fe_3O_4–Fe_2O_3, except for equilibrium 7.35 which was examined in conjunction with the Cu_2O–CuO oxygen buffer.

7.4.1 Metal chloride–CO_2 systems

Figures 7.7–7.9 illustrate the behaviour of salt components in the fluid in the presence of CO_2. Quantitatively, variation in the concentration of any component with composition of the system may be expressed in terms of

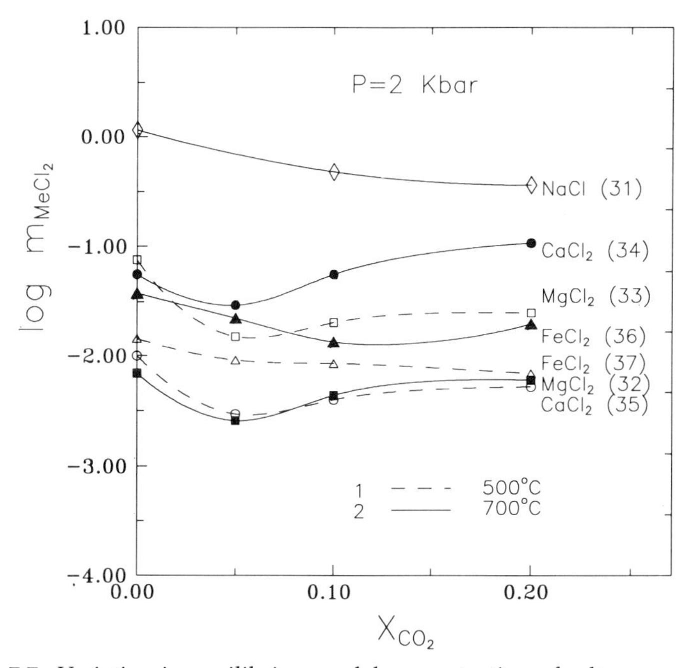

Figure 7.7. Variation in equilibrium molal concentration of salt components in the fluid as a function of the CO_2 content at 2 kbar. 1 – T = 500°C, 2 – T = 700°C; numbers represent values for total activity coefficients, numbers in parentheses indicate the equilibrium number.

the total activity coefficient. In this instance, the total activity coefficient, γ, is defined as: $\gamma = m^{I}_{MeCl_x} \cdot m^{II}_{MeCl_x}{}^{-1}$, where m^{I} is the equilibrium salt concentration in the pure aqueous fluid, and m^{II} the salt concentration in the presence of CO_2 or NaCl.

Figure 7.7 shows that at high temperatures, addition of CO_2 into the Na-bearing system (assemblage Ab–And–Qtz), with activities of Na_2O and HCl constant, produces a steady decrease in the equilibrium concentration of NaCl. $\gamma_{NaCl} = 3.2$ for $X_{CO_2} = 0.2$. The distinguishing feature of the behaviour of chlorides of the divalent elements Mg, Ca, and Fe at high temperatures is the extreme change in their equilibrium concentrations. At temperatures between 700 and 500 °C and 2 kbar pressure, the equilibrium salt concentration decreases sharply as X_{CO_2} increases from 0 to 0.05, while for X_{CO_2} in the range 0.05–0.2, it increases smoothly. As is apparent from Figure 7.8 for $CaCl_2$, a decrease in pressure from 2 to 1 kbar results first in the shift of this minimum in the salt concentration toward lower CO_2 contents, followed by its complete disappearance. Such behaviour of chloride salts at high temperatures in the presence of non-polar gas may be accounted for by their molecular hydration. In the case of divalent salts, discrete short- and long-range types of hydration spheres are assumed to exist, separated by an energy barrier. A decrease in the equilibrium salt concentration in the compositional range $0 < X_{CO_2} < 0.05$ is due to the effect produced by the long-range hydration sphere, while its increase at X_{CO_2} in the range 0.05–0.2 follows the law of mass action in accordance with the equilibrium constant

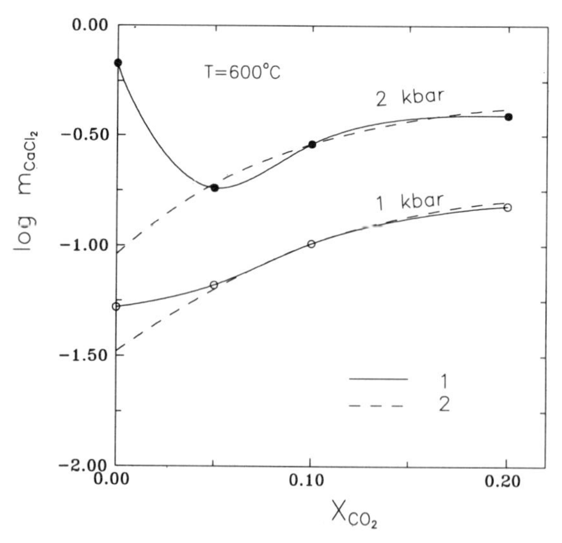

Figure 7.8. Variation in molal concentration of $CaCl_2$ in the fluid in equilibrium with the assemblage sphene–rutile–quartz as a function of the CO_2 content of the system at 600°C and 2 and 1 kbar. Line 1 is a fit to the experimental points, line 2 is extrapolated to $X_{CO_2} = 0$.

for unhydrated forms. It is known, however, that chloride salts are practically insoluble in pure CO_2 (the solubility of NaCl in pure CO_2 is $\sim 3.2 \times 10^{-5}$ mol · l^{-1} at 600°C, 2 kbar, from the author's experiments). Therefore, a further increase in the CO_2 content of the system should result in the development of another extreme state for the system due to the influence of the short-range hydration sphere. Such a hydration model predicts the existence of several extreme states responsible for changes in the equilibrium salt concentration over a wide range of CO_2 contents in the fluid. In the case of univalent salts, such as NaCl, two types of hydration are also possible; however, their energy-based relations are such that changing the role of one dominant hydration sphere for another occurs at much higher X_{CO_2} values, the change being of a gradual character. Previous studies (Fein and Walther, 1989; Walther and Orville, 1983) on the solubility of quartz and portlandite in the water–gas system also addressed this problem and suggested the existence of high-temperature molecular hydration.

At low temperatures (< 500°C), the change in the equilibrium concentration of metal chlorides follows a similar pattern to that observed at high temperatures. Under these conditions, however, the salts show increased dissociation, and at T < 500°C, the minimum in the equilibrium concentration shifts to higher X_{CO_2}. This is demonstrated by the data (Figure 7.9) for the system $MgCl_2$–H_2O–CO_2 at T < 500°C. The decrease in the equilibrium salt concentration at high X_{CO_2} is in this case a consequence both of decreasing the degree of $MgCl_2$ dissociation and the change in hydration state.

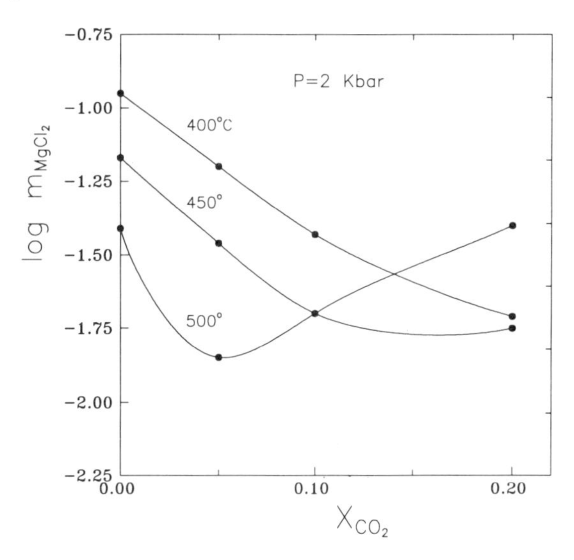

Figure 7.9. Variation in molal concentration of $MgCl_2$ in the fluid in equilibrium with $MgWO_4$–WO_3 as a function of the CO_2 content of the system at T between 500 and 400°C and P = 2 kbar.

7.4.2 Behaviour of salts in the presence of NaCl

The behaviour of metal chlorides in the presence of NaCl is shown in Figures 7.10–7.12. At high temperatures and fixed activities of HCl and of the cation in oxide form, the addition of NaCl into the system produces an increase in the metal chloride concentration in the fluid. A possible explanation is the formation of complex molecules such as $MeCl_2 \cdot nNaCl$. Analysis of the total activity coefficients in Figure 7.10 shows that in the NaCl-bearing system, the increase in the equilibrium concentrations of the chlorides investigated is in the order: $\gamma FeCl_2 < \gamma CaCl_2 < \gamma MgCl_2$ (where γ is defined as in 7.4.1), i.e. the interaction with NaCl increases in proportion to the molecular mass of the component.

At low temperatures (< 500°C), addition of NaCl causes a decrease in equilibrium metal chloride concentration. This is because of the higher degree of dissociation, which leads to a significant 'common ion effect'. $FeCl_2$ provides an exception (Figure 7.10), suggesting that $FeCl_2$ is fully associated at the P–T parameters considered in the present study (e.g. Hemley *et al.*, 1992).

Data for $MgCl_2$ (Figure 7.11) best demonstrate the effect of adding NaCl to a fluid in equilibrium with the assemblage talc–quartz. At 700°C, NaCl addition produces a uniform increase in the concentration of

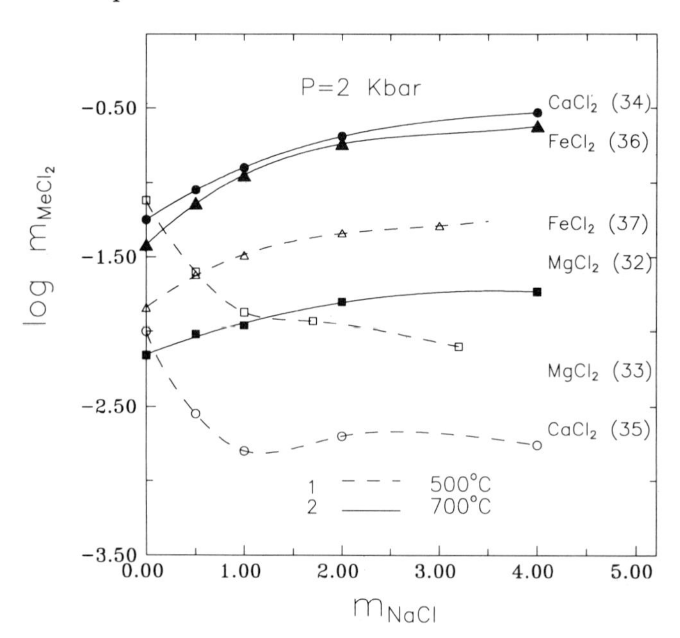

Figure 7.10. Variation in equilibrium molal concentration of salt components in the fluid as a function of the NaCl content of the system at 2 kbar. 1 – T = 500°C, 2 – T = 700°C; numbers in parentheses correspond to the equilibrium numbers used in the text.

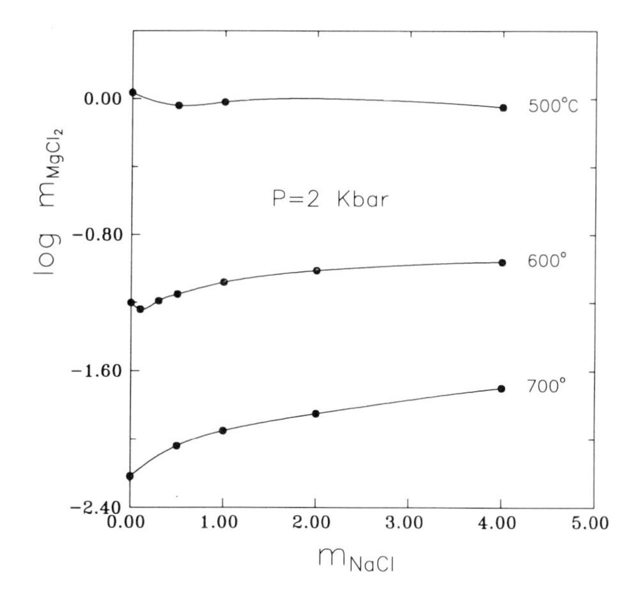

Figure 7.11. Variation in molal concentration of $MgCl_2$ in the fluid in equilibrium with the assemblage talc–quartz as a function of the NaCl content of the system at T in the range 700–500°C and P = 2 kbar.

$MgCl_2$. At 600 °C, $MgCl_2$ concentration exhibits a minimum, while at 500 °C, this dependence (with a particularly weak minimum) becomes non-diagnostic. Figure 7.12 also illustrates the behaviour of $MgCl_2$ in equilibrium with the assemblage $MgWO_4$–WO_3 at temperatures between 500 and 400 °C when NaCl is added to the system. At 400 °C, the equilibrium concentration of Mg decreases considerably as NaCl is added to the system, but this effect is absent at 500 °C. Differences in behaviour of $MgCl_2$ at 500°C between the two buffering assemblages are due to differences in intrinsic concentrations of $MgCl_2$ in equilibrium with each, which means that the degree of $MgCl_2$ dissociation is also different. A decrease in temperature expands the concentration range where the common ion effect is operative, as is apparent from Figure 7.12.

7.4.3 Additional experimental investigations

Further experiments considered the effect of T–P parameters and NaCl and CO_2 contents on $CaCl_2 : MgCl_2$ and $CaCl_2 : FeCl_2$ ratios, and the influence of CO_2 on the NaCl : HCl ratio in the hydrothermal fluids in equilibrium with univariant assemblages. Experiments were performed in the temperature–pressure range 400–700 °C and 1–2 kbar at up to 4 m NaCl and $X_{CO_2} < 0.2$. The following cation exchange equilibria were investigated:

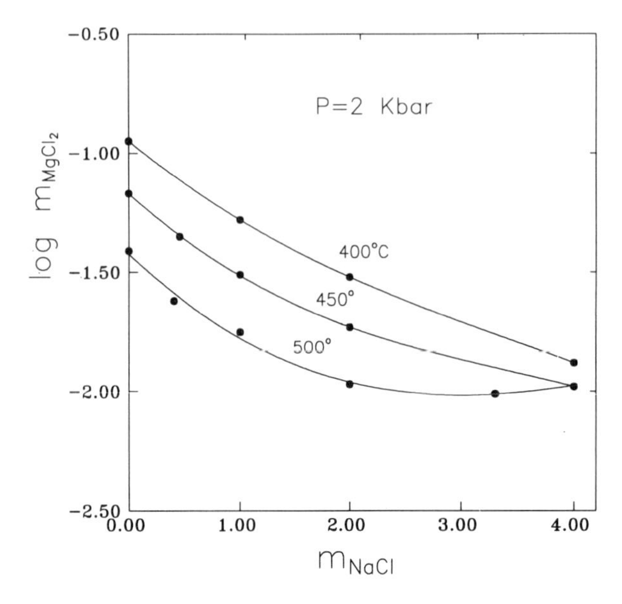

Figure 7.12. Variation in molal concentration of $MgCl_2$ in the fluid in equilibrium with $MgWO_4$–WO_3 as a function of the NaCl content of the system at T between 500 and 400°C and P = 2 kbar.

$$\underset{\text{wollastonite}}{2CaSiO_3} + MgCl_2 = \underset{\text{diopside}}{CaMgSi_2O_6} + CaCl_2, \tag{7.39}$$

$$\underset{\text{diopside}}{7CaMgSi_2O_6} + \underset{\text{quartz}}{2SiO_2} + 3MgCl_2 + 2H_2O = \underset{\text{tremolite}}{2Ca_2Mg_5(OH)_2[Si_8O_{22}]} + 3CaCl_2, \tag{7.40}$$

$$\underset{\text{hedenbergite}}{CaFeSi_2O_6} + FeCl_2 = \underset{\text{fayalite}}{Fe_2SiO_4} + \underset{\text{quartz}}{SiO_2} + CaCl_2, \tag{7.41}$$

$$0.5Na_2Ta_4O_{11} + HCl = Ta_2O_5 + NaCl + 0.5H_2O. \tag{7.42}$$

The investigations were carried out without using the Ag–AgCl buffer and therefore provide a test for the reliability of the earlier set of experiments on the behaviour of chloride salts in the presence of NaCl and CO_2.

Figures 7.13 and 7.14 show the influence of P and T on the ratios $CaCl_2 : MgCl_2$ and $CaCl_2 : FeCl_2$ in fluid in equilibrium with the assemblages of equilibria 7.39 and 7.41, respectively. Increase in T, or decrease in P result in an increase in $MgCl_2 : CaCl_2$ or $FeCl_2 : CaCl_2$. In addition, a fan-shaped variation in these ratios with total dissolved salts is

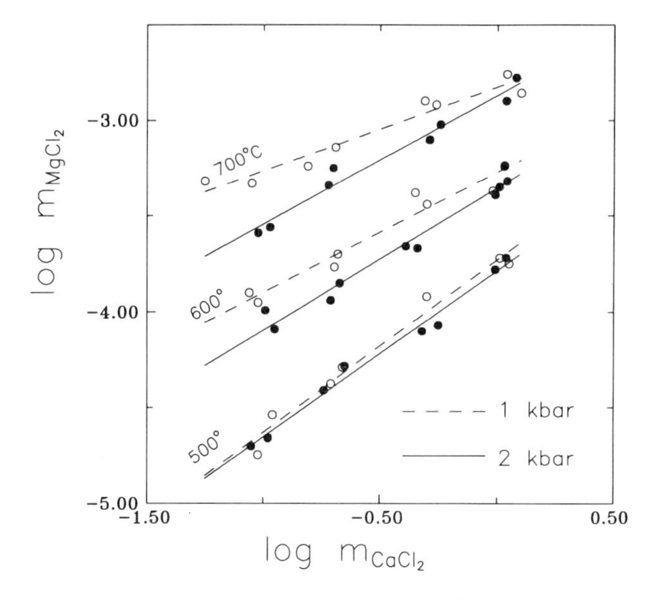

Figure 7.13. Concentration ratios of $CaCl_2$ and $MgCl_2$ in the fluid in equilibrium with the assemblage diopside–wollastonite at different T and P.

observed. For example, $\partial \log m_{FeCl_2} \cdot \partial \log m_{CaCl_2}^{-1}$ varies from 1.30 at 400 °C to 0.53 at 700 °C (see Korzhinskiy, 1986, 1987 for details).

Figures 7.15 and 7.16 depict the effect of NaCl and CO_2 concentrations on the divalent chloride ratios for fluids buffered by equilibria 7.40 and 7.41, respectively, at 600 °C and 2 kbar. The addition of either NaCl or CO_2 into equilibrium 7.41 is accompanied by increased $FeCl_2 : CaCl_2$ (Figure 7.16). By contrast, adding CO_2 to equilibrium 7.40 increases $MgCl_2 : CaCl_2$, while NaCl addition reduces it. This is in accordance with the conclusions of the preceding section, where it was found that adding NaCl increases the equilibrium concentration of divalent chlorides in proportion to their molecular masses. Thus, addition of NaCl into the system Ca–Mg produces a relative increase in the calcium concentration, while NaCl addition into the system Ca–Fe results in the relative increase of iron concentration.

The mechanism by which non-polar gas affects the equilibrium ratios of the chlorides is of a somewhat different character. Being present in the fluid, these salts are hydrated, and the degree of their hydration seems to be directly proportional to their ionic radii. According to Helgeson *et al.* (1981), relative ionic radii are: $r_{Ca^{2+}} > r_{Mg^{2+}}$ and $r_{Ca^{2+}} > r_{Fe^{2+}}$. Thus, adding CO_2 in these systems should reduce the equilibrium concentration of calcium relative to magnesium and iron. However, the information in the preceding section for the behaviour of their chlorides at 700 °C indicates that the predicted behaviour is valid for calcium and iron, but is invalid for calcium and magnesium. At 700°C, the decrease in the

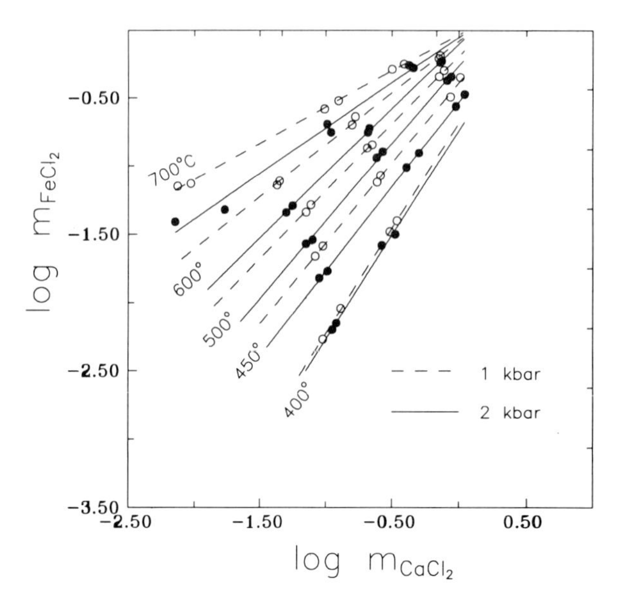

Figure 7.14. Concentration ratios of $CaCl_2$ and $FeCl_2$ in the fluid in equilibrium with the assemblage hedenbergite–fayalite–quartz at different T and P.

equilibrium concentration of $MgCl_2$ is greater than that of $CaCl_2$ for a given increase in X_{CO_2}. This may be a consequence of using different

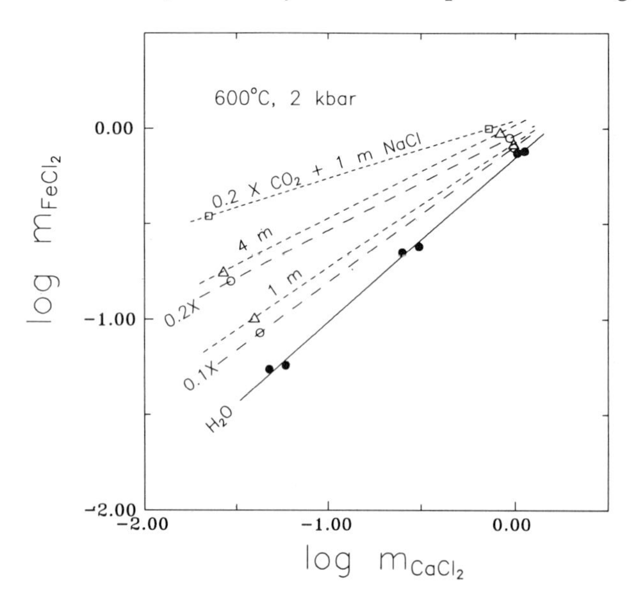

Figure 7.15. Variation in concentration ratios of $CaCl_2$ and $MgCl_2$ in the fluid in equilibrium with the assemblage tremolite–diopside–quartz as a function of the CO_2 and NaCl contents of the system at 600°C and 2 kbar.

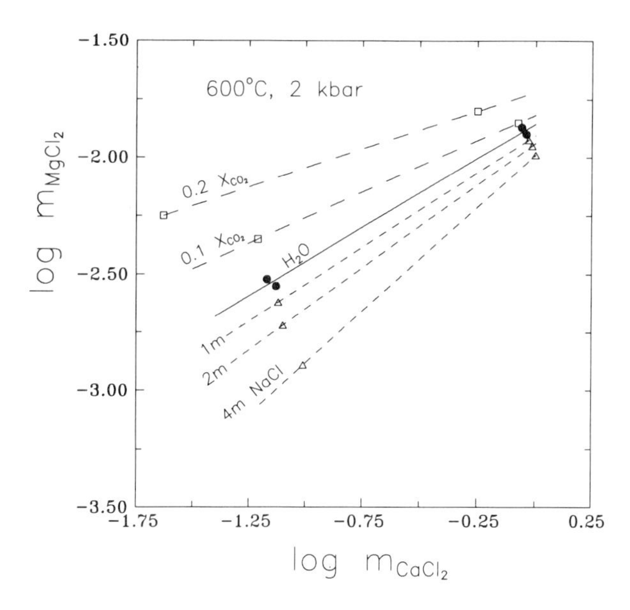

Figure 7.16. Variation in concentration ratios of $CaCl_2$ and $FeCl_2$ in the fluid in equilibrium with the assemblage hedenbergite–fayalite–quartz as a function of the CO_2 and NaCl contents of the system at 600°C and 2 kbar.

buffer assemblages. In particular, the use of talc in studies of $MgCl_2$ behaviour requires that the stoichiometric coefficient for water be equal to 1. 333, as follows from equilibrium 7.33. In the case of $CaCl_2$ (Equation 7.35), this coefficient equals 1.

Figure 7.17 displays the effect of CO_2 on the NaCl:HCl ratio in the fluid in equilibrium 7.42 with the tantalum-bearing assemblage $Na_2Ta_4O_{11}$–Ta_2O_5 at 600°C and 2 kbar. This assemblage was chosen because the phases applied display a higher stability with respect to T, P as opposed, for example, to the assemblage albite–andalusite–quartz, and because Ta_2O_5 is much less soluble than quartz. Introduction of CO_2 reduces the equilibrium ratio $m_{NaCl} \cdot m_{HCl}^{-1}$, i.e. in moderately concentrated solutions, the equilibrium concentration of the acid component increases by 0.7–0.8 log units compared to NaCl. At constant HCl content, the equilibrium concentration of NaCl reduces as CO_2 is added into the system. This is in good agreement with the data obtained using the assemblage albite–andalusite–quartz and the Ag–AgCl buffer (see Figure 7.7).

7.5 CONCLUDING DISCUSSION

The distinguishing feature of the experimental studies presented here is that they were performed at fixed activity of the acid component HCl,

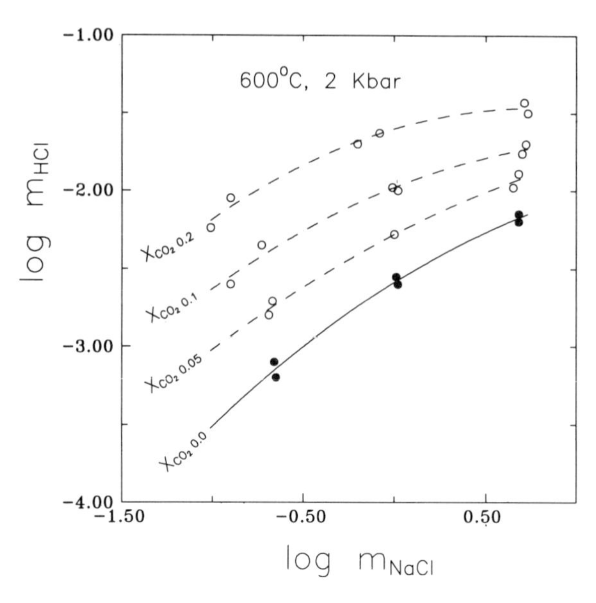

Figure 7.17. Variation in concentration ratios of NaCl and HCl in the fluid in equilibrium with the assemblage $Na_2Ta_4O_{11}$–Ta_2O_5 as a function of the CO_2 content of the system at 600°C and 2 kbar.

conjugate with chloride salt. By possessing the minimum degree of freedom, investigations of this sort permit estimation of ionic–molecular interactions between different species in fluid mixtures of complex composition and provide a basis for testing the available theories of the ionic–molecular state of salt substance dissolved in fluid mixture at elevated P and T. The ionic–molecular interactions may be expressed in terms of the 'realistic' composition of species dissolved in fluid, activity coefficients of these species, or as corresponding dissociation constants or molecular complex formation.

The experimental studies reported above show that the effect of background gas and salt on the behaviour of acid and salt components in a hydrothermal medium are fundamentally different. This is a consequence of the difference in their interaction with water at elevated T and P. Such interactions are negligible for acid and gas components, but molecular salt components are hydrated. HCl 'salts out' under the action of the background electrolyte, while salt components interact with it and with one another, forming complex molecules of the type $Na_xMe_yCl_z$.

Adding gas into the system reduces the volume concentration of water, while adding the electrolyte increases its concentration. Introducing additional components may be regarded as a decrease or increase in the internal pressure in the system at the chosen T and P. If the pressure dependence of a process active in a pure aqueous system is known, then, based on variation in the volume concentration of water in a given system, it may be possible to predict to a first approximation how the process will evolve when a gas or salt background component is added into the system.

The relative behaviour of components (i.e. variation in their concentration ratios at fixed activity ratios) may also be predicted when gas or salt is added into the system. This may be done on the basis of the molecular mass and ionic radius of given components. Our data show that the degree of molecular hydration of a component is in proportion to its ionic radius, whereas the degree of molecular complex formation varies with its molecular mass.

7.5.1 Geological implications

The experimental studies reported here may be applied to a wide range of hydrothermal processes. For example, they allow the evolution of postmagmatic dissolution and deposition of rock-forming and ore minerals to be predicted, and systematic changes in mineral assemblage to be understood. As shown above, the formation of chloride salt of an element or metal (Me) dissolved in a fluid phase may be described by equilibrium 7.3:

$$MeO_X + 2XHCl = MeCl_{2X} + XH_2O,$$

where MO_X represents a metal oxide or an oxide activity buffer. From the investigations presented above, equilibrium 7.3 may also be represented as:

$$MeO_X + 2XHCl + (n + m)H_2O + qNaCl =$$

$$MeCl_{2X} \cdot nH_2O \cdot mH_2O \cdot qNaCl, \qquad (7.43)$$

where n and m are the number of water molecules of short- and long-range hydration types, respectively, and q is the number of NaCl molecules participating in the molecular complex formation. Equilibrium 7.43 is not only P–T dependent. It is also strongly dependent on HCl activity, the content of the background electrolyte NaCl in the fluid phase, and on water activity in the system: hence it responds to the content of gas components. Our data show that the processes which increase the density of a hydrothermal fluid all favour an 'acid effect' of the fluid phase on mineral assemblages, i.e. cations are leached into solution. Such processes include temperature decrease, pressure increase, removal of CO_2, or an increase in the background salt concentration. The contrary processes favour the opposite 'alkaline effect' of fluid on surrounding country rocks.

Korzhinskiy (1958) identified early alkaline, acidic, and late alkaline stages of postmagmatic hydrothermal activity. The development of the early alkaline stage, at which fenites, Mg skarns, and feldspathoids were formed at the contact of magmatic and country rocks, is favoured by progressive heating, accumulation of CO_2 in the fluid, and by reduction in the density of the fluid phase. Decrease in temperature, removal of a gas component from the system (e.g. by phase separation), increase in fluid density, and increase in the concentration of chloride salts all favour the acid effect, responsible for the production of metasomatites, greisens, etc. The development of the late alkaline stage at which carbonate deposition and related ore mineralization occur is favoured by accumulation of a gas component in the fluid phase, total pressure drop and by a decrease in the overall salinity of solution, possibly as a result of dilution by surface waters. It should be stressed that this interpretation assumes a simple, homogeneous evolution of the system, and a moderate salinity fluid. Natural hydrothermal processes are of a dynamic nature and frequently operate under two-phase fluid conditions. In addition, the fluid may interact with wall rocks. All these factors must be taken into account in interpreting real natural systems.

In more detail, our experimental results enable the relative behaviour of the divalent elements Mg, Ca, and Fe to be examined. According to field observations (Zharikov, 1968), Mg skarns form at the highest temperature stages, and as the temperature is lowered, they are replaced by calcareous ones, then by Fe scarns. Our data show the Mg to Ca skarn transition could reflect an increase in temperature, decrease in pressure or in density of a hydrothermal fluid, accumulation of CO_2 in the system, or a decrease in the overall fluid salinity. The replacement of Ca-bearing assemblages by Fe-bearing ones is favoured by temperature decrease, increase in solution density, or reduction in the gas content or overall salinity of the fluid. They appear to provide a useful basis for the more detailed interpretation of hydrothermal rocks formation.

REFERENCES

Fein, J.B. and Walther, J.V. (1989) Portlandite solubilities in supercritical Ar–H_2O mixtures: Implications for quantifying solvent effects. *Amer. J. Sci.*, **289**, 975–93.

Frantz, J.D. and Eugster, H.P. (1973) Acid–base buffers: use of Ag + AgCl in the experimental control of solution equilibria at elevated pressures and temperatures. *Amer. J. Sci.*, **267**, 268–86.

Frantz, J.D. and Popp, R.K. (1979) Mineral–solution equilibria: (1) An experimental study of complexing and thermodynamic properties of aqueous $MgCl_2$ in the system MgO–SiO_2–H_2O–HCl. *Geochim. Cosmochim. Acta*, **43**, 1223–39.

Frantz, J.D. and Marshall, W.L. (1984) Electrical conductances and ionization constants of salts, acids, and bases in supercritical aqueous fluids: 1. Hydrochloric acid from 100 to 700°C and at pressures to 4 kbars. *Amer. J. Sci.*, **284**, 651–67.

Gehrig, M., Lentz, H. and Frank, E.U. (1983) Concentrated aqueous sodium chloride solutions from 200 to 600°C and 3 kbar. Phase equilibria and PVT–data. *Ber. Bunsenges. Phys. Chem.*, **87**, 597–600.

Helgeson, H.C., Kirkham, D.H. and Flowers, G.C. (1981) Theoretical prediction of the thermodynamic behaviour of aqueous electrolytes at high pressures and temperatures: IV. Calculation of activity coefficients, osmotic coefficients, and apparent molal and standard and relative partial molal properties to 600°C and 5 kb. *Amer. J. Sci.*, **281**, 1249–1516.

Hemley, J.J. (1959) Some mineralogical equilibria in the system K_2O–Al_2O_3–SiO_2–H_2O. *Amer. J. Sci.*, **257**, 241–70.

Hemley, J.J., Cygan, G.L., Fein, J.B., *et al.* (1992) Hydrothermal ore-forming processes in the light of studies in rock-buffered systems: I. Iron–copper zinc–lead–sulphide solubility relations. *Economic Geology*, **87**, (1), 1–22.

Hilbert, R. (1984) PVT–Daten von Wasser und von Wasserigen Natrium–Chlorid Losungen bis 873 K, 4000 bar und 25 Gewichtsprozent NaCl. *Dissertation Hochsch. Ulverlag.*, Greiburg.

Hiroshi, S. and Koichiro, F. (1989) Pressure dependence of mineral–water reaction equilibrium in the low pressure range. *Water–Rock Interaction. Proc. 6th International Symposium*, Malvern, U.K., Miles (ed.), c Balkema, Rotterdam. pp. 635–8.

Korzhinskiy, D.S. (1958) Hydrothermal acid–alkaline differentiation. *Doklady Akademii Nauk, SSSR*, **122** (2), 267–70 (in Russian).

Korzhinskiy, M.A. (1981) Investigations of the Ag–AgCl buffer in a range of low values of f_{H_2}, in *Contr. to Physico-chemical Petrology, IX* (eds V.A. Zharikov and V.V. Fed'kin) Nauka Press, Moscow (in Russian), pp. 41–51.

Korzhinskiy, M.A. (1986) Diopside–wollastonite equilibrium in a supercritical chloride fluid. *Geokhem. Int.*, **23** (3), 143–53.

Korzhinskiy, M.A. (1987) The calcium–iron ratio in a supercritical chloride fluid in equilibrium with skarn mineral assemblages. *Geokhem. Int.*, **24** (9), 39–55.

Korzhinskiy, M.A. (1992) Dissolved salt behaviour in chloride CO_2–H_2O fluids: the system H_2O–HCl–CO_2. *Geochem. Int.*, **29** (1), 1–13.

Lvov, S.N. and Wood, R.H. (1990) Equation of state of aqueous NaCl solutions over a wide range of temperatures, pressures and concentrations of fluid phase equilibria. *Fluid Phase Equilibria*, **60**, 273–87.

Marshall, W.L. (1970) Complete equilibrium constants, electrolyte equilibria, and reaction rates. *J. Phys. Chem.*, **74**, 346–55.

Mel'nik, Yu.P. (1978) *Thermodynamic Properties of Gases under Conditions of Depth Petrogenesis*. Naukova Dumka Press, Kiev (in Russian).

Pitzer, K.S. (1983) Dielectric constants of water at very high temperature and pressure. *Proc. Nat'L Acad Sci. USA*, **80**, 4575–6.

Plyasunov, A.V. (1989) Component thermodynamic activities in the NaCl–H_2O system in the homogeneous range up to 723 K and 1 kbar. *Geokhem. Int.*, **26** (4), 37–46.

Popp, R.K. and Frantz, J.D. (1979) Mineral solution equilibria – II. An experimental study of mineral solubilities and the thermodynamic properties of aqueous $CaCl_2$ in the system CaO–SiO_2–H_2O–HCl. *Geochim. Cosmochim. Acta*, **43**, 1777–90.

Robie, R.A., Hemingway, B.S. and Fisher, J.R. (1978) Thermodynamic properties of minerals and related substances at 298.15 K and 1 bar (10^5 pascals). Pressure at higher temperatures. *US Geol. Sur. Bull.*, **1452**, Washington.

Sato, M., Uematsu, M. and Watanabe, K. (1984) Proposal of the new skeleton tables for the thermodynamic properties of water and steam. *Proc. 10th International Conference on the Properties of Steam*, **1**. Mir Publishers, Moscow, pp. 71–88.

Shmonov, V.M. and Shmulovich, K.I. (1978) Measurement of P–V–T properties of the system H_2O–CO_2 at 500°C and to 5 kbar, in *Experiment i Tekhnika*

Vysokikh Gazovykh i Tverdofazovykh Davleniy. Nauka Press, Moscow (in Russian), pp. 133–7.

Shmulovich, K.I. and Shmonov, V.M. (1978) Tables of thermodynamic properties of gases and liquids, in *Carbon Dioxide*, **3**. Izdatelstvo Standartov, Moscow (in Russian),

Shmulovich, K.I. (1988) *Carbon Dioxide in High-temperature Mineral Formation Processes*. Nauka Press, Moscow (in Russian).

Sourirajan, S. and Kennedy, G.C. (1962) The system H_2O–NaCl at elevated temperatures and pressures. *Amer. J. Sci.*, **260**, 115–41.

Walther, J.V. and Orville, P.M. (1983) The extraction quench technique for determination of the thermodynamic properties of solute complexes: Application to quartz solubility in fluid mixtures. *Amer. Miner.*, **68**, 731–41.

Zharikov, V.A. (1968) Skarn deposits, in *Genesis of Endogenic Ore Deposits*. Nauka Press, Moscow (in Russian), pp. 220–302.

GLOSSARY OF SYMBOLS

α	activity
$\mathring{a}$	ion size parameter
A	Debye–Hückel parameter
B	Debye–Hückel parameter
f	fugacity
K	equilibrium constant
K_D	dissociation constant
m	molality
$\overline{m}$	mass
M	molecular mass
S	proportionality coefficient
v	specific volume
V	volume
V^E	excess volume
X	mole fraction
Z	compressibility factor
γ	mean activity coefficient
γ^*	fugacity coefficient of pure substance
ρ	density
$\tilde{\rho}$	density of water in a mixture
ϕ	activity coefficient in a gas mixture

CHAPTER EIGHT

Phase equilibria in fluid systems at high pressures and temperatures

Kirill I. Shmulovich, Sergey I. Tkachenko and Natalia V. Plyasunova

8.1 INTRODUCTION

Fluid inclusions of brines with daughter salt crystals are known from magmatic, metamorphic and metasomatic rocks, and have received considerable attention in the literature (e.g. Poty et al., 1974; Crawford *et al.*, 1979; Kreulen, 1980; Touret, 1981; Yardley and Bottrell, 1988). Not infrequently, they coexist with pure or almost pure carbonic inclusions. Such observations demonstrate that natural fluids must be considered as composed of at least 3 essential components: salt, CO_2 and H_2O, but there has been a paucity of experimental data on fluid systems of this complexity. Studies of metamorphic fluid evolution including the effects of fluid immiscibility by Bowers and Helgeson (1983), Trommsdorff *et al.* (1985), Trommsdorff and Skippen (1986) and Skippen (1988) have been based on the data of Gehrig (1980).

In this study we have investigated the topology of phase diagrams for fluid equilibria in salt-bearing fluid systems, based on experimental measurements to 2 kbar and 500°C, with extrapolations to higher T and P. It has become possible to extend the investigated range of parameters of state, using both phase-equilibria data obtained by direct sampling of experimental water–salt systems and by synthesis of fluid inclusions in the ternary H_2O–CO_2–salt systems.

Fluids in the Crust: Equlibrium and transport properties.
Edited by K.I. Shmulovich, B.W.D. Yardley and G. G. Gonchar.
Published in 1994 by Chapman & Hall, London. ISBN 0 412 56320 7

Binary systems involving water and a non-polar gas (CO_2, N_2, or CH_4) have been extensively studied (e.g. Takenouchi and Kennedy, 1964; Krader, 1985; Japas and Franck, 1985) and are not considered here.

Water–salt binary systems are less well understood. Since the pioneering work of Sourirajan and Kennedy (1962) and that of Ravich (1974), the system NaCl–H_2O, for example, has been studied experimentally in the near-critical region only (Bischoff and Rosenhauer, 1986). Natural processes, however, are frequently related to the cation exchange between solution and rock, for example, sodium-rich chloride solutions separating from crystallizing granitoid melt may alter the cation composition of fluid as a result of reaction with rock. This study therefore presents experimental data for liquid–vapour equilibria in the binary systems H_2O–NaCl, H_2O–KCl, H_2O–$CaCl_2$ and H_2O–$MgCl_2$, and investigates immiscibility field boundaries in the ternary systems H_2O–CO_2–$CaCl_2$ and H_2O–CO_2–NaCl, at 3 and 5 kbar, 500–700 °C.

8.2 BINARY H_2O–SALT SYSTEMS

8.2.1 Experimental technique

Experiments to investigate the composition of coexisting liquid and vapour in binary salt–H_2O systems were run in a gold-lined extraction–quench hydrothermal apparatus fitted with two Ti valves and a pressure transducer with minimal inner volume. This permits direct sampling of coexisting liquid 'L' and vapour 'G'. Details regarding the construction and use of this set-up can be found in Chapter 3 (Section 3.2.2 and Figure 3.2, this volume). Equilibration of the compositions of phases 'L' and 'G' was carried out when the reactor was positioned horizontally to provide the maximum surface for exchange; sampling was carried out when it was positioned vertically, with phases 'G' and 'L' sampled at the top and bottom of the reactor, respectively. The reactor volume was 120 cm^3, the volume of samples, 1–4 cm^3, whereas the volume of 'cold' sampling zones (that differ in temperature from the temperature of the run) was less than 50 mm^3. This is the smallest ratio of the volume of the cool zone through which the sample is extracted to the volume of the sample that has yet been achieved; therefore, the data on vapour compositions should be more reliable than estimates from experimental configurations with a larger cool zone.

8.2.2 Results

Liquid 'L'–vapour 'G' equilibria in the systems H_2O–NaCl (reference system) and H_2O–KCl have been investigated in the near-critical region at 400–700°C, and the systems H_2O–$MgCl_2$ and H_2O–$CaCl_2$ have been studied to 600°C; at higher temperatures, strong hydrolysis shifts the compositions of the systems away from the binary joins, and appreciable amounts of hydrolysis products like HCl and $Me(OH)_2$ appear. Preliminary results can be found in Tkachenko and Shmulovich (1992); here, tables summarizing all the coexisting phase compositions are presented.

Raw experimental data are summarized in Table 8.1. The accuracy of measurements is estimated to be: pressure, ± 1.5 bar; temperature, ± 0.5°C, salt concentration in solutions near saturation, < ± 2%. The accuracy of measured salt concentrations in phase 'G' is estimated as ± 20% relative.

Comparison of the data obtained in this work for the system H_2O–NaCl with those from the literature (e.g. Sourirajan and Kennedy, 1962; Urusova, 1974, Bischoff and Pitzer, 1989, Bodnar *et al.*, 1985) demonstrates good agreement between the authors' data and the most precise data of Bischoff and Rosenbauer (1986). However, in these runs, the concentrations of salt in phases 'G' and 'L' are, respectively, 1% lower and 1–2% higher than in the experiments by Bischoff and Pitzer (1989). This was to be expected because our apparatus has a minimal 'cold' zone for fluid sampling (Chapter 3, this volume) and should therefore produce more accurate determinations of the limiting gas and liquid compositions. The critical pressures for isothermal sections agree well in all available studies. Although a synthetic fluid inclusion approach is undeniably less accurate than the direct sampling method, the data obtained by Zhang and Frantz (1989) using synthetic fluid inclusions for the critical curve in the H_2O–$CaCl_2$ system agree with our data in Table 8.1 within the errors of measurement. From Table 8.1, it follows that the critical pressures in the investigated systems increase in the order: $P_{NaCl\,(KCl)} < P_{MgCl_2} < P_{CaCl_2}$. Figure 8.1 illustrates the experimental critical curves used in these systems, supplemented for the H_2O–$CaCl_2$ system by points taken from Zhang and Frantz (1989) at 600 and 700°C. At 600°C, the critical pressure for this system from Table 8.1 is in good agreement with the pressure estimated by Zhang and Frantz (1989).

Also included in Figure 8.1 is a curve for the granitic solidus (the Qtz–Ab–Or–H_2O eutectic, Ebadi and Johannes, 1991) at $X_{(H_2O)} = 1$. At pressures to 3 kbar, salt concentrations up to 20% by weight slightly raise the melting temperature, since $X^{fl}_{H_2O} > 0.9$. The critical curves in the H_2O–NaCl and H_2O–KCl systems intersect the granite solidus at a pressure of 1.3 kbar, while the critical curve for the H_2O–$CaCl_2$ system intersects it at 1.7 kbar. At lower pressures than these, a water–salt fluid may unmix into two phases in the subsolidus region, or may separate from the melt as two phases, while at higher pressures, the fluid remains homogeneous or, alternatively, melting initiates.

Table 8.1. Experimental data for phase compositions on the surface of liquid–vapour equilibria in water–salt systems. All phase compositions are in % by weight of salts

400°C			500°C			600°C			650°C			700°C		
P, bar	wt % vapour	NaCl liquid	P, bar	wt % vapour	NaCl liquid	P, bar	wt % vapour	NaCl liquid	P, bar	wt % vapour	NaCl liquid	P, bar	wt % vapour	NaCl liquid
282.4	–	2.08	580.0	–	10.12	922.5	–	26.50	–	–	–	–	–	–
281.7	–	1.20	576.7	–	20.19	919.1	13.80	29.30	–	–	–	–	–	–
281.1	–	4.50	573.0	5.87	–	915.0	–	30.20	–	–	–	–	–	–
281.0	0.95	–	568.8	–	23.29	911.3	10.20	–	–	–	–	–	–	–
280.7	0.78	–	565.0	4.21	–	894.0	7.60	–	–	–	–	–	–	–
280.3	–	5.24	560.8	–	25.36	859.2	4.55	–	–	–	–	–	–	–
279.5	0.70	–	557.1	3.13	–	816.9	2.73	–	–	–	–	–	–	–
278.3	–	6.17	551.1	–	27.70	798.7	2.26	–	–	–	–	–	–	–
276.6	–	7.50	546.1	2.27	–	–	–	–	–	–	–	–	–	–
276.4	0.50	–	541.9	2.14	–	–	–	–	–	–	–	–	–	–
275.1	0.46	–	538.8	1.92	–	–	–	–	–	–	–	–	–	–
274.6	–	8.23	535.3	–	30.68	–	–	–	–	–	–	–	–	–
274.2	0.42	–	525.5	1.51	–	–	–	–	–	–	–	–	–	–
272.8	–	9.06	–	–	–	–	–	–	–	–	–	–	–	–
270.1	–	10.74	–	–	–	–	–	–	–	–	–	–	–	–
265.3	0.28	–	–	–	–	–	–	–	–	–	–	–	–	–
264.2	–	13.89	–	–	–	–	–	–	–	–	–	–	–	–
261.2	0.26	–	–	–	–	–	–	–	–	–	–	–	–	–
257.4	–	16.71	–	–	–	–	–	–	–	–	–	–	–	–
256.1	0.22	–	–	–	–	–	–	–	–	–	–	–	–	–
252.2	0.18	–	–	–	–	–	–	–	–	–	–	–	–	–
248.3	0.13	–	–	–	–	–	–	–	1091.2	–	30.01	1254.0	–	31.33
244.1	0.11	–	–	–	–	–	–	–	1084.8	15.31	–	1246.7	19.05	–
239.7	0.08	–	–	–	–	–	–	–	1076.7	–	37.24	1231.4	–	39.41
234.1	0.06	–	–	–	–	–	–	–	1070.7	12.28	–	1206.8	10.46	–
230.2	–	27.99	–	–	–	–	–	–	1064.2	–	39.96	1193.3	–	43.20
226.7	0.045	–	–	–	–	–	–	–	1056.5	9.72	–	1177.1	8.21	–
–	–	–	–	–	–	–	–	–	1043.8	8.48	–	1162.4	7.69	–
–	–	–	–	–	–	–	–	–	1027.9	7.21	–	–	–	–

Table 8.1 (Contd)

400°C			500°C			600°C			650°C			700°C		
P, bar	wt % vapour	$CaCl_2$ liquid	P, bar	wt % vapour	$CaCl_2$ liquid	P, bar	wt % vapour	$CaCl_2$ liquid	P, bar	wt % vapour	$CaCl_2$ liquid	P, bar	wt % vapour	$CaCl_2$ liquid
311.3	3.88	–	800.0	–	25.60	1324.1	19.74	–	–	–	–	–	–	–
311.1	–	12.08	794.4	12.64	–	1323.9	–	30.61	–	–	–	–	–	–
308.8	–	14.51	787.6	11.70	–	1313.5	16.01	–	–	–	–	–	–	–
308.7	2.17	–	782.8	–	29.83	1307.0	–	35.41	–	–	–	–	–	–
307.3	1.70	–	770.8	–	31.73	1298.7	13.31	–	–	–	–	–	–	–
305.6	–	16.07	769.9	10.81	–	1297.3	–	36.35	–	–	–	–	–	–
304.9	1.32	–	760.6	9.52	–	1288.1	11.90	–	–	–	–	–	–	–
303.4	–	17.53	751.5	–	34.27	1285.9	–	39.18	–	–	–	–	–	–
302.5	1.00	–	747.4	–	34.53	1277.3	10.92	–	–	–	–	–	–	–
300.4	0.77	–	741.1	8.14	–	1262.0	10.43	–	–	–	–	–	–	–
300.0	–	19.00	732.8	–	36.28	1260.0		43.70	–	–	–	–	–	–
297.2	0.63	–	723.9	6.89	–	1252.7	9.16	–	–	–	–	–	–	–
295.3	0.56	–	712.0	6.10	–	1235.9	7.78	44.73	–	–	–	–	–	–
288.9	–	23.54	–	–	–	1216.7	6.44	–	–	–	–	–	–	–
288.5	0.47	–	–	–	–	–	–	–	–	–	–	–	–	–
284.8	–	25.03	–	–	–	–	–	–	–	–	–	–	–	–
282.4	–	25.57	–	–	–	–	–	–	–	–	–	–	–	–
278.9	0.39	–	–	–	–	–	–	–	–	–	–	–	–	–
271.0	0.35	–	–	–	–	–	–	–	–	–	–	–	–	–
P, bar	wt % vapour	$MgCl_2$ liquid	P, bar	wt % vapour	$MgCl_2$ liquid	P, bar	wt % vapour	$MgCl_2$ liquid	P, bar	wt % vapour	$MgCl_2$ liquid	P, bar	wt % vapour	$MgCl_2$ liquid
305.8	2.72	–	700.3	–	21.44	1037.3	23.54	–	–	–	–	–	–	–
303.0	1.86	10.03	699.0	9.49	–	1029.7	17.59	–	–	–	–	–	–	–
301.5	1.33	–	691.3	–	23.73	1024.7	–	32.80	–	–	–	–	–	–
300.8	1.19	–	689.5	8.64	–	1021.6	16.18	–	–	–	–	–	–	–
300.4	–	11.51	687.2	7.60	–	1018.2	–	33.47	–	–	–	–	–	–
300.2	1.14	–	680.4	6.73	–	1007.7	–	36.29	–	–	–	–	–	–
298.6	1.00	–	680.1	–	27.23	1002.5	13.45	–	–	–	–	–	–	–
297.8	–	12.62	671.7	–	27.65	984.9	12.03	–	–	–	–	–	–	–
295.6	–	13.50	666.2	5.31	–	984.3	–	40.10	–	–	–	–	–	–
290.8	–	15.86	655.7	–	30.12	974.9	–	40.72	–	–	–	–	–	–
290.0	0.60	–	653.7	4.90	–	967.2	9.73	–	–	–	–	–	–	–
286.8	–	17.32	642.3	–	31.05	963.6	–	42.19	–	–	–	–	–	–
285.0	–	18.09	637.1	3.93	–	954.0	–	42.95	–	–	–	–	–	–

Table 8.1 (Contd)

400°C			500°C			600°C			650°C			700°C		
P, bar	wt % vapour	$MgCl_2$ liquid	P, bar	wt % vapour	$MgCl_2$ liquid	P, bar	wt % vapour	$MgCl_2$ liquid	P, bar	wt % vapour	$MgCl_2$ liquid	P, bar	wt % vapour	$MgCl_2$ liquid
281.3	0.37	–	633.4	–	32.00	948.7	8.60	–	–	–	–	–	–	–
276.6	0.33	–	627.6	3.71	–	939.6	7.04	–	–	–	–	–	–	–
276.4	–	19.99	620.0	3.14	–	925.6	6.50	–	–	–	–	–	–	–
–	–	–	617.1	–	32.28	–	–	–	–	–	–	–	–	–
–	–	–	612.6	3.11	–	–	–	–	–	–	–	–	–	–
P, bar	wt % vapour	Kcl liquid	P, bar	wt % vapour	Kcl liquid	P, bar	wt % vapour	Kcl liquid	P, bar	wt % vapour	Kcl liquid	P, bar	wt % vapour	Kcl liquid
285.0	1.84	4.35	581.0	–	20.42	904.4		27.04	–	–	–	–	–	–
283.8	1.13	–	578.0	7.84	–	902.8	14.87	–	–	–	–	–	–	–
282.3	–	5.69	575.1	6.62	–	896.3	11.29	–	–	–	–	–	–	–
282.0	0.81	–	574.5	–	23.58	893.3		30.95	–	–	–	–	–	–
281.7	–	6.90	571.3	–	24.58	882.1		32.89	–	–	–	–	–	–
281.1	0.71	–	570.5	5.47	–	876.9	7.88	–	–	–	–	–	–	
280.8	0.65	–	567.4	4.78	–	853.0	6.17	–	–	–	–	–	–	–
278.9	0.44	–	566.1	–	25.62	839.3	4.87	–	–	–	–	–	–	–
276.5	–	10.26	561.6	4.20	–	817.2	3.83	–	–	–	–	–	–	–
271.3	0.27	–	557.5	–	27.85	–	–	–	–	–	–	–	–	–
269.4	–	14.34	556.6	3.49	–	–	–	–	–	–	–	–	–	–
266.8	–	15.40	544.7	2.65	–	–	–	–	–	–	–	–	–	–
264.3	0.19	–	530.6	2.07	–	–	–	–	–	–	–	–	–	–
252.6	0.11	–	513.7	1.39	–	–	–	–	–	–	–	–	–	–
250.6	–	24.05	–	–	–	–	–	–	–	–	–	–	–	–
247.5	0.10	–	–	–	–	–	–	–	–	–	–	–	–	–
241.9	0.08	–	–	–	–	–	–	–	–	–	–	–	–	–
–	–	–	–	–	–	–	–	–	1063.5	–	26.20	1204.3	–	31.80
–	–	–	–	–	–	–	–	–	1057.8	15.91	–	1199.8	21.41	–
–	–	–	–	–	–	–	–	–	1048.9	–	31.47	1185.5	15.95	–
–	–	–	–	–	–	–	–	–	1043.6	12.11	–	1182.3	–	35.57
–	–	–	–	–	–	–	–	–	1033.8	–	34.32	1163.5	–	37.63
–	–	–	–	–	–	–	–	–	1029.1	10.49	–	1157.2	12.04	–
–	–	–	–	–	–	–	–	–	1019.3	–	37.82	1143.1	10.60	–
–	–	–	–	–	–	–	–	–	1014.0	8.69	–	1119.8	8.39	–
–	–	–	–	–	–	–	–	–	997.8	6.92	–	–	–	–

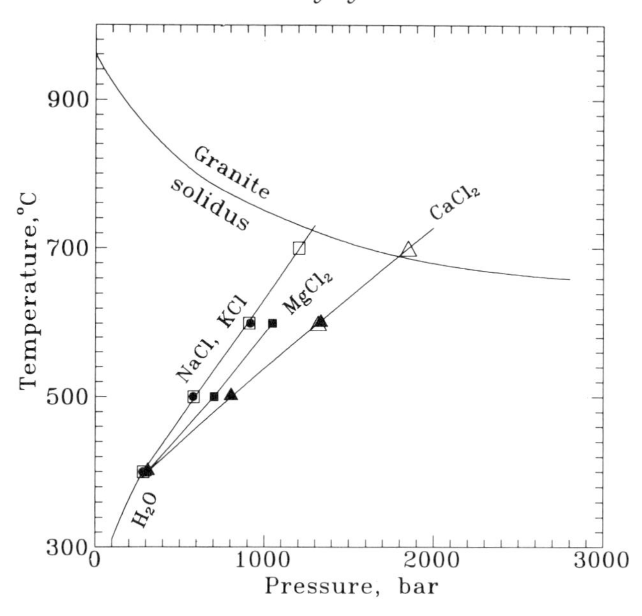

Figure 8.1. Critical curves in the H_2O–salt systems. Large open triangles are for the system H_2O–$CaCl_2$ from Zhang and Frantz (1989). Granitic solidus is from Ebadi and Johannes (1991). Critical parameters for the systems H_2O–NaCl and H_2O–KCl coincide within the scale of the diagram.

Surfaces of the 'L'–'G' phase equilibria in the systems H_2O–NaCl and H_2O–KCl look similar. At 400 °C, the critical pressures in the H_2O–KCl system are only 3 bar higher than in the H_2O–NaCl system; between 500 and 600 °C, an inversion is observed, and at 700 °C, the critical pressures in the H_2O–NaCl system are about 50 bar higher than in the H_2O–KCl system. Therefore, the location of phase boundaries is not markedly affected by Na $\rightleftharpoons$ K exchange between fluid and rock, and NaCl and KCl may be treated as one component in the study of fluid phase equilibria. The (Na, K $\rightleftharpoons$ Mg, Ca) exchange process raises critical pressures and hence may cause an initially homogeneous fluid to unmix into two phases. For instance, reaction of an NaCl-rich fluid that separated from granitoid melt with carbonate rocks may cause fluid to unmix into 'L' and 'G' phases at constant P and T, at pressures a little above 1–3 kbar.

8.3 TERNARY H_2O–CO_2–SALT SYSTEMS

Ternary systems have so far been studied over the limited compositional range and at pressures less than 3 kbar (Takenouchi and Kennedy, 1965; Gehrig, 1980; Zhang and Frantz, 1989). In this region, Henry's law is met, and CO_2 solubility in phase 'L' increases with pressure. Data for the system H_2O–CH_4–NaCl (Krader, 1985) show a reversal in this pressure-dependence of the solubility of the gas phase in salt water, and suggest that a similar reversal may occur in systems involving CO_2.

8.3.1 Experimental techniques

We have experimentally investigated phase equilibria in ternary H_2O–CO_2 salt systems at 5 kbar using the synthetic fluid inclusion method of Bodnar and Sterner (1987). The details of the experiment and tables summarizing all the experimental runs done for the study are given in Plyasunova and Shmulovich (1991) and Shmulovich and Plyasunova (in press); here, we give a review of the latest data obtained. A noteworthy feature of the technique is the way by which CO_2 was introduced into the capsule: the gas was made to freeze into it, rather than being added as oxalic acid or oxalate salt. Thus, there were no additional components beyond those investigated, their hydrolysis products, and quartz.

8.3.2 Results

The results from our studies of the H_2O–CO_2–NaCl and H_2O–CO_2–$CaCl_2$ systems at 5 kbar, obtained using the synthetic fluid inclusion technique, are presented in Figures 8.2–8.4. At 5 kbar, in the system involving $CaCl_2$, there are several discrepant points; checking revealed that the fluid from these runs contained up to 20% N_2, which is believed

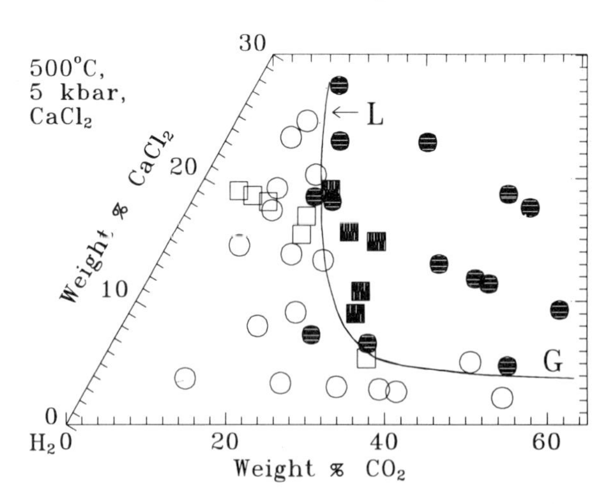

Figure 8.2. Phase diagram for the system H_2O–CO_2–$CaCl_2$ at 5 kbar and 500°C. Open symbols refer to a homogeneous fluid, solid symbols to a two-phase one, 'L' represents a water–salt phase, 'G', a CO_2-rich one. Squares indicate the revised methods by which the capsules were prepared after finding N_2 by Raman spectroscopy. The uncertainty associated with drawing the two-phase field boundary is ± 2%.

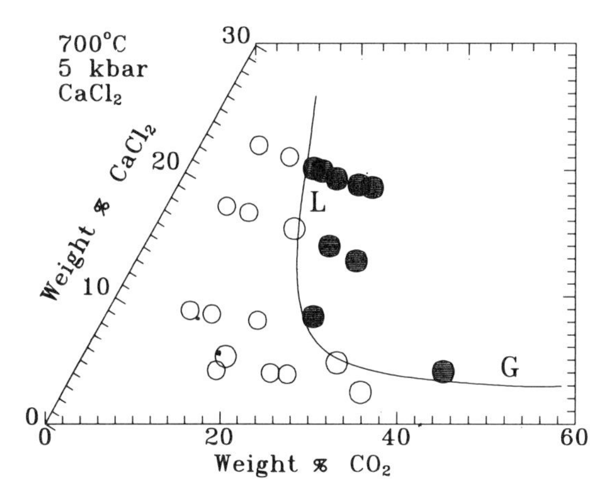

Figure 8.3. The same system as that in Figure 8.2, but at 700°C.

to have condensed into the capsule while it was being cooled with CO_2 at liquid N_2 temperature. A set of additional runs was therefore performed, using a modified CO_2-loading method to ensure that the temperature of the capsule was not low enough for this to happen. The data from these runs are shown in Figure 8.2 as squares. The data for the H_2O–CO_2–$CaCl_2$ system at 500 °C and 3 kbar are not shown graphically, since they are indistinguishable from the data of Zhang and Frantz (1989). This join has received further study because CO_2 solubility in phase 'L' at 5 kbar proved to be appreciably lower than that at 3 kbar. Experiments at 5 kbar and both 500 and 700 °C (Figures 8.2 and 8.3) show no differences in the location of the miscibility field boundaries. In

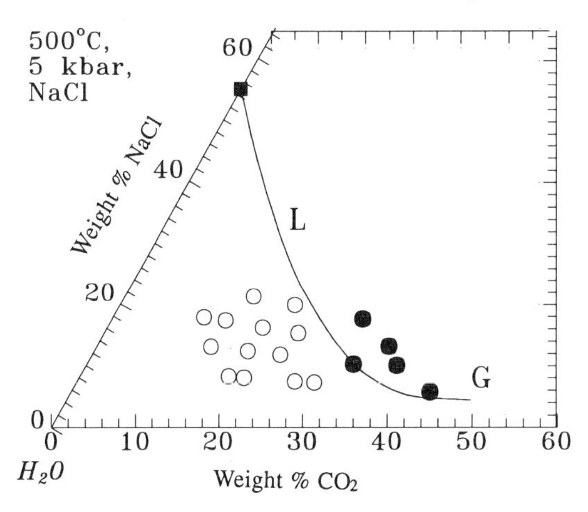

Figure 8.4. Phase diagram for the system H_2O–CO_2–NaCl at 5 kbar and 500°C. Square represents a limit of NaCl (solid) solubility at 500°C and atmospheric pressure.

other words, at a pressure of 5 kbar or higher, variations of temperature do not have a detectable influence on the composition of phase 'L' (for phase 'G', few data are available), and the 500 °C results may be used for higher temperatures.

A summary diagram illustrating limits for the occurrence of a homogeneous fluid at 500°C in the system H_2O–CO_2–$CaCl_2$ (NaCl) is presented in Figure 8.5, where CO_2 concentrations are recalculated to mol% units commonly used by petrologists. The observed shift in the immiscibility field boundaries at 500 °C with increasing pressure is significantly larger than the experimental uncertainties and has considerable petrological significance (below). It follows from Figures 8.2–8.5 that for a fluid with $CO_2 \cdot (CO_2 + H_2O)^{-1} > 0.3$, salt concentrations less than 10% by weight NaCl are sufficient for its composition to lie in the two-phase region. Direct compositional measurements of coexisting phases in the system H_2O–CO_2–$CaCl_2$ at 5 kbar and 500 °C have yielded a salt concentration of about 1.5–2% by weight in phase 'G' at $X_{(CO_2)} = 0.5$ (Shmulovich and Plyasunova, in press).

The two major results of the phase-equilibria studies in the ternary systems are the recognition of an inversion of the pressure dependence of CO_2 solubility in phase 'L' between 3–5 kbar and the determination that only very low salt concentrations may cause H_2O–CO_2 fluids to unmix.

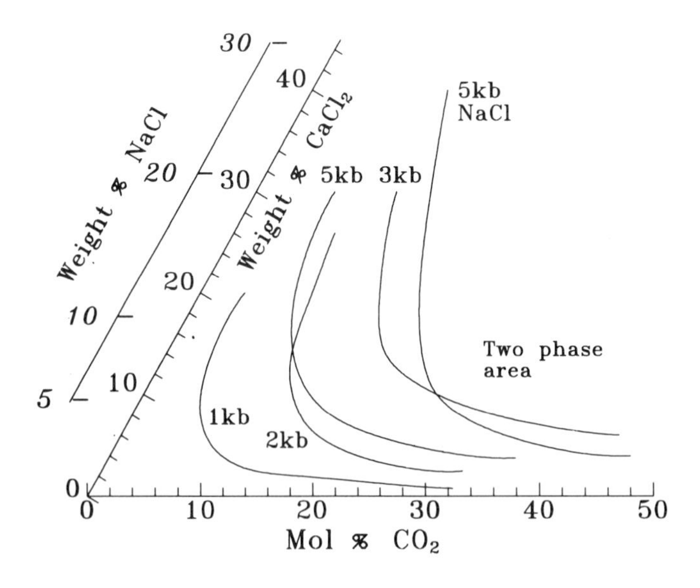

Figure 8.5. Summary diagram for phase equilibria at 500°C. The miscibility gap boundaries for the system H_2O–CO_2–$CaCl_2$ at 1 and 2 kbar are from Zhang and Frantz (1989); at 3 kbar, from the authors' data, which agree perfectly with Zhang and Frantz's (1989) data. Salt's axes are combined for equal molalities.

8.4 FLUID IMMISCIBILITY AND MAGMATIC CRYSTALLIZATION

Fluid immiscibility in the fluid that separates during the crystallization of granitoid intrusions has been considered by Ryabchikov (1975) for binary H_2O–NaCl fluids. As a granitoid melt with an initial ratio $NaCl \cdot (NaCl + H_2O)^{-1} \approx 0.05$–$0.1$ cools and crystallizes, it first becomes saturated with respect to a concentrated water–salt solution, i.e. salt is partitioned into the fluid more strongly than water. As crystallization proceeds, the concentration of salt in fluid reduces, and the following alternative scenarios are possible. At pressures above 1.4–1.8 kbar (see Figure 8.1), the concentration of water in the separating fluid increases steadily, and crystallization culminates in the separation of an aqueous fluid (95% H_2O or higher). At lower pressures, the solidus line intersects a miscibility gap, and two fluid phases, 'L' and 'G' (Figure 8.1), may separate from the crystallizing melt in the final stages of the process. In natural granites therefore, the 'granitic melt'–'L'–'G' equilibrium is likely at $P < 1.3$ kbar, since NaCl and KCl are the dominant salt components of granitoids. The phase diagram for a model H_2O–CO_2 binary fluid separating from the crystallizing melt is topologically similar to that for the system granite–H_2O–NaCl at $P > 1.4$ kbar.

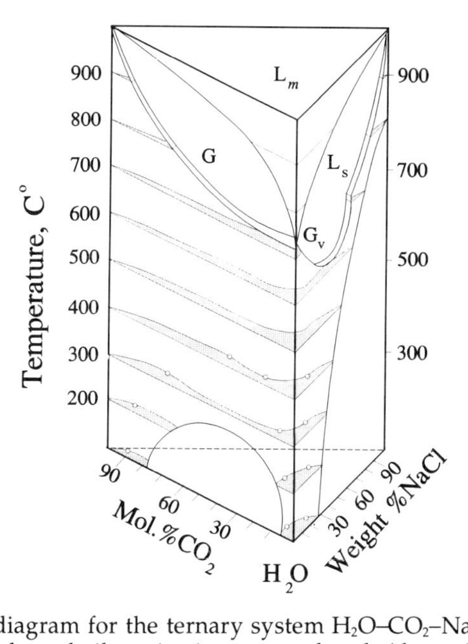

Figure 8.6. T–X diagram for the ternary system H_2O–CO_2–NaCl at 1 kbar. Surface L_m indicates the volatile ratios in saturated melt (dependent on the initial composition), L_S refers to a water–salt phase, G stands for a CO_2-rich phase, while G_V denotes vapour. The region of homogeneous fluid existence on isothermal sections is hatched. Open circles on the diagram at 100–400°C are from Gehrig (1980).

An isobaric phase diagram for three-component H_2O–CO_2–NaCl fluid separating from crystallizing granitic melt at 1 kbar may be represented as a prism (Figure 8.6). On the diagram, the portion of isothermal sections where a homogeneous fluid exists is hatched. 'L_m' represents a surface of fluid component concentrations in a saturated silicate solution, 'L_s' stands for a water–salt phase, 'G' refers to a CO_2-rich phase, 'G_V' to water vapour. From the diagram, it may be observed that only a narrow composition field is occupied by homogeneous (one-phase) fluid. Also, from the preceding section (see Figure 8.5), it follows that replacement of some portion of NaCl by $CaCl_2$ narrows this field further. The possibilities for fluid composition evolution within the homogeneous field are extremely limited; therefore, the two-phase fluid state must commonly accompany magmatic distillation.

8.5 FLUID IN METAMORPHISM AND ANATECTIC MELTING

As was noted in the Introduction, salt contents near, or greater than, saturation at room conditions, i.e. about 25% equivalent weight NaCl, are quite widespread in metamorphic fluids. Indeed, a salt-absent fluid is a theoretical abstraction, although very low-salinity fluids are also known from metamorphic settings. In order to gain insight into the influence of salt on the phase state of fluid, consider a projection of the isothermal section for the diagram in Figure 8.5 on to the P–C (salt) plane, where C denotes any compositional variable, not necessarily mole fraction (X). Also, consider the pseudobinary system CO_2–H_2O with 30% by weight $CaCl_2$ (a solution with 20% by weight NaCl has the same molality), which is below saturation at room temperature. Figure 8.7 is a phase diagram based on Gehrig (1980) for a pressure < 2 kbar and Zhang and Frantz (1989) for the H_2O–CO_2–$CaCl_2$ system at 3 kbar, as well as extrapolation of CO_2 solubility in salt solutions to 10 kbar from the data illustrated in Figures 8.2–8.5. The compositions of phase 'G' (CO_2-rich) are determined rather crudely, with uncertainty in the measurements being up to ± 10 mol% CO_2. The concentration of salt in phase 'G' is much lower than in phase 'L', i.e. the diagram is a projection rather than a section. Nevertheless, this diagram clearly demonstrates a wide field in which two fluid phases coexist under high-grade metamorphic conditions.

Widespread fluid inclusions consisting of pure or almost pure CO_2 correspond in composition to phase 'G' (Figure 8.7), especially when it is considered that the inclusions with 25–30 mol% H_2O may appear as pure CO_2 when viewed under the optical microscope, depending on their shape. The composition of much less abundant water–salt inclusions with daughter salt crystals matches the composition of phase 'L'. Thus,

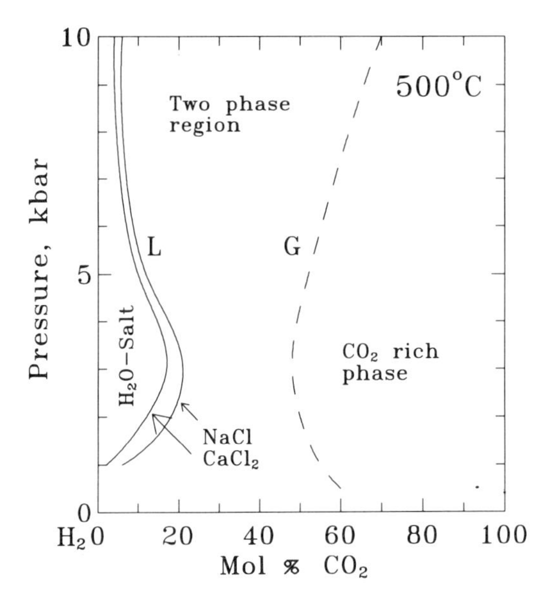

Figure 8.7. Isothermal (500°C) projection of the system H_2O–CO_2–salt (30 wt% $CaCl_2$, or 20 wt% NaCl in water) on to the P–X plane with extrapolation up to 10 kbar. Dashed boundary for the CO_2-rich phase is shown very crudely.

the occurrence of two fluid-inclusion series of markedly distinct composition (Ramboz *et al.*, 1982; Crawford *et al.*, 1979) do not necessarily reflect the evolution of fluid in time. Rather, their occurrence may indicate that they were formed simultaneously. As the fraction of $CaCl_2$ in the salt composition of a metamorphic fluid becomes larger with pressure (Vapnik, 1988), the miscibility gap expands, and the CO_2 content in phase 'L' further decreases at $P > 5$ kbar.

The effect of intracrystal inclusion migration under conditions of thermal gradient and/or stress (Antony and Cline, 1972) will result in the preferential loss of 'L' inclusions (due to the higher silicate solubility in phase 'L'). Hence the occurrence of a series of CO_2-rich fluid inclusions in minerals from metamorphic rocks does not necessarily demonstrate a predominantly CO_2-rich fluid composition during metamorphism.

Anatectic melting in amphibolites and granulites also falls in the range of conditions under which a two-phase fluid may occur. Figure 8.8 presents a projection of the isobaric section for the system H_2O–CO_2–NaCl with 7–10 mol% NaCl in H_2O on to the T–$X_{(CO_2)}$ plane at 7.5 kbar, along with curves representing the partitioning of H_2O and CO_2 between a granitic melt (the Qtz–Ab–Or eutectic, Ebadi and Johannes, 1991) and a fluid. The compositions of the water and CO_2-rich phases have very different salt concentrations, i.e. the line for the CO_2-rich phase is projected from the section plane for a very low salt concentration, less than

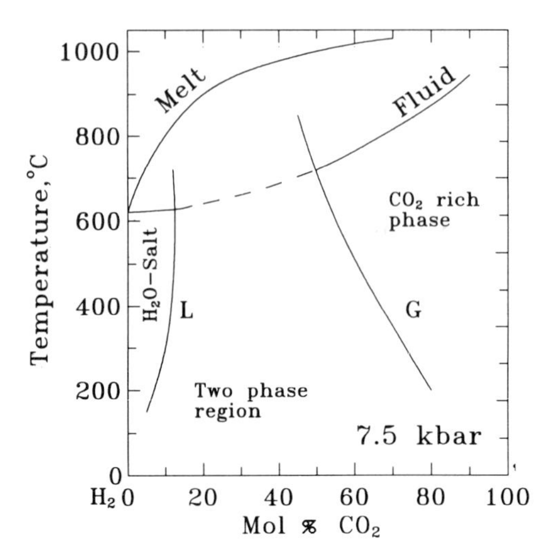

Figure 8.8. Isobaric (7.5 kbar) T – X projection for the system H_2O–CO_2–NaCl (20 wt% NaCl in phase 'L'). CO_2–water ratio in the saturated granitic melt and coexisting fluid is from Ryabchikov (1975).

10% equivalent weight NaCl, whereas the H_2O phase is salt-rich. It can be seen from the diagram that melting of the granitic eutectic may occur in the presence of either or both the water–salt and CO_2-rich fluids. Note that the left portion of the diagram in Figure 8.8 is constructed from the data on melting in pure water; the presence of salt at high pressure (solution density) will reduce the activity of H_2O (Shmulovich, 1988) and somewhat raise melting temperatures. Anatectic melting may take place at any bulk fluid composition, since the temperatures do not go beyond 'reasonable' estimates even at the CO_2-rich end. The actual fluid state however, is controlled by a phase diagram on which bulk compositions may plot within the immiscible field. In terms of thermodynamics, this implies that activity coefficients of components are so high that small CO_2 concentrations in phase 'L' come into equilibrium with 50–70 mol % CO_2 concentrations in phase 'G'. Similarly, low salt concentrations in phase 'G' tend to equilibrate with water–salt solution 'L', although equilibrium equations are in the latter case more complex because of the salt dissociation. It follows that activity estimates of components, retrieved from petrological data, are ambiguous. For example, a high value of $a_{(CO_2)}$ may indicate that a mineral assemblage is in equilibrium not only with a CO_2-rich phase, but with phase 'L' as well, in which CO_2 concentration does not exceed 10–15 mol %, while the activity coefficient exceeds 5.

As with other high-level granitoids, when anatectic melts (Figure 8.6) crystallize, the first fluid portions are enriched in CO_2 ('G'); only at the final stage does a liquid phase separate.

8.6 PHENOMENA ACCOMPANYING THE TWO-PHASE FLUID STATE

The existence of two coexisting fluid phases at high P and T implies partitioning of components between them. Partitioning of the major components, H_2O, CO_2 and salts, is characterized by the phase diagrams themselves (Figures 8.2–8.5). The saline portion is dominantly in phase 'L', while the non-polar gas is in phase 'G'. The thermodynamic consequences of fractionation of water between the two phases, both of which contain it in appreciable quantities, are more subtle. Fluid phase equilibria define the equality of component activities, i.e. $(a_{L,H_2O}) = (a_{G,H_2O})$. In salt solutions, $a_{(H_2O)} < X_{(H_2O)}$ at $P > 4$ kbar (Shmulovich, 1988), thereby reducing the concentration of water in phase 'G' for which only positive deviations from Raoult's law are known. The partitioning of water between fluid phases leads to the situation where water activities in coexisting phases may become much lower than its bulk concentration in the system, this being, in general, consistent with low values of a_{H_2O} at high metamorphic grades.

Consider some other consequences of the coexistence of two equilibrium fluid phases.

8.6.1 Fractionation of hydrolysis products

In water–salt solutions, hydrolysis reactions of the type:

$$MeCl_x + H_2O = Me(OH)_x + x\ HCl$$

take place. These reactions are generally shifted strongly to the left; at room temperature, the concentrations of Na(OH) and of HCl in salt solutions are negligible; the NaCl solution has a near neutral pH. At 'L'–'G' equilibrium, however, the hydrolysis reaction is accompanied by reactions to redistribute the components between coexisting phases. Because of the low dielectric constant of phase 'G' (be it a low-density water–vapour or a CO_2-rich phase), electrolyte dissociation in this phase is small, and the concentrations of components in phase 'G' are dependent on their own vapour pressures. As a result, phase 'G' becomes enriched in HCl, while phase 'L' undergoes complementary enrichment in $Me(OH)_x$, while maintaining equilibrium between the two fluids. Acidic fractionation between fluid phases in this way has been experimentally demonstrated by Vakulenko *et al.* (1989) on NaCl solutions. Figure 8.9 illustrates the pH values of the solutions obtained by quenching phases 'G' and 'L' from different experimental temperatures. Recall that under the experimental conditions $a^+_{H(G)} = a^+_{H(L)}$, however, the weaker dissociation of HCl in phase 'G' means that a much higher total HCl concentration is required to match the pH of coexisting aqueous liquid in which the HCl is more strongly dissociated. In our experiments in

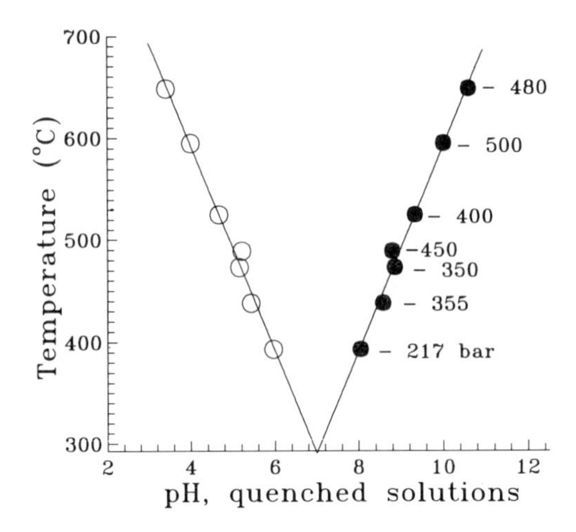

Figure 8.9. Quench solution pHs in the system H_2O–NaCl, with liquid (L, ●) and vapour (G, O) sampled by separate condensation. The run pressure is shown on the right. Reproduced with permission from the authors (Vakulenko *et al.*, 1989)

the H_2O–$CaCl_2$ and H_2O–$MgCl_2$ systems, conducted using the experimental set-up described above (Chapter 3, Section 3.2.2), we obtained differences in pH between phases 'G' and 'L' of up to 10 units. Spatial phase separation upon fluid infiltration through rocks in the field of temperature–pressure gradients will result in the production of chemically active acidic and alkaline solutions as the two fluids are no longer able to interact with each other. Such a mechanism for the formation of strong acids and alkalis seems to be potentially one of the most effective in nature, and is well known from geothermal fields, where acid gases are fractionated into the steam phase (Giggenbach, 1980). A similar process governs separation of acidic magmatic gases ('G') from melts ('L'); however, due to the small values for the water to silicate ratio, the residual melt does not become markedly more alkaline.

8.6.2 Fluid overpressure

Steinberg *et al.* (1984) pointed out that when two fluid phases of different density separate in the gravitational field, the lighter one will rise, and cause a fluid overpressure to develop. In a closed system, when the compressibility of the dominant, liquid phase 'L' is small, gravitational differentiation will cause pressure to increase as the rising vapour bubble attempts to expand. In the ideal case of $V_{syst} = \text{const}$, $\frac{dV}{dP} = 0$, and there is no chemical interaction between phases 'L' and 'G'; a rise in pressure during floating of even one phase 'G' bubble will result in a pressure increase by:

$$\Delta P = \rho_{(L)} \times g \times l,$$

where $\rho_{(L)}$ is the density of phase 'L', g the acceleration of gravity, and l the height of the pathway traversed by the phase 'G' bubble.

Under natural conditions, the compressibility of both phases and the pressure dependence of their mutual solubility must be taken into account in the equation for ΔP. At higher pressures (2–4 kbar, depending on the composition of the system), the CO_2-rich phase may be denser than the water–salt one ('L'); therefore, it is phase 'L' that floats up. In any event, spatial phase separation in the gravitational field will increase fluid pressure, and if $P_{(tot)} = P_{(fl)}$ before the process commences, then fluid overpressure will develop producing a network of interconnected pores, in addition to encouraging mass transport and recrystallization.

8.6.3 Phase separation

Aside from gravitational separation, phase separation upon infiltration of a two-phase fluid through rocks is well known from petroleum geology, and has also been experimentally investigated in the systems H_2O–N_2 and H_2O–Ar (Koshemchuk, 1993) as well as being known from synthetic fluid inclusion studies (Antony and Cline, 1972). The extent of separation is a function of the pressure gradient along the flow path: the lower the pressure gradient, the greater the extent of separation. Depending on rock type, porosity, permeability, and the relative proportions of tne two phases in the initial fluid mixture, different scenarios of component separation are possible on infiltration. These include cases where phase 'G' leaves behind phase 'L' or vice versa, or where the two phases infiltrate jointly without being separated from one another. A theoretical model for component separation upon infiltration, developed for two phases in the simple system H_2O–NaCl (Balashov *et al.*, 1989), has shown that water is transported mainly by phase 'G', because of its low viscosity, whereas NaCl is transported primarily by phase 'L'. The salt content in the filter (i.e. the rock) increases until a steady state is attained.

8.6.4 Mineral solubility in a two-phase fluid

The influence of two-phase state on mineral dissolution–deposition is poorly understood. The boiling front may be a geochemical barrier, where mineral supersaturation and crystallization are a consequence of boiling of the solvent (water). An opposite process is, however, also possible; it was detected in the study of PbS and ZnS solubilities in solutions of strongly hydrolysing salts such as $MgCl_2$ and NH_4Cl (Petrenko *et al.*, 1989). At constant temperature and autoclave volume, in the region of two-phase fluid state, the dissolution of sulphides in salt solution showed an inverse dependence on the initial amount of the solvent. That is, the joint action of Hemley *et al.*'s (1986) pressure effect on ore mineral solubilities and of the reaction of hydrolysis products with sulphides, produced an increase in the mineral solubility in the system as the amount of the solvent was reduced. The importance of this phenomenon and its geochemical consequences are not clear at present.

8.6.5 Stable isotope fractionation

Oxygen–hydrogen isotope fractionation between liquid and vapour along the boiling curve of pure water decreases with temperature to be close to zero at the critical temperature of water, as the difference between phases disappears. In water–salt solutions, critical phenomena occur at higher temperatures, and isotopic fractionation at 400–700°C is unknown. In the study of water–salt solutions (section 8.2); some portions of the samples were analysed for the ratios D:H and ^{18}O:^{16}O in phases 'L' and 'G'. The measurements in the system H_2O–NaCl at 400–600°C were suggestive of differences in the isotope composition of the phases; however, fractionation coefficients are large at room temperature, and there was a several-month delay between extraction of the samples and their analysis; as a result, there was a large scatter in the data, preventing the effect from being evaluated quantitatively. It is the subject of continuing research. The system H_2O–KCl yielded reproducible results, the difference in the ratio D:H between phases 'L' and 'G' ($10^3 \ln\alpha$) = 12 with ^{18}O:^{16}O constant and = 2.4 for the ^{18}O:^{16}O with D:H constant. The difference in the isotope composition of the phases is temperature independent in the range 400–600°C and is very close to the value of the salt effect at $T < 100$°C.

High-temperature isotopic fractionation between phases 'L' and 'G' is due to two mechanisms: the partitioning of isotopes between phases of different density and the salt isotope effect. Because phase 'G' is almost pure water, and critical phenomena above the critical temperature of H_2O are due to the presence of salt; there is no way of separating the contribution from these two mechanisms.

Equilibration of muscovite with water and with KCl solution yielded almost identical values for the salt isotope effect to those obtained by direct equilibration of liquid and gas phases (Suvorova and Shmulovich, 1992; Tkachenko and Shmulovich, 1992). The difference in δD between muscovite and pure water (650°C, 2 kbar) amounted to 20.5%., whereas the same value for 30% by weight KCl solution proved to be only 9‰. This difference of 11.5‰. δD between muscovite in equilibrium with pure water and with 30% by weight KCl solution of the same isotopic composition practically coincides with that retrieved from the 'L'–'G' equilibrium.

Isotopic fractionation on the phase-equilibria surface may simulate the occurrence of 'juvenile' and/or 'meteoric' fluid flows in the absence of any exchange between a fluid and its environment. The values of $10^3 \ln\alpha$ for H_2 and O_2 are equal to one-half the difference between the average isotope composition of oceanic water (SMOW) and that of water in magmatic melts from island arcs. That is, Rayleigh distillation makes it possible to produce one composition from the other (without buffering the isotopic ratios of oxygen by fluid–rock interaction). The 'L'–'G' phase equilibrium assumes isotopic equilibrium as well, and the isotope compositions of minerals are therefore independent of which of the phases is a crystal growth medium. As was noted above however, different physical properties of fluid phases are responsible for the mechanisms of their spatial separation. The movement of solutions in gradient fields (pressure, temperature, gravity, infiltration through a thin-pore medium) disturbs the solution–rock equilibria. In this case, it is the crystal growth medium that determines the isotope composition of minerals. Unfortunately, isotopic fractionation in the ternary H_2O–CO_2–salt systems has not as yet been investigated; preliminary results are available for the binary H_2O–salt systems alone.

8.6.6 Two-phase fluids at regional metamorphic pressures

At high pressures (> 5 kbar), mass transfer is probably more limited because of low rock permeability, and metasomatic reactions are also of limited occurrence. If the bulk fluid composition lies in the two-phase region, fluid phase equilibria buffer the activities of H_2O and CO_2. Their activities respond to the release of these components during the course of metamorphic devolatilization reactions. Component fractionations partially account for the low value of $a_{(H_2O)}$ as well. Fluid with the bulk composition 85 mol% H_2O, 10 mol% CO_2, and 5 mol% NaCl is two-phase at 7 kbar and both 500 and 700 °C. Phase 'G' contains 60–70 mol% CO_2, 40–30 mol% H_2O, and a negligible salt concentration. Activity coefficients in phase 'G' are close to, or little more, than 1, and hence $a_{(H_2O)} = 0.3$–0.4. From the equality of water activities in phases 'L' and 'G', it follows that $a_{(H_2O)} \approx 0.3$–0.4 for phase 'L' even at $X_{(H_2O)} > 0.9$. This value is again 2–3 times greater than $a_{(H_2O)}$ measured in granulites (Aranovich *et al.*, 1987). However, salt concentrations in phase 'L' are normally much higher (> 5 mol%), and the activity of water may also be reduced at high pressures by the pore size effect (Belonozhko and Shmulovich, 1987).

8.7 CONCLUSIONS

Quantitative calibration of phase equilibria in fluid systems has shown that the field of immiscibility spans a wide compositional range under conditions that include crystallization of near-surface intrusions, high-grade regional metamorphism, and anatectic melting. Homogeneous solutions occur in the very confined compositional range, low in salt and/or CO_2, and much narrower than might be expected. We were surprised to find that for a model natural fluid (the H_2O–CO_2–salt system), although a pressure increase from surface conditions to 3 kbar enlarges the compositional range of homogeneous fluid, i.e. increases the mutual solubility of components, their mutual solubility decreases at higher pressures. In practice, this leads to the situation where only those solutions whose compositions are close to the H_2O–CO_2 and H_2O–salt boundary systems may exist as single phases. This accounts for the widespread utility of binary models for natural fluids and points to the fact that the petrologic results obtained on this basis may be considered reasonable.

The two-phase fluid state is a rather common state in the process of magmatic melt degassing (magmatic distillation). Only when the crystallizing melt becomes highly depleted in both chlorine and non-polar gases (CO_2 or N_2) can the separating fluid avoid unmixing into two phases, with consequent fractionation. The melt–fluid equilibrium may likewise be treated in the same way as the 'L'–'G' equilibrium, with its own limits to immiscibility and fractionation coefficients. However, the fractionation processes themselves and their consequences should be similar. The major problem is to establish criteria to determine whether two-phase 'L1' (melt)–'G' or three-phase 'L1' (melt)–'L2' (salt solution)–'G' equilibria are attained during the process of degassing (including emission of volcanic gases). Thermodynamically, the two cases are similar ($a_{(i)} = \text{const}$), but geochemical consequences of spatial phase separation are different because 'L2' is an alkaline phase, which may, for example, be responsible for albitization.

ACKNOWLEDGEMENTS

Fluid-inclusion studies were started in the laboratory of Prof. P. Metz, Tubingen University, Germany. Discussions of immiscibility phenomena in natural fluids with Bruce Yardley were extremely helpful.

REFERENCES

Antony, T.R. and Cline, H.E. (1972) The thermomigration of biphase vapor–liquid droplet in solid. *Acta Metallurgica*, **20**, 247–55.

Aranovich, L.Ya., Shmulovich, K.I. and Fed'kin, V.V. (1987) The H_2O and CO_2 regime in regional metamorphism. *Intern. Geolog. Review*, **29**, 1379–1401.

Balashov, V.N., Korotaev, M.Yu. and Zaraisky, G.P. (1989) Stationary infiltration of H_2O–NaCl solutions under heterogenization conditions. *Geokhem. Int.*, **26** (6), 58–67.

Belonozhko, A.B. and Shmulovich, K.I. (1987) A molecular-dynamics study of a dense fluid in micropores. *Geokhem. Int.*, **24** (6), 1–12.

Bischoff, J.L. and Rosenbauer, R.J., (1986) The system NaCl–H_2O: Relations of vapor–liquid near the critical temperature of water and of vapor–liquid – halite from 300 to 500 °C. *Geochim. Cosmochim. Acta*, **50**(7), 1437–44.

Bischoff, J.I. and Pitzer, K.S. (1989) Liquid–vapour relations for the system NaCl–H_2O: Summary of the P–T–X surface from 300 to 500 °C. *Amer. J. Sci.*, **289**, 217–48.

Bodnar, R.J., Burnham, C.W. and Sterner, S.M. (1985) Synthetic fluid inclusions in natural quartz: III. Determination of phase equilibrium properties in the system H_2O–NaCl to 1000 °C and 1500 bars. *Geochim. Cosmochim. Acta*, **49**, 1861–73.

Bodnar, R.J. and Sterner, S.M. (1987) Synthetic fluid inclusions, in *Hydrothermal Experimental Techniques* (eds G.C. Ulmer and H.L. Barnes) Wiley-Intersciences, New York, N.Y., pp. 423–57.

Crawford, M.L., Kraus, D.W. and Hollister, L.S. (1979) Petrologic and fluid inclusions study of calc–silicate rocks, Prince Rupert, British Columbia. *Amer. J. Sci.*, **279**, 1135–59.

Ebadi,A. and Johannes, W. (1991) Beginning of melting and composition of first melts in the system Qtz–Ab–Or–H_2O–CO_2. *Contrib. Mineral. Petrol.*, **106**, 286–95.

Gehrig, M. (1980) Phasengleichgewichte und PVT–Daten ternarer Mischungen ausWasser, Kohlendioxid und Natriumclorid bis 3 kbar und 550 °C. Dissertation, Univer. Karlsruhe, Hochschulverlag, Freiburg, 212 pp.

Giggenbach, W.F. (1980) Geothermal gas equilibria. *Geochim. Cosmochim. Acta*, **44**, 2021–32.

Hemley, J.J., Cygan, G.L. and D'Angelo, W.M. (1986) Effect of pressure on ore mineral solubilities under hydrothermal conditions. *Geology*, **14**, 377–79.

Japas, M.L. and Franck, E.U. (1985) High pressure phase equilibria and PVT–data of water–nitrogen system to 673 K and 250 MPa. *Ber. Bunsenges. Phys. Chem.*, **89**, 793–800.

Koshemchuk, S.K. (1993) The investigation of regularities of two-phase filtration of the water–gas system through porous media. Ph.D. thesis, Moscow, Vernadsky Institute of Geochemistry and Analytical Chemistry, 28 pp.

Krader, T. (1985) Phasengleichgewichte und kritische Kurven des ternaren Systems H_2O–CH_4–NaCl bis 250 MPa und 800 K. Dissertation Univer. Karlsruhe, 107 pp.

Kreulen, R. (1980) CO_2-rich fluids during regional metamorphism on Naxos (Greece). Carbon isotopes and fluid inclusions. *Amer. J. Sci.*, **280**, 745–71.

Petrenko, G.V., Malinin, S.D. and Arutyunyan, L.A. (1989) An experimental investigation of solubility of heavy metal sulfides in the heterogeneous systems chloride solution–vapor phase. *Geokhimia*, **6**, 882–97 (in Russian).

Plyasunova, N.V. and Shmulovich, K.I. (1991) Phase equilibria in the system $CaCl_2$–H_2O at 5 kbar and 500 °C. *Doklady Academii Nauk SSSR*, **319** (3), 738–41 (in Russian).

Poty, B.P., Stalder, H.A. and Weisbrod, A.M. (1974) Fluid inclusions studies in quartz from fissures of Western and Central Alps. *Schweiz. Mineral. Petrogr. Mitt.*, **54**, 717–52.

Ramboz, C., Pichavant, M. and Weisbrod, A. (1982) Fluid immiscibility in natural processes: Use and misuse of fluid inclusion data. *Chemical Geology*, **37**, 29–48.

Ravich, M.I. (1974) *Water–salt Systems at Elevated Temperatures and Pressures.* Nauka Press, Moscow (in Russian).

Ryabchikov, I.D. (1975) *The Thermodynamic of Fluid Phase of Granitic Magmas.* Nauka Press, Moscow (in Russian).

Skippen, G. (1988) Phase relations in model fluid systems. *Rendiconti Della Societa Italiana di Mineralogia e Pterologia*, **43**, 7–14.

Shmulovich, K.I. (1988) *Carbon Dioxide in High-temperature Processes of Mineral Formations*. Nauka Press, Moscow (in Russian).

Shmulovich, K.I. and Plyasunova, N.V. (in press) Phase equilibria in the ternary system H_2O–CO_2–salt ($CaCl_2$, NaCl) at high temperatures and pressures. *Geochem. Int.*

Sourirajan, S. and Kennedy, G.C. (1962) The system H_2O–NaCl at elevated temperatures and pressures. *Amer. J. Sci.*, **260**, 115–41.

Steinberg, G.S., Steinberg, A.S. and Merzhanov, A.G. (1984) The fluid mechanism of increasing pressure in magmatic systems. *Doklady Academii Nauk*, **279** (5), 1081–4 (in Russian).

Suvorova, V.A. and Shmulovich, K.I. (1992) *High temperature isotope salt effect*. In Proceed. IV Intern. Sympos. Experim. Miner. Petrol. and Geochem. Clermont-Ferrand, p. 43.

Takenouchi, S. and Kennedy, G.C. (1964) The binary system H_2O–CO_2 at high temperatures and pressures. *Amer. J. Sci.*, **262**, 1055–67.

Takenouchi, S. and Kennedy, G.C. (1965) The solubility of carbon dioxide in NaCl solutions at high temperatures and pressures. *Amer. J. Sci.*, **263**, 445–64.

Tkachenko, S.I. and Shmulovich, K.I. (1992) Liquid–vapour equilibria in the systems water–salt (NaCl, KCl, $CaCl_2$, or $MgCl_2$) at 400–600 °C. *Doklady Akademii Nauk SSSR*, **326** (6), 1055–9 (in Russian).

Touret, J. (1981). Fluid inclusions in high grade metamorphic rocks. *Mineral. Assoc. Canada Short Handbook*, **6**, 182–208.

Trommsdorff, V., Skippen, G. and Ulmer, P. (1985) Halite and sylvite assolid inclusions in high-grade metamorphic rocks. *Contrib. Mineral. Petrol.*, **89**, 24–9.

Trommsdorff, V. and Skippen, G. (1986) Vapour loss (Boiling) as a mechanism for fluid evolution in metamorphic rocks. *Contrib. Miner. Petrol.*, **94**, 317–22.

Urusova, M.A. (1974) Phase equilibria and thermodynamic characteristics of solutions in the systems NaCl–H_2O and NaOH–H_2O at 350–550 °C. *Geochem. Int.*, **11** (5), 944–50.

Vakulenko, A.G., Alekhin Yu.V. and Razina, M.V. (1989) *Solubility and thermodynamic properties of alkali chlorides in steam*. In Proc. II Intern. Sympos. 'Properties of water and steam', Prague, pp. 395–401.

Vapnik, Eu.A. (1988) The fluid regime of anatexis and diatexis at different metamorphic conditions. Ph.D. thesis, Leningrad Univers., 23 pp. (in Russian).

Yardley, B.W.D. and Bottrell, S.H. (1988) Immiscible fluids in metamorphism: Implications of two-phase flow for reaction history. *Geology*, **16**, 199–202.

Zhang, Yi-G. and Frantz, J.D. (1989) Experimental determination of the compositional limits of immiscibility in the system $CaCl_2$–H_2O–CO_2 at high temperatures and pressures using synthetic fluid inclusions. *Chemical Geology*, **74**, 289–308.

CHAPTER NINE

Diffusion of electrolytes in hydrothermal systems: free solution and porous media

Victor N. Balashov

9.1 INTRODUCTION

Metasomatic and metamorphic assemblages are generated during the long-term chemical evolution of rocks. According to present-day views, material transport by a fluid phase that fills a porous space is crucial to this process. Transfer of material by the fluid phase may be the result of both diffusion and infiltration (Korzhinskiy, 1970). Within any given system, transfer over relatively short distances and times is dominated by diffusion, irrespective of the presence of a convection component.

Diffusion coefficients for components in solid phases from 200 to 700 °C are of the order of 10^{-20}–10^{-12} $cm^2.s^{-1}$ (Hofmann *et al.*, 1974; Hart, 1981). Therefore, transport by diffusion over distances in excess of 10 cm is unlikely even within time intervals of geological magnitude (~ 3–4 Ma or ~10^{14} s). Under standard laboratory conditions, values for diffusion coefficients measured in aqueous solutions are 10^{-5} $cm^2.s^{-1}$ and tend to increase with increasing temperature (Robinson and Stokes, 1959; Erdey-Gruz, 1974). Consequently, diffusional transport of material in geochemical processes becomes more important if diffusion proceeds in fluids filling pores and fractures of the rocks. Since many crustal fluids are essentially water–salt solutions, diffusion in aqueous electrolyte solutions is of particular interest.

Diffusion in a pore solution commonly differs from that in a free solution. The chief factor is electrical interaction between phases. A jump

Fluids in the Crust: Equlibrium and transport properties.
Edited by K.I. Shmulovich, B.W.D. Yardley and G. G. Gonchar.
Published in 1994 by Chapman & Hall, London. ISBN 0 412 56320 7

in electric potential occurring at the interface between two phases results in the partitioning of charges between the phases. As a consequence, an electric double layer (EDL) forms (Shchukin *et al.*, 1982).

For the density of a diffusion flux through a cross-section of the homogeneous, isotropic porous medium, the following equation holds true:

$$J_i = -D_i^* \nabla C_i, \tag{9.1}$$

where C_i is the concentration of the ith component per unit volume of pore fluid and D_i^* the effective diffusion coefficient of the ith component through the porous medium.

Let us assume that all the above effects on D_i^* may be reduced to the form:

$$D_i^* = F æ_i D_i, \tag{9.2}$$

where F is the formation factor of the porous medium, D_i the diffusion coefficient of component i in free solution and $æ_i D_i$ the diffusion coefficient of component i in pore solution (Balashov, 1992).

9.2 THE DIFFUSION OF THE ELECTROLYTE $A_{\nu_1} B_{\nu_2}$ IN DILUTE SOLUTIONS

9.2.1 Stoichiometric and activity relations

Consider the general diffusion of the electrolyte $A_{\nu_1} B_{\nu_2}$. All possible ionic species can be designated as $A_{\nu_{1i}} B_{\nu_{2i}}^{z_i}$, i = 1, 2,. . ., n. Let z_1 be the charge of ions of type A and z_2 the charge of ions of type B, then:

$$z_1\nu_1 + z_2\nu_2 = 0, \tag{9.3}$$

$$z_i = \nu_{1i} z_1 + \nu_{2i} z_2.$$

For example, for ions of salts A and B,

$$\nu_{11} = 1, \nu_{21} = 0 : A^{z_1}; \tag{9.4}$$

$$\nu_{12} = 0, \nu_{22} = 1 : B^{z_2}$$

The molality of the ith species will be designated as m_i.

Since the rate of ionic reactions is high, it may be assumed that during diffusion dissociation reactions are governed by local chemical equilibrium (LCE),

$$\ln K_i + \ln(a_i) = \nu_{1i}\ln(a_1) + \nu_{2i}\ln(a_2), \tag{9.5}$$

where K_i is the thermodynamic equilibrium constant.

The mean electrolyte activity, *a*, and the mean stoichiometric electrolytic activity coefficient, γ, are defined by the relationships:

$$a = \gamma m, \tag{9.6}$$

where m is electrolyte molality,

$$a = (a_1^{\nu_1} a_2^{\nu_2})^{\frac{1}{\nu}} = (\alpha_1^{\nu_1} \alpha_2^{\nu_2})^{\frac{1}{\nu}}.(\gamma_1^{\nu_1} \gamma_2^{\nu_2})^{\frac{1}{\nu}}.m, \tag{9.7}$$

$$\gamma = (\alpha_1^{\nu_1} \alpha_2^{\nu_2})^{\frac{1}{\nu}}.(\gamma_1^{\nu_1} \gamma_2^{\nu_2})^{\frac{1}{\nu}}, \tag{9.8}$$

where

$$\nu = \nu_1 + \nu_2, \tag{9.9}$$

$$\alpha_i = m_i \cdot m^{-1}. \tag{9.10}$$

9.2.2 General expression for the total diffusion coefficient of electrolyte

Further statements are valid for the reference system related to solvent (H_2O). It should be emphasized that for dilute solutions, the diffusion coefficients in the solvent reference system are practically identical with those in Fick's conventional reference system (Anderson, 1981).

The diffusional flux of charged species i in units of mol. cm^{-2}. s^{-1} is determined by:

$$J_i = \sum_{j=i}^{n} l_{ij}^0 X_j, \tag{9.11}$$

where n is the total number of electrolyte species and l_{ij}^0 the phenomenological transport coefficients. The generalized forces X_j satisfy the relationship (Miller, 1966):

$$X_j = -\nabla\mu_j - z_j \nabla\phi, \tag{9.12}$$

where μ_j is the chemical potential of the jth species and

$$\phi = \phi^D F, \tag{9.13}$$

where ϕ^D is the diffusion potential in volts and F the Faraday constant. For dilute solutions, we may set the off-diagonal terms $l_{ij}^0 = 0, i \neq j$ and define the phenomenological coefficient for diffusion of component in its own chemical potential field as (Anderson, 1981; Miller, 1966):

$$l_{ii}^0 = \frac{D_i^0 C_i}{1000 \cdot RT}, \tag{9.14}$$

where D_i^0 is the diffusion coefficient at infinite dilution in units of $cm^2 \cdot s^{-1}$, C_i the molar concentration ($mol \cdot l^{-1}$), 1000 the normalized multiplier corresponding to the density of the diffusion flux in units of mol. cm^{-2}. s^{-1} from Equation 9.11. For dilute solutions,

$$C_i \cong \rho_{H_2O} \cdot m_i, \tag{9.15}$$

thus,

$$l_{ii}^0 = \frac{D_i^0}{RT} \rho\, m_i, \tag{9.16}$$

where

$$\rho = \frac{\rho_{H_2O}}{1000}. \tag{9.17}$$

The flow of the ith component is determined by the relationship:

$$J_i = -D_i^0\, m_i\, \rho(\nabla \ln(a_i) + z_i\, \nabla\phi/RT). \tag{9.18}$$

The flow of component A and that of the electrolyte are expressed as:

$$J_A = \sum \nu_{1i}\, J_i, \tag{9.19}$$

$$J_e = J_A \cdot \nu_1^{-1}.$$

From the condition of LCE (Equation 9.5), it follows that:

$$\nabla \ln(a_i) = \nu_{1i}\, \nabla \ln(a_1) + \nu_{2i}\, \nabla \ln(a_2). \tag{9.20}$$

To simplify the calculations, here introduce the designation:

$$D_i = D_i^0\, m_i \tag{9.21}$$

and the condition of zero electric current:

$$\sum z_i\, J_i = 0, \tag{9.22}$$

or

$$-\sum z_i\, D_i\, \nabla \ln(a_i) - \sum D_i\, z_i^2\, \nabla\phi/RT = 0.$$

This yields:

$$\nabla\phi/RT = -\frac{\sum z_i\, D_i\, \nabla \ln(a_i)}{\sum D_i\, z_i^2} \tag{9.23}$$

for the diffusion potential gradient, which may be simplified by introducing the designation:

$$\tilde{Z}_1 = \sum D_i\, z_i^2. \tag{9.24}$$

Substituting Equation 9.18 in Equation 9.19 and utilizing Equations 9.20, 9.23 and 9.24 gives:

$$\tilde{Z}_1\, J_A \cdot \rho^{-1} = -f_1\, \nabla \ln(a_1) - f_2\, \nabla \ln(a_2), \tag{9.25}$$

where

$$f_1 = \sum_{i,j} (\nu_{1i}^2\, D_i)\,(D_j\, z_j^2) - \sum_{i,j} (\nu_{1i}\, D_i\, z_i)\,(\nu_{1j}\, D_j\, z_j) \tag{9.26}$$

$$= \sum_{i \neq j} (\nu_{1i}^2 D_i)(D_j z_j^2) - \sum_{i \neq j} (\nu_{1i} D_i z_i)(\nu_{1j} D_j z_j),$$

$$f_2 = \sum_{i,j} (\nu_{1i} \nu_{2i} D_i)(D_j z_j^2) - \sum_{i,j} (\nu_{1i} D_i z_i)(\nu_{2j} D_j z_j) \tag{9.27}$$

$$= \sum_{i \neq j} (\nu_{1i} \nu_{2i} D_i)(D_j z_j^2) - \sum_{i \neq j} (\nu_{1i} D_i z_i)(\nu_{2j} D_j z_j).$$

Transformations in Equations 9.26 and 9.27 are made using the identity:

$$\sum_{i,j} A_i A_j = \sum A_i^2 + \sum_{i \neq j} A_i A_j. \tag{9.28}$$

The function f_1 can be expressed by:

$$f_1 = \nu_1^2 Y_1, \tag{9.29}$$

where Y_1 is written as:

$$Y_1 = \sum_{i \neq j} \left(\frac{\nu_{1i}}{\nu_1} \right)^2 D_i D_j z_j^2 - \sum_{i \neq j} \frac{\nu_{1i}}{\nu_1} D_i z_i \frac{\nu_{1j}}{\nu_1} D_j z_j. \tag{9.30}$$

Substituting the ratio $\nu_{1i}.\ \nu_1^{-1}$ in the form given by Equation 9.3:

$$\frac{\nu_{1i}}{\nu_1} = \frac{z_i}{z_1 \nu_1} + \frac{\nu_{2i}}{\nu_2}, \tag{9.31}$$

yields the following expression for Y_1:

$$Y_1 = \sum_{i \neq j} \left(\frac{\nu_{2i}}{\nu_2} \right)^2 D_i D_j z_j^2 - \sum_{i \neq j} \frac{\nu_{2i}}{\nu_2} D_i z_i \frac{\nu_{2j}}{\nu_2} D_j z_j, \tag{9.32}$$

which is identical to Equation 9.30.

The function f_2 may have the form:

$$f_2 = \nu_1 \nu_2 Y_2, \tag{9.33}$$

where Y_2 is determined by:

$$Y_2 = \sum_{i \neq j} \frac{\nu_{1i}}{\nu_1} \frac{\nu_{2i}}{\nu_2} D_i D_j z_j^2 - \sum_{i \neq j} \frac{\nu_{1i}}{\nu_1} D_i z_i \frac{\nu_{2j}}{\nu_2} D_j z_j. \tag{9.34}$$

It is clear that by substituting $\nu_{2i}.\ \nu_2^{-1}$ in the form given by Equation 9.3:

$$\frac{\nu_{2i}}{\nu_2} = \frac{z_i}{z_2 \nu_2} + \frac{\nu_{1i}}{\nu_1}, \tag{9.35}$$

into Equation 9.34, Y_2 rearranges to Equation 9.30 for Y_1. Substitution of $\nu_{1i} \cdot \nu_1^{-1}$ via Equation 9.31 changes Y_2 to Equation 9.32. Hence,

$$Y_1 = Y_2 = Y. \tag{9.36}$$

By using Equations 9.29, 9.33, and 9.36, Equation 9.25 can be written as:

$$\tilde{Z}_1 J_e \cdot \rho^{-1} = -\nu Y \nabla \ln(a). \tag{9.37}$$

Thus,

$$J_e = J_A \cdot \nu_1^{-1} = -\tilde{D}_e^0 \rho\, m \nabla \ln(a) = -\tilde{D}_e^0 \rho \left(1 + \frac{\partial \ln(\gamma)}{\partial \ln(m)}\right) \nabla m. \tag{9.38}$$

As a consequence, for the total diffusion coefficient of the electrolyte, taking account of complex formation in solution, the following expression is valid:

$$D_e = \tilde{D}_e^0 \rho \left(1 + \frac{\partial \ln(\gamma)}{\partial \ln(m)}\right). \tag{9.39}$$

For the generalized diffusion mobility of the electrolyte given by Equations 9.36 and 9.37, we obtain:

$$\tilde{D}_e^0 = \nu \left(\frac{Y}{m}\right) \cdot \tilde{Z}_1^{-1}, \tag{9.40}$$

where Y is given by one of the identical expressions 9.30, 9.32 or 9.34, and $\tilde{Z}_1$ is defined in Equation 9.24.

As a specific example, consider the diffusion of $CaCl_2$. The stoichiometric composition of its complexes and their charges are given in Table 9.1. Substitution of Equations 9.10 and 9.21 in expression 9.40 gives the following expression for the diffusion mobility of $CaCl_2$:

Table 9.1. Stoichiometric composition of ion complexes and their charges for $CaCl_2$

Species	z_i	i	ν_{1i}	ν_{2i}
Ca^{2+}	+2	1	1	0
Cl^-	−1	2	0	1
$CaCl^+$	+1	3	1	1
$CaCl_2^0$	0	4	1	2

$$\tilde{D}^0_{CaCl_2} = 3\left(\frac{D_1^0 D_2^0 \alpha_1 \alpha_2 + D_1^0 D_3^0 \alpha_1 \alpha_3 + D_2^0 D_3^0 \alpha_2 \alpha_3}{4D_1^0 \alpha_1 + D_2^0 \alpha_2 + D_3^0 \alpha_3} + D_4^0 \alpha_4\right), \tag{9.41}$$

where D_1, etc. may be identified with specific species utilizing Table 9.1. At complete dissociation $\alpha_1 = 1$, $\alpha_2 = 2$, $\alpha_3 = 0$, and $\alpha_4 = 0$; thus

$$\tilde{D}^0_{CaCl_2} = \frac{3D_1^0 D_2^0}{2D_1^0 + D_2^0}, \tag{9.42}$$

which is the classic Nernst equation.

Applin and Lasaga (1984) derived specific particular cases of Equations 9.30, 9.39, and 9.40 for diffusion of $MgSO_4$ and Na_2SO_4 electrolytes when dealing with Mg^{2+} and SO_4^{2-} and an ion pair $MgSO_4^0$ in the former case and with Na^+ and SO_4^{2-} and an ion pair $NaSO_4^-$ in the latter one.

For the case of a certain molality electrolyte, the values of α_i and their derivatives $\partial\alpha_i . \partial m^{-1}$ are unequivocally given by the set of mass action equations (Equation 9.5), with two linear equations added: the electroneutrality equation and the equation for the total molality of the electrolyte. Therefore, given the thermodynamic equilibrium constants for Equation 9.5, infinite dilution diffusion coefficients, and functions γ_i of the electrolyte molality, it seems possible to calculate the dependence of D_e on the electrolyte concentration.

9.3 P–T–m DEPENDENCIES OF DIFFUSION COEFFICIENTS IN ELECTROLYTE SOLUTIONS

9.3.1 Tracer diffusion coefficients for ions and ion pairs

The infinite dilution diffusion coefficient for the ion, D_i^0, is related to the limiting equivalent conductance, λ_i^0, by the Nernst–Einstein equation:

$$D_i^0 = \frac{\lambda_i^0 RT}{| z_i | F^2}, \tag{9.43}$$

where F is Faraday constant. Electrical conductances of a series of high-temperature solutions of alkaline and calc–alkaline salts, and of some acids (HCl, H_2SO_4) have been the subject of a number of experimental studies (Franck, 1956; Frantz and Marshall, 1982; Frantz and Marshall, 1984; Quist and Marshall, 1965, 1968, 1969; Ritzert and Franck, 1968).

There is a linear correlation between the limiting equivalent conductance, λ^0, and the standard partial molal entropy of ion at infinite dilution, S^0,

$$\lambda_{ij}^0 = a_{ij} + b_{ij} S_{ij}^0, \tag{9.44}$$

where the subscript i refers to a single ion, and the subscript j to a group of ions (cations and anions of like charge). The values of λ_i^0, and hence D_i^0, for various ions (Oelkers and Helgeson, 1988a), calculated using a range of experimental data and Equation 9.44, superseded Nigrini's (1970) model.

Developing ideas by Wishaw and Stokes (1954), Nigrini (1970) proposed a method to calculate the tracer diffusion coefficients for ion pairs from Ds for individual species. In that method, an ion pair is represented as a spherical ellipsoid with an equatorial diameter equal to the largest species diameter, and a major axis equal in length to the sum of the species diameters. Let D_1^0 and D_2^0 be the diffusion coefficients for the species (we suppose $D_2^0 > D_1^0$), then:

$$D_{as}^0 = \frac{D_1^0}{\left(1 + \left(\frac{D_1^0}{D_2^0}\right)^3\right)^{1/3} sf}, \tag{9.45}$$

where D_{as}^0 is the diffusion coefficient of the ion pair, and sf stands for the spherical factor defined as:

$$sf = \frac{(1 - \beta^2)^{1/2}}{\beta^{2/3} \ln\left(\frac{1 + (1 - \beta^2)^{1/2}}{\beta}\right)}. \tag{9.46}$$

For β, the following relationship is valid:

$$\beta = \frac{1}{\left(1 + \frac{D_1^0}{D_2^0}\right)}. \tag{9.47}$$

Table 9.2 lists diffusion coefficient values for a series of complexes, computed at 25°C and 1 atm according to Equation 9.45, and a set of estimates from Applin (1987) and Applin and Lasaga (1984). Table 9.2 also includes standard partial molal electrolyte volumes, which in principle should correlate with the partial molal volume of an ion pair. In the work by Applin (1987) attention is given to the fact that ion pairs with positive partial volumes exhibit diffusion coefficients of around $2 \cdot 10^{-5}$ $cm^2 . s^{-1}$, whereas those with negative partial molar volumes show diffusion coefficients of $\leqslant 10^{-5}$ $cm^2 . s^{-1}$. A possibly important factor omitted in the calculation of the diffusion coefficients of ion pairs from Equation 9.45 is the extent to which electrical interaction between ions reduces their electrostatic effect on water. This effect should cause a change in the size of the hydrate shells. The results in Table 9.2, however, suggest that Equation 9.45 can be satisfactorily applied provided the partial molal volumes are negative. Because the standard partial molal volumes of electrolytes become predominant (Helgeson and Kirkham, 1974) as the temperature rises (at least up to pressures of 3 kbar), we have used Equation 9.45 for the estimation of the diffusion coefficients for ion pairs.

The data on the tracer diffusion coefficients for HCl, NaCl, KCl, NaOH, $MgCl_2$ and $CaCl_2$ complexes, calculated via Equation 9.45 at T from 25 to 1000 °C and P = 100, 300 and 500 MPa using ion data (Oelkers and Helgeson, 1988b), are compiled in Table 9.3 (a,b,c). These data are relatively close to values of the H_2O liquid lattice self-diffusion coefficient calculated by Labotka (1991).

Table 9.2. Tracer diffusion coefficients of ion-pairs in aqueous solutions at 25°C and 1 atm from Applin (1987) and according to Equation 9.45.

Species	$\overline{V}_e^0$, $cm^3.mol^{-1}$	$D^0.10^5$, Eqn. 9.45	$cm^2.sec^{-1}$
KCl^0	26.8	1.52	2.16
$NaCl^0$	16.6	1.2	1.99
$CaSO_4^0$	- 3.8	0.69	0.65
$MgSO_4^0$	- 7.2	0.64	0.8
$CaCO_3^0$	- 21.6	0.65	0.6
$MgCO_3^0$	- 22.9	0.61	0.58

9.3.2 Thermodynamic dissociation constants in electrolyte solutions

Marshall and Franck (1981) showed that values for the thermodynamic dissociation constant of water, K_w, from 25 to 1000°C at 0.01 to 10 kbar are well described by the polynomial (molality units):

$$\log K_w = \sum_{i=0}^{n_A} A_i \left(\frac{1}{T}\right)^i + \left(\sum_{i=0}^{n_B} B_i \left(\frac{1}{T}\right)^i\right) . \log \rho_{H_2O}, \qquad (9.48)$$

where $n_A = 3$, and $n_B = 2$. Frantz and Marshall (1982, 1984) showed that, within the errors of measurement, the P–T dependence of the first and second dissociation constants of $CaCl_2$, $MgCl_2$, and HCl is described with Equation 9.48 at $n_A = 1$, $n_B = 0$. Oelkers and Helgeson (1988b) calculated thermodynamic dissociation constants of alkali metal halides at temperatures from 400 to 800°C and pressures from 50 to 400 MPa using available conductance data (Franck, 1956; Quist and Marshall, 1965, 1968, 1969; Ritzert and Franck, 1968). We have processed those data numerically with the aid of polynomial 9.48 using a least squares approach. Standard deviation values indicate that in this case, too, polynomial 9.48 is good for approximation. To approximate the dissociation constants of NaOH and HCl, the data from Plyasunov (1988, 1989) have been used. Polynomial parameters, the number of approximation data points, and standard deviations for different compounds are listed in Table 9.4.

9.3.3 A model for electrolyte activity

For dilute electrolyte solutions, the conventional electrostatic Debye–Hückel theory is applicable (Robinson and Stokes, 1959; Erdey-Gruz, 1974; Helgeson, *et al.*, 1981). The activity coefficient of the ith ion

Table 9.3. Tracer diffusion coefficients, $D.10^5$, $cm^2 \cdot sec^{-1}$, for HCl, NaCl, KCl, NaOH, $MgCl_2$ and $CaCl_2$ species, calculated via Equation 9.45 at T from 25 to 1000°C and P = 100, 300 and 500 MPa using ion data (Oelkers and Helgeson, 1988a)

Table 9.3a. P = 100 MPa

Species	25°C0.1 MPa	100°C	150°C	200°C	250°C	300°C	350°C	400°C	500°C	600°C	700°C	800°C	900°C	1000°C
HCl^0	2.02	6.31	10.1	14.3	18.9	23.6	28.3	32.9	41.3	48.6	54.7	59.9	64.6	68.5
$NaCl^0$	1.2	4.03	6.7	9.8	13.4	17.3	21.3	25.4	33.9	42.3	50.3	58.0	65.1	71.9
KCl^0	1.52	4.5	7.2	10.4	14.0	18.0	22.0	26.1	34.2	42.0	48.8	55.1	61.6	67.0
$MgCl^+$	0.69	2.8	4.7	6.9	9.5	12.2	15.1	18.1	24.6	31.4	38.4	45.3	51.5	56.2
$MgCl_2^0$	0.68	2.7	4.5	6.6	9.1	11.6	14.3	17.2	23.2	29.5	35.8	42.1	47.7	52.1
$CaCl^+$	0.77	2.9	4.9	7.3	9.8	12.6	15.5	18.5	24.6	30.7	36.5	42.0	47.2	52.1
$CaCl_2^0$	0.75	2.8	4.7	6.9	9.3	11.9	14.7	17.5	23.2	28.9	34.4	39.5	44.4	48.9
$NaOH^0$	1.32	4.5	7.5	11.1	15.1	19.5	23.8	28.2	36.8	44.7	51.8	58.0	63.4	68.3

Table 9.3b. P = 300 MPa.

Species	25°C	100°C	150°C	200°C	250°C	300°C	350°C	400°C	500°C	600°C	700°C	800°C	900°C	1000°C
HCl^0	2.09	5.8	9.0	12.5	16.2	20.0	23.8	27.6	34.7	41.2	47.0	52.1	56.7	60.7
$NaCl^0$	1.25	3.8	5.9	8.5	11.3	14.2	17.3	20.4	26.6	32.5	38.2	43.5	48.5	53.0
KCl^0	1.48	4.2	6.5	9.1	11.9	14.9	17.9	21.0	27.1	32.7	38.0	42.9	47.2	50.9
$MgCl^+$	0.78	2.6	4.1	5.9	7.9	9.9	12.1	14.2	18.5	22.7	26.6	30.3	33.7	36.8
$MgCl_2^0$	0.76	2.5	3.9	5.6	7.5	9.4	11.5	13.5	17.6	21.5	25.2	28.7	31.9	34.8
$CaCl^+$	0.78	2.7	4.4	6.1	8.2	10.3	12.5	14.7	19.2	23.4	27.5	31.3	34.8	38.1
$CaCl_2^0$	0.76	2.6	4.2	5.8	7.8	9.8	11.8	13.9	18.1	22.1	25.9	29.5	32.7	35.8
$NaOH^0$	1.39	4.2	6.8	9.7	13.0	16.4	19.9	23.4	30.2	36.7	42.6	47.9	52.7	57.0

Table 9.3c. P = 500 MPa.

Species	25°C	100°C	150°C	200°C	250°C	300°C	350°C	400°C	500°C	600°C	700°C	800°C	900°C	1000°C
HCl^0	2.03	5.62	8.53	11.7	15.0	18.4	21.7	25.1	31.4	37.3	42.6	47.4	51.7	55.5
$NaCl^0$	1.21	3.58	5.62	7.91	10.4	13.0	15.6	18.3	23.5	28.6	33.3	37.7	41.8	45.6
KCl^0	1.49	4.02	6.13	8.48	11.0	13.6	16.2	18.8	23.9	28.7	33.0	36.8	40.0	42.9
$MgCl^+$	0.74	2.45	3.89	5.50	7.22	9.02	10.9	12.7	16.4	19.9	23.2	26.2	29.0	31.6
$MgCl_2^0$	0.72	2.36	3.73	5.26	6.89	8.60	10.3	12.1	15.6	18.9	22.0	24.8	27.4	29.9
$CaCl^+$	0.77	2.58	4.09	5.77	7.55	9.41	11.3	13.2	17.0	20.6	24.0	27.2	30.2	32.9
$CaCl_2^0$	0.75	2.47	3.91	5.49	7.17	8.92	10.7	12.5	16.1	19.5	22.7	25.6	28.3	30.9
$NaOH^0$	1.33	4.04	6.41	9.10	12.0	15.0	18.1	21.2	27.3	33.0	38.3	43.0	47.4	51.5

Table 9.4. Parameters of polynomial 9.48 for calculation of thermodynamic dissociation constants in electrolyte solutions

Species	A_i			B_i		*Number of appr points*	*Stand. deviation*
	i = 0	1	2	i = 0	1		
HCl^0	– 3.307	2893.6	$-5.207 \cdot 10^5$	26.653	– 10097.5	64	0.13
$LiCl^0$	3.349	– 3864.0	$1.345 \cdot 10^6$	19.78	– 8159.5	38	0.05
$NaCl^0$	3.019	– 4487.7	$1.552 \cdot 10^6$	16.08	– 6888.9	53	0.03
KCl^0	4.829	– 6686.8	$2.301 \cdot 10^6$	18.17	– 8857.2	41	0.06
$RbCl^0$	3.997	– 5662.0	$1.799 \cdot 10^6$	17.82	– 9065.5	51	0.08
$CsCl^0$	4.75	– 7650.8	$2.715 \cdot 10^6$	15.77	– 8147.7	34	0.09
$MgCl^+$	– 4.8	3415.0	0.0	18.2	0.0	—	—
$MgCl_2^0$	– 3.21	2407.0	0.0	9.6	0.0	—	—
$CaCl^+$	– 5.05	3112.0	0.0	15.5	0.0	—	—
$CaCl_2^0$	– 3.21	2407.0	0.0	9.6	0.0	—	—
$NaOH^0$	0.230	319.656	– 102489	33.092	– 13858	22	0.08
KOH^0	– 1.687	764.543	0.0	6.28	0.0	12	0.12

possessing a charge, z_i is expressed as (molality units):

$$\log \gamma_i = \frac{z_i^2 A\sqrt{I}}{1 + B\sqrt{I}}, \tag{9.49}$$

where I stands for the ionic strength of the solution expressed as:

$$I = \frac{1}{2} \sum_i m_i z_i^2. \tag{9.50}$$

Coefficients A and B are functions of temperature, density and dielectric constant of H_2O; coefficient B is directly proportional to the ion size parameter for solution $\hat{\mathring{a}}$ [Å]. All these values have been calculated according to Helgeson and Kirkham (1974) and Helgeson *et al.* (1981).

9.3.4 Results of calculations of concentration and temperature dependencies of diffusion coefficients

Calculated and experimental data at 25°C and 1 atm for KCl, NaCl, and and $CaCl_2$ solutions in the concentration range 0–0.1m are compared in Figure 9.1. As can be seen, the data are in good agreement. A slight deviation of the calculated diffusion coefficient from experimental measurements at m > 0.05 is due to the electrophoretic effect (Miller, 1966), not taken into account in the calculation.

We have calculated the diffusion coefficients of HCl, NaCl, KCl, $MgCl_2$, $CaCl_2$ and NaOH at temperatures from 100 to 600°C and a pressure of 100 MPa in the aqueous solution concentration range 0–0.1 m (the

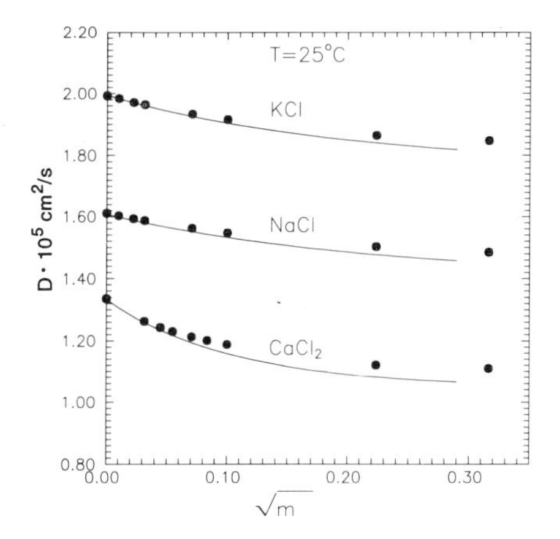

Figure 9.1. Concentration dependence of diffusion coefficients in aqueous solutions at 25°C. The abscissa indicates the square root of the solution molality.

dependence at 25 °C corresponds to P = 0.1 MPa). The results of the calculations for NaCl and $CaCl_2$ are shown in Figures 9.2 and 9.3.

For all the compounds under study, the concentration dependence of the diffusion coefficients in dilute solutions becomes more pronounced with increasing temperature (Figures 9.2 and 9.3). This is due to the increase in the degree of association with temperature, resulting in a transition to a region, in which diffusion transfer is dominated by ion pairs with lower D^0 values. Association for HCl, NaOH, $MgCl_2$ and $CaCl_2$ is most pronounced at 600 °C, whereas major variations in diffusion coefficients occur at high temperatures between 0 and 0.005 m. In the range 0.01–0.1 m the De's vary slightly at temperatures below 400 °C and remain practically constant at 500–600 °C.

Another set of calculations related to the temperature dependence of the infinite dilution diffusion coefficients and 'experimental' concentration of 0.05 m at 100, 300 and 500 MPa are given in Figure 9.4 (NaCl), Figure 9.5 (KCl), Figure 9.6 ($MgCl_2$), and Figure 9.7 ($CaCl_2$). An increase in pressure strongly hinders association, and at P = 500 MPa ions predominate in solution even at 600 °C. In the temperature range 100–600 °C, an increase in pressure produces a decrease in the diffusion coefficients of the salts considered, whereas between 25 and 100 °C the diffusion coefficients increase with pressure. Qualitatively, these correlations manifest themselves at higher concentrations, with D_e values becoming lower (Figures 9.4–9.7).

The experimental measurements of Balashov *et al.* (1983) and Zaraisky *et al.* (1986) appear to be in good agreement with predicted values, also shown in Figures 9.4–9.7. An appreciable increase in the D_{KCl} coefficient at 400 and 500 °C with respect to the 0.05 m predicted curve may be accounted for by weak thermal convection disregarded in the run (Figure

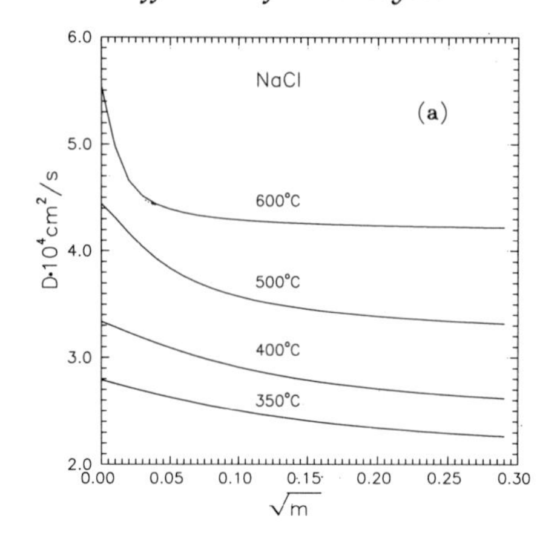

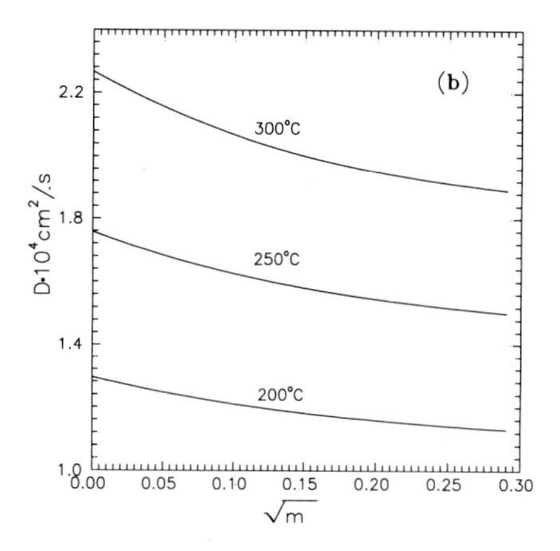

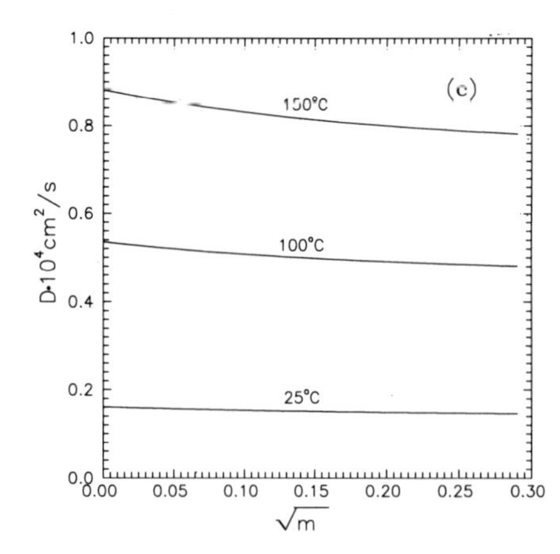

Figure 9.2. Concentration dependencies of the NaCl diffusion coefficient in hydrothermal solutions at P = 100 MPa.

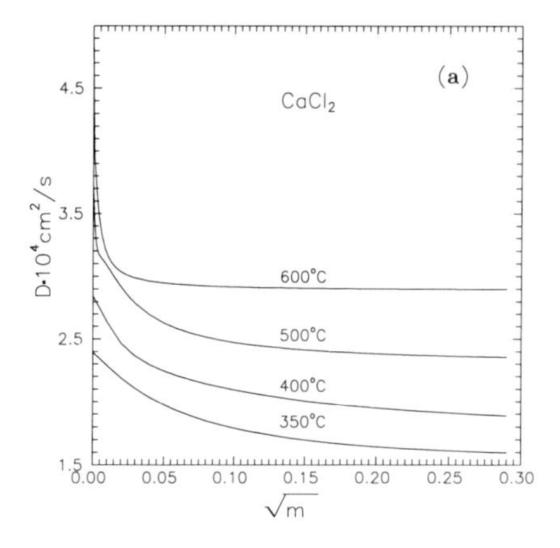

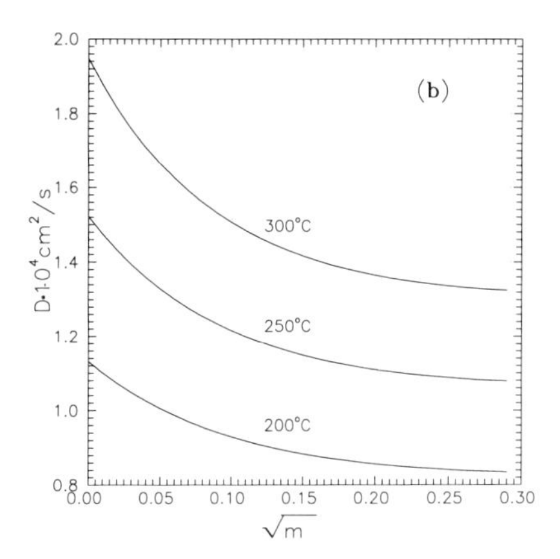

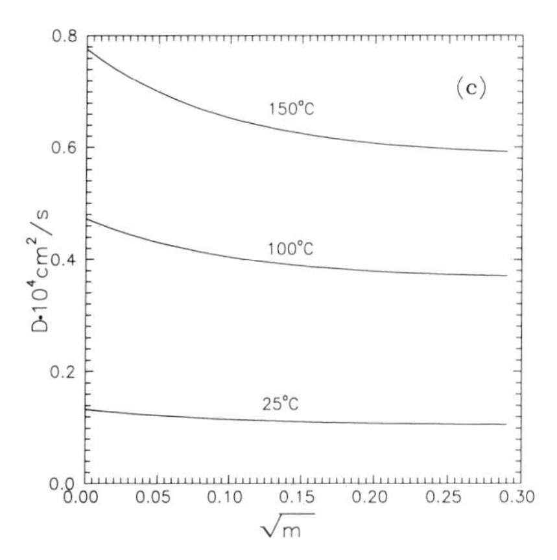

Figure 9.3. Concentration dependencies of the $CaCl_2$ diffusion coefficient in hydrothermal solutions at P = 100 MPa.

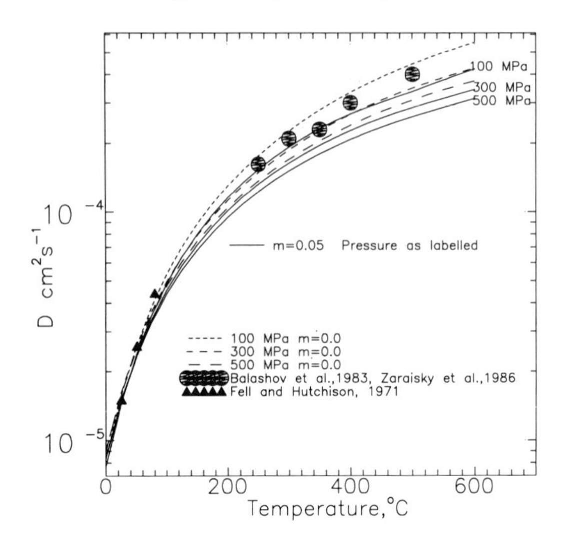

Figure 9.4. Temperature dependence of the NaCl diffusion coefficient at infinite dilution and 0.05 m concentration at different effective pressures: theoretical and experimental data.

9.5) (Zaraisky *et al.*, 1986). Finally, Figure 9.8 (m = 0.05) illustrates the temperature dependence of the diffusion of HCl, NaOH, KOH, CsCl and LiCl at P = 1 bar and 500 MPa. These data are of interest in that they seem to determine the lower (LiCl) and upper (HCl) boundaries separating hydrochloric acid, the alkalis and the alkali metal halides considered.

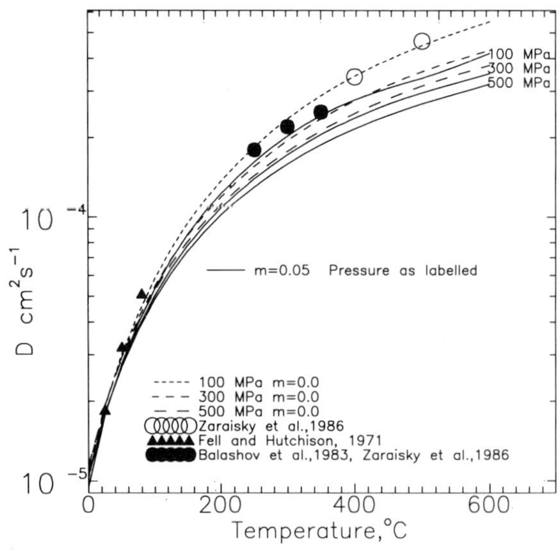

Figure 9.5. Temperature dependence of the KCl diffusion coefficient at infinite dilution and 0.05 m concentration at different pressures: theoretical and experimental data.

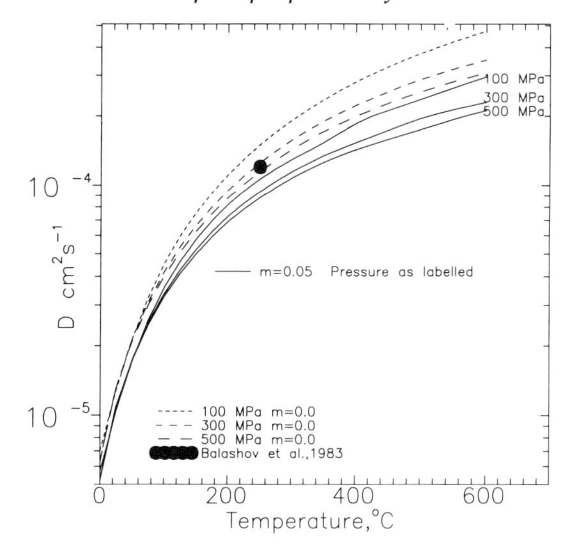

Figure 9.6. Temperature dependence of the $MgCl_2$ diffusion coefficient at infinite dilution and 0.05 m concentration at different pressures: theoretical and experimental data.

In order of increasing diffusion coefficients, alkali metal halides are arranged in the following standard sequence: LiCl, NaCl, KCl, RbCl, CsCl; it corresponds to the increase in the crystallographic cation radius. Overall, the rate of diffusion of NaOH and KOH is somewhat higher than that of CsCl. The diffusion of HCl is undoubtedly most rapid over the entire range of temperatures.

9.4 THEORETICAL CALCULATIONS OF TRANSPORT PROPERTIES OF ROCKS, BASED ON THE DATA ON FUNCTIONS OF MICROCRACK SIZE DISTRIBUTIONS

9.4.1 General theoretical model

The formation factor, F, defined by Equation 9.2, is of utmost importance for predicting values of effective diffusion coefficients in rocks. Its role in calculating diffusion through porous media is analagous to that of permeability in infiltration calculations. Our investigations of thermal decompaction of rocks (Zaraisky and Balashov, Chapter 10, this volume) revealed a significant dependence of F on temperature. At high P–T parameters, however, the F value is somewhat difficult to measure directly, therefore, in this case theoretical predictions are important.

This section of the paper considers one of the possible approaches to the solution of this problem; it was developed (Balashov and Zaraisky, 1982) on the basis of percolation theory (Shante and Kirkpatrik, 1971).

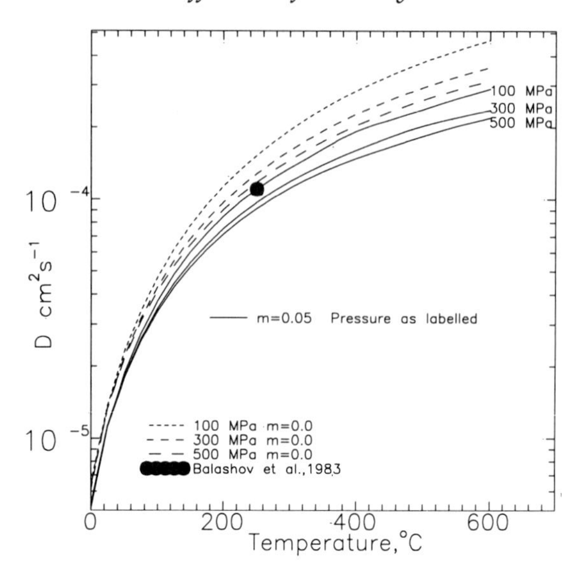

Figure 9.7. Temperature dependence of the $CaCl_2$ diffusion coefficient at infinite dilution and 0.05 m concentration at different pressures: theoretical and experimental data.

The intergranular planes *a priori* constitute a structure which penetrates a rock. When a critical probability of opening a single intergranular boundary is reached, such a system becomes conductive. The conductance of a single microcrack is a function of its width (diameter).

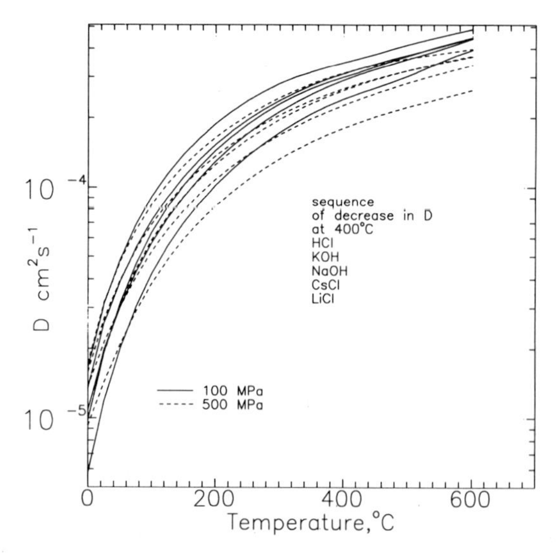

Figure 9.8. Temperature dependence of the HCl, KOH, NaOH, LiCl, and CsCl diffusion coefficients in solutions of m = 0.05 at different pressures: calculated data.

Percolation theory is mostly needed if the diameters (conductances) are characterized by a broad scatter of values. In these characteristics the distribution of pore space in rocks is similar to an exponentially non-uniform distribution and, therefore, corresponds to the conditions under which percolation theory is applicable. For values of diffusion and filtration resistance of a given set of microcracks (designated as $\tilde{p}_1$) with diameters in the range $\delta_1 < \delta < \delta_{max}$, the following equations are valid:

$$R_d^* = \int_{\mathcal{F}(\delta_1)}^{\mathcal{F}(\delta_{max})} \frac{d\mathcal{F}}{(1 - \mathcal{F}(\delta_1))\,\delta} = \frac{1}{p_1} \int_{1-p_l}^{1} \frac{d\mathcal{F}}{\delta}, \tag{9.51}$$

and

$$R_f^* = \int_{\mathcal{F}(\delta_1)}^{\mathcal{F}(\delta_{max})} \frac{d\mathcal{F}}{(1 - \mathcal{F}(\delta_1))\,\delta^3} = \frac{1}{p_1} \int_{1-p_l}^{1} \frac{d\mathcal{F}}{\delta^3}, \tag{9.52}$$

where $\mathcal{F}(\delta)$ is the function of microcrack size distribution, $p_l = 1 - \mathcal{F}(\delta_l)$. In general there is an infinite array of conductive sets $\tilde{p}_l$; the spatial density of the conductive cluster p_l is $p_1 P(p_1)$, where $P(p_1)$ is a percolation function of the intergranular planes structure.

In our approach to the calculation of transport properties, integration of the conductance function $1/R^*(p_1)$ over the cluster density variable $p_l\,P(p_l)$ should result in the estimation (up to a constant coefficient) of the integral conductance of the system:

$$L = \int_{p_c}^{1} \frac{dp_1\,P(p_1)}{R^*(P_1)}, \tag{9.53}$$

where by R^* is meant either R_d^* or R_f^*. The use of integral (9.53) satisfies the principle of maximum conductance and the other earlier assumptions made.

The proposed theoretical model has been experimentally tested for residual permeability of the New Ukrainian granite samples, which have experienced thermal decompaction (Zaraisky and Balashov, Chapter 10, this volume). The attempted theoretical investigation indicates that estimation of F and K^0 is possible if the structural characteristics of the system of microcracks, and their relationships to crack size distribution are assumed. In practice, the function $\mathcal{F}(\delta)$ is, unfortunately, not readily accessible; therefore, general assumptions for the nature of microcracks' size distributions are of particular importance.

9.4.2 Internally consistent calculation of structure-transport characteristics for a simplified media

The following calculation applies to media in which the microcrack widths have a log-uniform distribution. The maximum width to minimum width ratio for microcracks,

$$r = \frac{\delta_{max}}{\delta_{min}} \gg 1, \tag{9.54}$$

provides a major parameter for calculating the transport properties of such a rock. If the distribution is uniform, then the value of δ will be:

$$\ln(\delta) = \ln(\delta_{min}) + a\,\mathcal{F}, \tag{9.55}$$

where

$$a = \ln(r). \tag{9.56}$$

Expressing the percolation function $P(p)$ in terms of the polynomial in $(p{-}p_c)$ and integrating Equation 9.53 gives:

$$\begin{aligned} L_n &= L_n^{(1)} + L_n^{(2)}, \\ L_n^{(1)} &= na\delta_c^n\left(-C_1 + e^{-(p_f - p_c)na}\sum_i C_i\,(p_f - p_c)^{i-1}\right), \\ L_n^{(2)} &= \delta_c^n\left(\frac{1}{na} + p_f\right)e^{-(p_f - p_c)na} - \delta_{min}^n\left(\frac{1}{na} + 1\right), \end{aligned} \tag{9.57}$$

where coefficients C_i are functions of both the polynomial $(P(p - p_c))$ coefficients and P_c and na, n = 1 corresponds to the formation factor, and n = 3 to permeability. As seen from Equation 9.57, a critical diameter δ_c, ratio parameter a, and curve percolation parameters p_c and p_f, are of essential significance for calculation. The percolation term p_c is the critical percolation probability, and p_f is given by $P(p_f - p_c) = 1$. The formation factor and permeability of a uniform media of dense packed grains with diameters, d_{gr}, and tortuosity, k_{tr}, may be expressed as:

$$F = \frac{2}{d_{gr}k_{tr}}\,L_1, \tag{9.58}$$

$$K^0 = \frac{1000}{6d_{gr}k_{tr}}\,L_3, \tag{9.59}$$

where K^0 is in mD, L_1 and d_{gr} in μm, and L_3 in μm^3.

For the granite samples which have experienced hydrothermal decompaction, three types of experimental data have been utilized: the residual relative elongation, permeability and formation factor as a function of temperature (Balashov and Zaraisky, Chapter 10, this volume). The above simplified model for transport properties, based on the log-uniform crack-size distribution, enabled the temperature dependence of ln(r) for granite to be estimated. By assuming this function to be independent of effective pressure (the difference between the lithostatic and pore fluid pressures) and using available experimental data, it is possible to calculate the temperature dependencies of the formation factor and of the mean physical microcrack diameter, δ_F, for granite under hydrothermal conditions and different effective pressures. The value of δ_F is determined by:

$$\delta_F^2 = \frac{12\,K^0}{F}. \tag{9.60}$$

The results of the calculations are given in Figures 9.9–9.11. Initial data were true excess extensions at zero effective pressure (Balashov and Zaraisky, 1982) and permeability values *in situ* at 20 and 50 MPa (Zonov *et al.*, 1989).

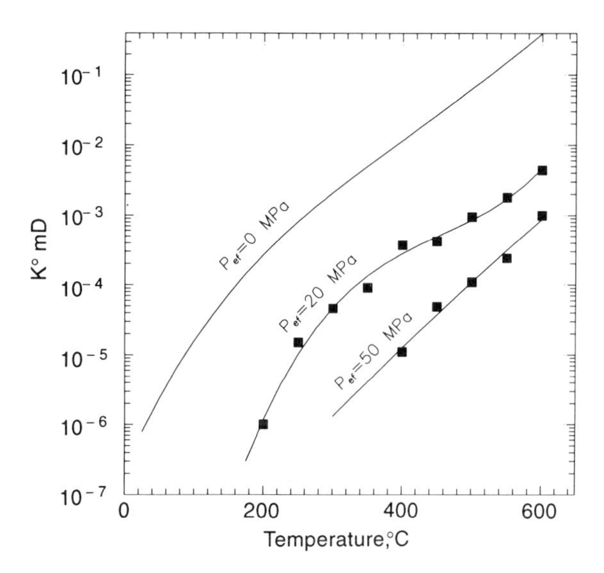

Figure 9.9. Temperature dependencies of the New Ukrainian granite permeability at different effective pressures.

At $P_{eff} > 0.0$ – experimental data of Zonov *et al.* (1989); at $P_{eff} = 0.0$ – model calculation based on the data for the granite samples which experienced hydrothermal decompation.

9.5 DIFFUSION OF ELECTROLYTES IN PORE SOLUTIONS: CORRECTION FOR THE ELECTRIC DOUBLE LAYER (EDL) EFFECT

The contact of rock minerals and aqueous solution is characterized by a jump in electric potential, Ψ_0, with the formation of the EDL. If

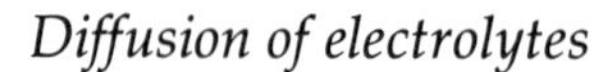

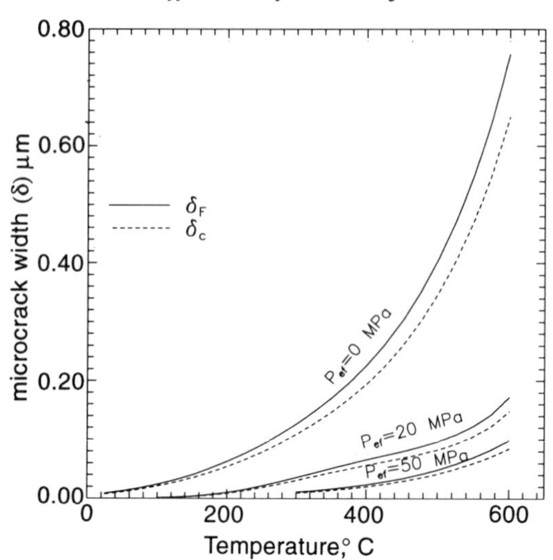

Figure 9.10. Model temperature dependences of characteristic microcracks diameters at different effective pressures. δ_c – critical percolation diameter.

the minerals are in contact with an electrolyte solution containing two kinds of ions with charges equal in magnitude but opposite in sign, + z and – z, at $zF\Psi_0/RT \ll 1$ or $\Psi_o \ll 25$ mV, then the surface charge value will be $\sigma \approx \varepsilon_0 \varepsilon \chi \Psi_0$, C.m^{-2} (where ε is the dielectric constant of solution and ε_0 is the dielectric constant of a vacuum). The electric potential (Ψ) drop in solution as a function of normal distance (x) to the surface will thus be given by $\Psi = \Psi_0 \exp(-\chi x)$. The EDL effective thickness χ^{-1} is the distance into the solution at which the potential value is $e \cong 2.718$

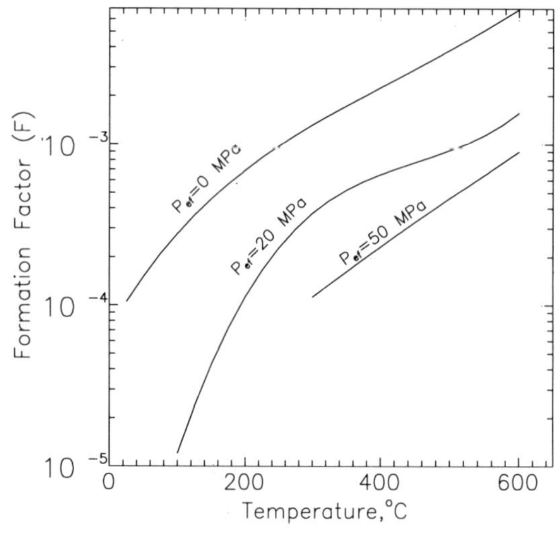

Figure 9.11. Model temperature dependencies of the formation factor at different effective pressures. Relative excess thermal elongation.

times less than that on the surface. At a given temperature, it is governed by the solution properties:

$$\frac{1}{\chi}=\left(\frac{\varepsilon_0 R}{2000z^2F^2}\right)^{\frac{1}{2}}\cdot\left(\frac{\varepsilon T}{\alpha C}\right)^{\frac{1}{2}}=1988.10^{-12}\left(\frac{\varepsilon T}{z^2\alpha C}\right)^{\frac{1}{2}},\ \mathrm{m}, \tag{9.61}$$

where α represents the degree of dissociation of a 1:1 electrolyte and C stands for the concentration in $\mathrm{mol}\cdot 1^{-1}$.

At a temperature of 25°C, a completely dissociated solution of a 1:1 aqueous electrolyte at concentrations of 10^{-3}, 10^{-2}, and 10^{-1} M yields the following values for χ^{-1}: 96, 30, 9.6 Å. At T = 500°C and P = 125 MPa, aqueous NaCl solutions of the same concentrations yield the following values for χ^{-1}: 59, 22, and 10.2 Å calculated with parameters from Helgeson and Kirkham (1974). Thus, if the pores are small (< 0.01 μm in width), partitioning of electric charges between pore walls and electrolyte solution can produce a significant effect on transfer processes. The problem simplifies considerably if transfer in the pore solution is considered in terms of electric parameters, charge and potential, averaged over the pore volume (Alekhin *et al.*, 1982b; Lakshtanov and Kopylov, 1985). It should be emphasized that such an approach is doubtless valid as electric layers overlap one another in fairly small pores; furthermore, it should somewhat enhance effects in large pores. Nevertheless, such assessment of the effect of the surface–solution partitioning of electric charges on the transfer process seems in large part quite acceptable.

9.5.1 Diffusion flux of the dissociated electrolyte in pores

Consider the diffusion of the completely dissociated electrolyte $A_{\nu_a}B_{\nu_b}$ in pores ($z_a > 0$). From Equation 9.3, we obtain:

$$\frac{z_a}{\nu_b}=-\frac{z_b}{\nu_a}=\bar{z}. \tag{9.62}$$

Let σ be the surface charge density in units of $\mathrm{C.\,cm^{-2}}$ and z_χ the charge on surface complexes (centres). The surface charge 'molality' averaged over the pore volume may then be determined from the following expression describing a flat slot-like channel ($m_\chi > 0$):

$$m_\chi=\frac{2\,|\sigma|\,S^P}{\rho V^P F}=\frac{2\,|\sigma|}{\rho\delta^P F}, \tag{9.63}$$

where $2S^P$ is the inner surface, V^P the volume of such a slot-like pore, δ^P its width. In our further calculations, we will take:

$$\delta^P=\delta_F. \tag{9.64}$$

The condition of electric neutrality relates the molalities of ions to that of the surface charge as follows:

$$z_a m_a + z_b m_b + z_\chi m_\chi = 0. \tag{9.65}$$

When an ion diffuses in pore solution, it is essential that the effect of diffusion (membrane) potential, $\phi = \phi^M F$, be taken into account, by analogy with diffusion in free solution. A radically new feature is the electro-osmotic 'drift' (convection transfer) of the charged pore liquid, caused by the gradient in membrane potential. The value directly responsible for this migration is a ζ (zeta) potential (Shchukin *et al.*, 1982).

For a diffusion flux in the pore solution, the following equation is valid:

$$\frac{J_\alpha}{\rho} = -D_\alpha^0 (m_\alpha \nabla \ln(a_\alpha) + m_\alpha z_\alpha \frac{1}{RT} \nabla \phi) + m_\alpha O \nabla \phi, \tag{9.66}$$

where $\alpha = a,b$. The value of O responsible for the electro-osmotic transfer may be expressed in terms of the ζ potential as follows:

$$O = 10^4 \frac{\varepsilon\varepsilon_0\zeta}{\eta F}, \text{ cm}^2\text{.equiv. s}^{-1}\text{. J}^{-1}, \tag{9.67}$$

where ζ stands for the electrokinetic zeta potential, ε and η are the dielectric constant and viscosity of the pore solution, respectively.

As in the case of homogeneous continuum, the condition of zero electric current,

$$z_a J_a + z_b J_b = 0, \tag{9.68}$$

holds true for diffusion in a porous medium.

From Equations 9.62 and 9.68, it follows that the electrolyte flow in pores may be determined by:

$$J_{A_{\nu_a} B_{\nu_b}} = J_e = J_a \cdot \nu_a^{-1} = J_b \cdot \nu_b^{-1}. \tag{9.69}$$

We will now rewrite J_α from Equation 9.66 as:

$$J_\alpha \cdot \rho^{-1} = -D_\alpha \nabla \ln(a_\alpha) - \overline{D}_\alpha z_\alpha (RT)^{-1} \nabla \phi, \tag{9.70}$$

where

$$D_\alpha = D_\alpha^0 m_\alpha, \tag{9.71}$$

$$\overline{D}_\alpha = D_\alpha (1 + \Delta_\alpha), \tag{9.72}$$

$$\Delta_\alpha = -ORT.\, D_\alpha^{0^{-1}}.\, z_\alpha^{-1}. \tag{9.73}$$

Substituting Equation 9.70 in Equation 9.68 and using the resulting equation gives the expression:

$$(RT)^{-1} \nabla \phi = -\frac{\sum D_\alpha z_\alpha \nabla \ln(a_\alpha)}{\sum \overline{D}_\alpha z_\alpha^2} \tag{9.74}$$

for the electric potential gradient. By substituting Equation 9.74 in Equation 9.70 and using Equation 9.69 as the basis, we can determine the electrolyte flow as:

$$\frac{J_e}{\rho} = \frac{J_\alpha}{\rho \cdot \nu_\alpha^{-1}} = -\frac{D_\alpha}{\nu_\alpha \nabla \ln a_\alpha} + \frac{\overline{D}_\alpha z_\alpha}{\nu_\alpha} \frac{\sum D_\alpha z_\alpha \nabla \ln a_\alpha}{\sum \overline{D}_\alpha z_\alpha^2}. \tag{9.75}$$

The mean electrolyte activity in pores is determined routinely from ionic activities in pore solution (Equation 9.7). Rearranging Equation 9.75 to include the mean electrolyte activity term, a little algebraic manipulation yields the following intermediate form:

$$J_e \overline{\overline{Z}}_1 \cdot \rho^{-1} = -\overline{z}^2 (D_a \overline{D}_b \nabla \ln(a_a^{\nu_a}) + \overline{D}_a D_b \nabla \ln(a_b^{\nu_b})), \tag{9.76}$$

where $\overline{z}$ is given by Equation 9.62, and

$$\overline{\overline{Z}}_1 = \sum \overline{D}_\alpha z_\alpha^2 = \overline{D}_a z_a^2 + \overline{D}_b z_b^2. \tag{9.77}$$

By making use of Equation 9.72, we can express Equation 9.76 in the form of interest:

$$J_e \cdot \rho^{-1} = -\frac{D_a D_b z^{-2} \nu}{\overline{\overline{Z}}_1} (\nabla \ln(a) + \nu^{-1} \nabla \ln(a_a^{\nu_a \Delta_b} a_b^{\nu_b \Delta_a})), \tag{9.78}$$

or in alternative form:

$$J_e \cdot \rho^{-1} = -\frac{D_a D_b z^{-2} \nu}{\overline{\overline{Z}}_1} \left(1 + \nu^{-1} \frac{d(\ln(a_a^{\nu_a \Delta_b} a_b^{\nu_b \Delta_a}))}{d(\ln(a))}\right) \nabla \ln a, \tag{9.79}$$

where, as before, $\nu = \nu_a + \nu_b$, and a is the mean electrolyte activity. Equation 9.79 expresses the diffusion flux of the electrolyte in pore solution in terms of the mean electrolyte activity gradient.

9.5.2 Comparison of diffusion in pore and free solutions

Our main interest is to compare diffusion occurring in pore solutions with that in free solutions. We may therefore make comparisons using values for the mean electrolyte activity and mean activity gradient. This may be done by setting the mean electrolyte activity in pore solution equal to that in free solution,

$$a = a_f, \tag{9.80}$$

where the subscript f symbolizes the free solution. Thus, the free solution in equilibrium with the pore solutions is chosen for comparison; in both cases, the standard state is taken to be that of an infinitely dilute free solution. Note that Equation 9.80 is equivalent to the Donnan equilibrium.

The difference in ion–ion interaction between the pore and free solutions is taken into account by including the ionic strength of the pore solution,

$$I = \frac{1}{2}(m_a z_a^2 + m_b z_b^2 + m_\chi z_\chi^2). \tag{9.81}$$

In reality, the charge and potential of pores are governed by equilibrium between ions and charged complexes contained in pore solution and adsorption complexes and centres located on the inner pore surface (Ivanova, 1989; Shchukin *et al.*, 1982). The calculation of such equilibrium requires information on the appropriate thermodynamic constants and knowledge of possible patterns that adsorption processes could follow. Only a limited number of such data is at present available, and these require more profound and thorough analysis than that possible within the scope of the present study. Therefore, for our estimates, we have simplified the problem by singling out two extreme patterns of the system behaviour: (a) where the charge is constant, and (b) where the mean potential of pores is constant. It should be immediately noted that case (a) corresponds to the real behaviour of the whole group of ion exchange materials displaying high electrosurface activities, i.e. ionites. Analysis of experimental data on rocks and related materials, a group of substances which exhibit relatively low electrosurface activities, (Alekhin *et al.*, 1982b; Ivanova, 1989) indicates that case (b) is most likely applicable for dilute solutions ($< 10^{-2}$m), when the surface charge is essentially a function of the electrolyte concentration, whereas case (a) is more realistic for more concentrated solutions ($> 10^{-2}$m). The boundaries enclosed in parentheses are, of course, arbitrary.

Let a_f be the mean electrolyte activity in the free solution of molality m_f. Then the equilibrium relation 9.80 may be written as:

$$\nu \ln a_f = \nu_a \ln\gamma_a + \nu_b \ln\gamma_b + \nu_a \ln(m_a) + \nu_b \ln(m_b), \tag{9.82}$$

where $\gamma_{a,b}$ and $m_{a,b}$ are the activity coefficients and molalities of ions in pore solution, respectively. The condition of constant charge density of pore walls, $|\sigma| = \text{const}$, is equivalent to $m_\chi = \text{const}$ at characteristic δ_F, as follows from Equation 9.63. In this case, equilibrium equation 9.82, electroneutrality equation 9.65, the equation for determining the ionic strength of the pore solution 9.81 and that for calculating activity coefficients 9.49 make up a consistent system, which uniquely defines m_a and m_b in the pore solution provided m_f and a_f are known.

If the condition of constant mean potential of a pore, $\tilde{\Psi} = \tilde{\Psi}_v F$, is fulfilled, m_χ will no longer be a constant value. The above set of equations is complemented by two more equations relating $m_{a,b}$ and $\tilde{\psi}$:

$$m_\alpha = \nu_\alpha \tilde{m}^0 \exp(-z_\alpha \tilde{\Psi}(RT)^{-1}). \tag{9.83}$$

Two new unknowns are m_χ and $\tilde{m}^O$. Note that if Equation 9.83 is used, it is a simple matter to reduce Equations 9.82 and 9.65 to the forms, which incorporate m_χ and $\tilde{m}^O$ only, respectively:

$$\nu_a \ln \gamma_a + \nu_b \ln \gamma_b + \nu \ln(\tilde{m}^0) = \nu \ln(a_f) - \nu_a \ln \nu_a - \nu_b \ln \nu_b, \tag{9.84}$$

$$\tilde{m}^0 \left(\sum \nu_\alpha z_\alpha \exp(-z_\alpha \tilde{\Psi}(RT)^{-1})\right) + z_\chi m_\chi = 0. \tag{9.85}$$

Thus, in this case too ($\tilde{\Psi} = \text{const}$), ion concentrations in the pore solution $m_{a,b}$ are unequivocally governed by the molality (m_f) of the electrolyte in the free solution in equilibrium with the pore solution.

To make our comparison somewhat more complete, assume that solution contains a neutral electrolyte complex with an activity coefficient of 1. From the pore solution–free solution equilibrium, it follows that:

$$a_{0f} = a_0,$$

or alternatively,

$$m_{0f} = m_0 \tag{9.86}$$

where m_{0f} and m_0 are the molalities of the neutral complex in the free and pore solutions, respectively.

For a diffusive electrolyte flux in the pore solution, the following expression is valid, when the neutral ion pair is included:

$$J_e \cdot \rho^{-1} = -D_e^{00} m_0 \nabla \ln a_0 + m_0 O \nabla \phi - \tilde{D} \nabla \ln a, \tag{9.87}$$

where $\tilde{D}$ is given by Equation 9.79, and $m_0 O \nabla \phi$ corresponds to electro-osmotic transfer of the neutral complex. By applying relation 9.20 to pore solution, Equation 9.87 changes to:

$$J_e \cdot \rho^{-1} = -\left(D_e^{00} m_0 \nu - \frac{m_0 O d\phi}{d \ln a} + \tilde{D}\right) \nabla \ln a. \tag{9.88}$$

For this case above (ions and neutral complex), $\tilde{D}_e^0$, Equation 9.38, is given by:

$$\tilde{D}_e^0 = D_e^{00} (1 - \alpha) \nu + D_e^0 \alpha, \tag{9.89}$$

where D_e^{00} is the self-diffusion coefficient of the neutral ion pair, D_e^0 the Nernst diffusion coefficient of the electrolyte, and α the degree of dissociation. Then, for a diffusion flux of the electrolyte of molality m_f in free solution, Equation 9.38, the following expression holds true:

$$J_e^f \cdot \rho^{-1} = -(D_e^{00}(1-\alpha)m_f\nu + D_e^0\alpha_{m_f})\nabla\ln a_f. \tag{9.90}$$

We define the coefficient æ as a ratio of the diffusive flux in pore solution to that in free solution. Assuming that Equation 9.80 holds and that the mean activity gradients are equal, i.e.:

$$\nabla\ln a = \nabla\ln a_f, \tag{9.91}$$

gives

$$ae = \frac{J_e}{J_e^f} = \frac{D_e^{00}\, m_0\nu - \dfrac{m_0 O d\phi}{d\ln a} + \tilde{D}}{D_e^{00}\,(1-\alpha)m_f\nu + D_e^0\,\alpha_{m_f}}. \tag{9.92}$$

Thus, for any value for the mean electrolyte activity, æ is strongly dependent on both the derivatives entering into $\tilde{D}$ (Equation 9.79):

$$\frac{d(\ln(a_a^{\nu_a\Delta_b}\, a_b^{\nu_b\Delta_a}))}{d(\ln(a))} = \nu_a\Delta_b \frac{d\ln a_a}{d\ln a} + \nu_b\Delta_a \frac{d\ln a_b}{d\ln a}, \tag{9.93}$$

and the value for the membrane potential derivative, $d\phi \cdot d\ln a^{-1}$, with respect to the logarithm of the mean electrolyte activity.

9.6 DIFFUSION OF NaCl IN PORE SOLUTIONS OF GRANITE TO 600°C

This section applies the theoretical model developed here to the diffusion of NaCl dilute solutions into and through the New Ukrainian biotite granite at $P_{fl} = 100$ MPa. First of all, we need a theoretical temperature model for a porous space in granite. i.e. values of F(T) and $\delta_F(T)$ at P_{eff} of 0, 20, and 50 MPa (Figures 9.9–9.11). By conventionally defining functions $m_\chi(T)$ or $\tilde{\Psi}(T)$, we then calculate parameters for diffusion in pores under these conditions. Following the above model, we then take into account ion–ion interaction and association. The calculation assumes that absolute mobility and viscosity are unchanged, as one passes from free solution to pore solution.

The density and viscosity of water have been calculated over a wide temperature range in accordance with Helgeson and Kirkham (1974) and Dudziak and Frank (1966).

9.6.1 Estimation of purely temperature characteristics of the EDL model: $m_\chi(T)$, $\tilde{\psi}(T)$, $\zeta(T)$ and $\Delta_\alpha(T)$

For the case of diffusion at constant charge (m_χ = const), the temperature dependence of m_χ was calculated from δ_F using Equation 9.63; in each case, $|\sigma| = 2.10^{-6}$ C. cm^{-2} (Alekhin *et al.*, 1982a; Ivanova, 1989).

For the case of diffusion at constant mean potential of pores ($\tilde{\Psi}$ = const), the value of $\tilde{\Psi}_v$ was calculated from $|\sigma| = 2.10^{-6}$ C. cm^{-2} and $\delta_F(T)$ using Equations 9.65 and 9.83 at 25°C and $\tilde{m}^O = 0.01$m.

The calculations utilized $\zeta(T)$ generated by linearly extrapolating experimental data (Alekhin *et al.*, 1984; Ivanova, 1989). At each temperature, the values of Δ_{Na^+,Cl^-} were calculated from the ζ potential using Equations 9.67 and 9.73.

9.6.2 Discussion and conclusion: A calculated example

Comparisons of the temperature dependencies of the NaCl diffusion coefficient in free solutions at m = 10^{-3}, 10^{-2}, and 5.10^{-2} with diffusion coefficient values (= æD) in pore solution at different effective pressures are illustrated in Figures 9.12–9.14. Figure 9.12 depicts retardation of diffusion in pores, which clearly increases to become particularly evident at m_χ = const. In this case, an increase in the effective pressure due to decrease of characteristic pore diameters causes this effect to grow in importance.

An increase in the electrolyte concentration to 0.01m causes charge and potential effects to draw closer together, as seen from Figure 9.13. At this concentration, the regime of constant potential produces a substantial increase in retardation of diffusion, while the regime of constant charge is characterized by a slight increase in diffusion (at P_{eff} = 0, T above 100 °C; at P_{eff} = 20 MPa, T above 300 °C). At a NaCl concentration of

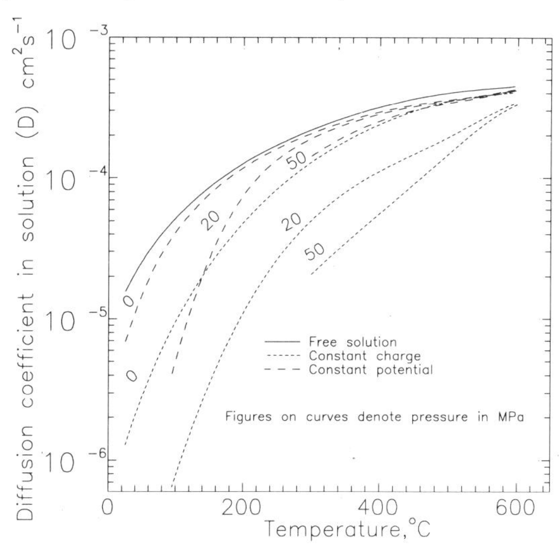

Figure 9.12. Temperature dependencies of the NaCl diffusion coefficient in free and pore solutions with reference to the thermal granite model for diffusion regimes of constant surface charge and constant effective surface potential at different effective pressures (P_{fl} = 100 MPa) and a concentration of free solution in equilibrium with a pore one of 0.001 m.

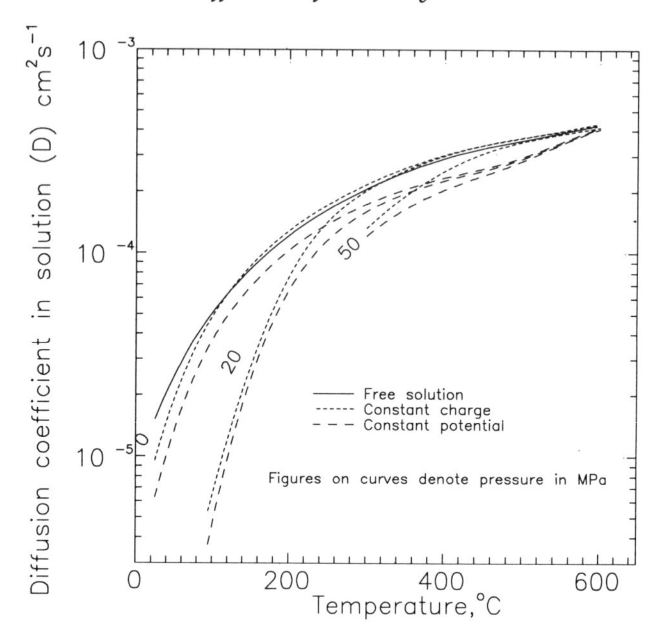

Figure 9.13. Temperature dependencies of the NaCl diffusion coefficient in free and pore solutions with reference to the thermal granite model for diffusion regimes of constant surface charge and constant effective surface potential at different effective pressures (Pfl = 100 MPa) and a concentration of free solution in equilibrium with a pore one of 0.01 m.

0.05m and temperature above 150 °C, the effect of pore diffusion is conserved to become somewhat larger under the regime of constant potential (Figure 9.14). At constant charge a very minor increase in diffusion occurs in this temperature region at all effective pressures.

The next problem to address is the combined influence of structural and EDL effects on diffusion using a model granite at m_{NaCl} of 10^{-3}, 10^{-2}, and 5.10^{-2} (Figures 9.15–9.17). Solid lines represent functions corresponding to the product of F(T) and D(T,m), i.e. they reflect the effect of the formation factor on diffusion only. All the figures clearly display the leading role of the formation factor, whose effect becomes particularly pronounced with increasing solution concentration. The curves are seen to fall into three groups according to effective pressure. An increase in the effective pressure from 0 to 50 MPa causes a 10–times retardation of diffusion through granite. By and large, if the charge is constant, diffusion may be retarded by more than one order of magnitude as a result of the EDL effect, with the retardation being particularly evident at an effective pressure of 50 MPa. For the case of constant potential, the maximum retardation of diffusion by about a factor of 3 is observable in the most concentrated solution studied (0.1m). An increase in the effective pressure causes the observed effects to shift systematically to higher temperatures (Figures 9.15–9.17).

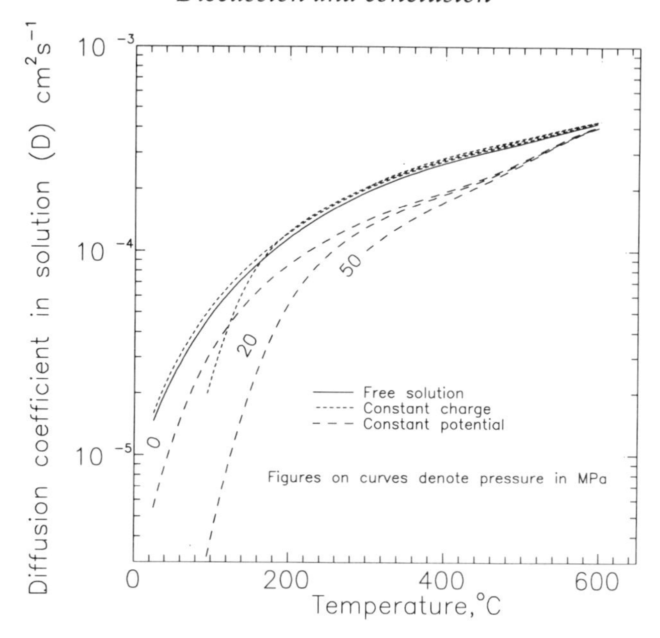

Figure 9.14. Temperature dependencies of the NaCl diffusion coefficient in free and pore solutions with reference to the thermal granite model for diffusion regimes of constant surface charge and constant effective surface potential at different effective pressures (P_{fl} = 100 MPa) and a concentration of free solution in equilibrium with a pore one of 0.05 m.

If we assume that the regime of constant potential is a characteristic of dilute solution, then a 2–3-times retardation of diffusion in dilute solutions (~ 10^{-3}m) at effective pressures to 20 MPa and temperatures to 300 °C is to be expected. This retardation will be enhanced by increased effective pressure or solution concentration, but reduced by increasing temperature.

The investigation indicates that theoretical prediction of diffusion in fluid-saturated porous media of rocks is possible over a wide range of hydrothermal conditions, while the formation factor of the porous medium plays a leading role in controlling diffusion through consolidated rocks. Electrosurface properties of the porous medium have significant effects on diffusion in pores containing dilute electrolyte solutions, and will be especially important for low porosity rocks with small pores.

ACKNOWLEDGEMENTS

Creative discussions with Georgiy Zaraisky, Yurii Alekhin and Leonid Lakshtanov are gratefully acknowledged. Theoretical investigations of the effect of the electrical charge of the pore surface on diffusion transfer

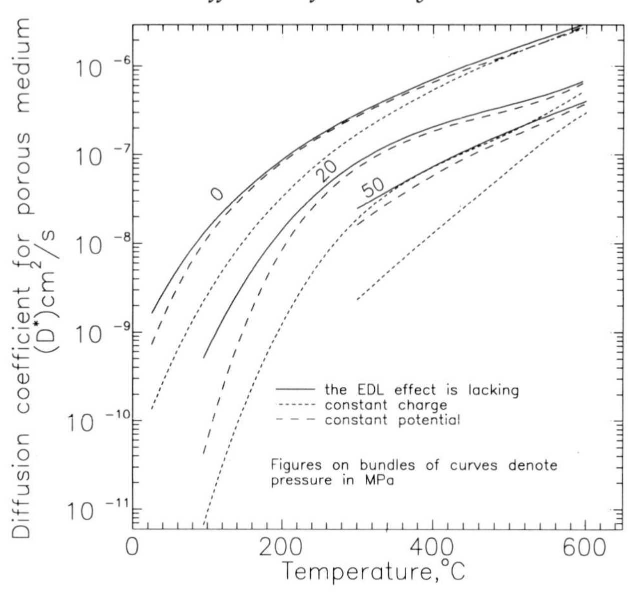

Figure 9.15. Temperature dependencies of the NaCl diffusion coefficient through a porous medium with reference to the thermal granite model at different effective pressures ($P_{f1} = 100$ MPa) if the EDL effect is lacking, as well as for diffusion regimes of constant surface charge and constant effective surface potential at a concentration of free solution in equilibrium with a pore one of 0.001 m.

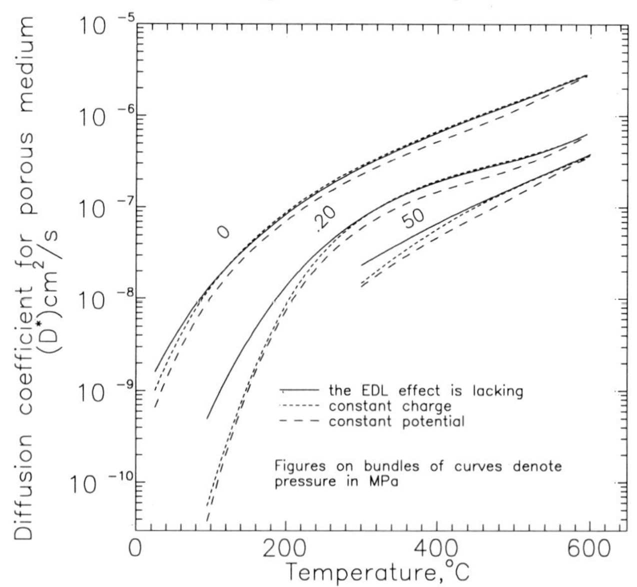

Figure 9.16. Temperature dependencies of the NaCl effective diffusion coefficient through a porous medium with reference to the thermal granite model at different effective pressures ($P_{f1} = 100$MPa) if the EDL effect is lacking, as well as for diffusion regimes of constant surface charge and constant effective surface potential at a concentration of free solution in equilibrium with a pore one of 0.01 m.

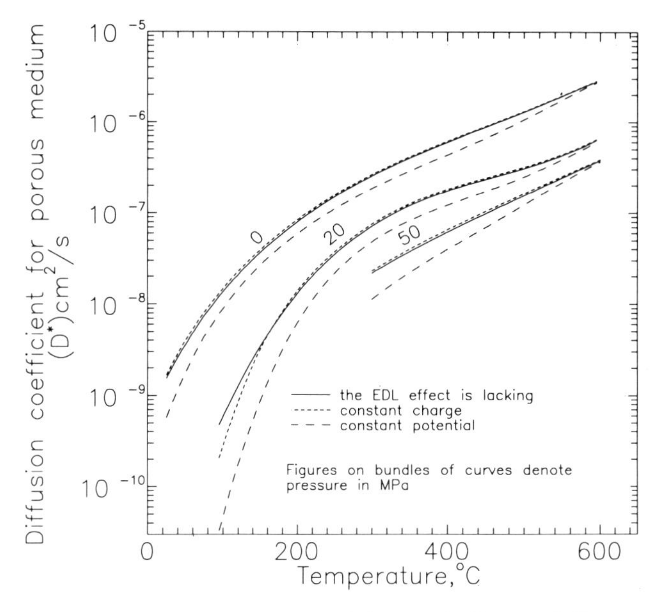

Figure 9.17. Temperature dependencies of the NaCl effective diffusion coefficient through a porous medium with reference to the thermal granite model at different effective pressures (P_{fl} = 100 kbar) if the EDL effect is lacking, as well as for diffusion regimes of constant surface charge and constant effective surface porential at a concentration of free solution in equilibrium with a pore one of 0.05 m.

were initiated by works of Yu. Alekhin, L. Lakshtanov, A. Vakulenko and P. Kopylov. I take the opportunity to thank Vladimir Traskin who was the first to call my attention to possible application of percolation theory to mass transfer in rock porous media.

The research described in this chapter is supported by the Russian Fund of Fundamental Investigation under Grant No. 93–05–8848.

REFERENCES

Alekhin, Yu.V., Vakulenko, A.G. and Lakshtanov, L.Z. (1982a). Methods for studying transport phenomena during isothermal filtration into porous media. In *Contr. to Physico-chemical Petrology*, **X**, pp. 45–68. Nauka Press, Moscow (in Russian).

Alekhin, Yu.V., Vakulenko, A.G. and Lakshtanov, L.Z. (1982b). Filtration effect and its bearing on the convection and diffusion mass transfer in porous media. In *Dynamic Models in Physical Geochemistry*, pp. 144–62. Nauka Press, Novosibirsk (in Russian).

Alekhin, Yu.V., Zharikov, V.A. Ivanova, L.I., and Lakshtanov, L.Z. (1984). Electrosurface properties of natural and synthetic porous media under hydrothermal conditions. *Dokl. Acad. Nauk SSSR*, **274**, N6, pp. 1454–7 (in Russian).

Anderson, D.E. (1981). Diffusion in electrolyte mixtures. In *Reviews in Mineralogy* (eds Lasaga A.C. & Kirkpatrick R.J.), vol. **8** (Kinetics of Geochemical Processes), chapter 6, pp. 211–60.

Applin K.R. and Lasaga A.C. (1984). The determination of SO_4^{2-}, $NaSO_4^-$, and $MgSO_4^0$ tracer diffusion coefficients and their application to diagenetic flux calculations. *Geochim. Cosmochim. Acta*, **48**, N10, pp. 2151–62.

Applin, K.R. (1987). The diffusion of dissolved silica in dilute aqueous solution. *Geochim. Cosmochim. Acta*, **51**, pp. 2147–51.

Balashov, V.N. and Zaraisky, G.P. (1982). Experimental and theoretical investigation of thermal decompaction of rocks on heating. In *Contr. to Physicochemical Petrology*, **X**, pp. 69–109. Nauka Press, Moscow (in Russian).

Balashov, V.N., Zaraisky, G.P., Tikhomirova, V.N. and Postnova, L.E. (1983). Diffusion of rock-forming components in pore solutions at 200 °C and 1 kbar. *Geokhimia*, **1**, pp. 30–42 (in Russian).

Balashov, V.N. (1992). Diffusion on mass transfer in hydrothermal systems. Ph.D. thesis, Moscow (in Russian).

Dudziak, K.H. and Franck, E.U. (1966). Messungen der Viscositat des Wassers bis 560 °C und 3500 bar. *Ber. Bunsenges. Physik. Chem.*, **70**, N9/10, pp. 1120–8.

Erdey-Gruz (1974). *Transport Phenomena in Aqueous Solutions*, 592 pp. Akademiai Kiado, Budapest.

Fell, C.J.D. and Hutchison, H.P. (1971). Diffusion coefficients for sodium and potassium chlorides in water at elevated temperatures. *J. Chem. Eng. Data*, **16**, N4, pp. 427–9.

Franck, E.U. (1956). Hochverdichter Wasserdampf III. Ionendissoziation von KCl, KOH und H_2O in uberkritischem wasser. *Zeitschr. fur Physik. Chemie*, **8**, pp. 192–206.

Frantz, J.D. and Marshall, W.L. (1982). Electrical conductances and ionization constants of calcium chloride and magnesium chloride in aqueous solutions at temperatures to 600 °C and pressures to 4000 bars. *Amer. J. Sci.*, **282**, pp. 1666–93.

Frantz, J.D. and Marshall, W.L. (1984). Electrical conductances and ionization constants of salts, acids and bases in supercritical aqueous fluids. I. Hydrochloric acid from 100 °C to 700 °C and at pressures to 4000 bars. *Amer. J. Sci.*, **284**, pp. 651–67.

Hart, S.R. (1981). Diffusion compensation in natural silicates. *Geochim. Cosmochim. Acta*, **45**, N3, p. 279.

Helfferich, F. (1959). *Ionenaustauscher (Grundlagen Struktur Herstellung Theorie)*. Verlag Chemie GMBH Weiheim Bergstr.

Helgeson, H.C. and Kirkham, D.H. (1974). Theoretical prediction of the thermodynamic behavior of aqueous electrolytes at high pressures and temperatures. I. Summary of the thermodynamic–electrostatic properties of the solvent. *Amer. J. Sci.*, **274**, pp. 1089–1198.

Helgeson, H.C., Kirkham, D.H. and Flowers, G.C. (1981). Theoretical prediction of the thermodynamic behavior of aqueous electrolytes at high pressures and temperatures. IV. Calculation of activity coefficients, osmotic coefficients and apparent molal and standard and relative partial molal properties to 600 °C and 5 kb. *Amer.J.Sci.*, **281**, N10, pp. 1249–1516.

Hofmann, A.W., Giletti, B.J., Yoder, H.S.Jr., Yund, R.A. (eds) (1974) *Geochemical Transport and Kinetics*. Carnegie Institution of Washington, publication 634.

Ivanova, L.I. (1989). Electrosurface properties of aluminium and silicon oxides in electrolyte solutions over a wide temperature range. Ph.D. thesis, Leningrad (in Russian).

Korzhinskiy, D.S. (1970). *Theory of Metasomatic Zoning*. Clarendon Press, Oxford.

Labotka, T.C. (1991). Chemical and physical properties of fluids. In *Reviews in Mineralogy* (ed. Kernick, D.H.), vol. 26 (Contact Metamorphism), pp. 43–104.

Lakstanov, L.Z. and Kopylov, P.N. (1985). Separation of electrolyte solutions upon filtration through porous media. *Khimia i Tekhnologia Vody*, **7**, N4, pp. 8–11 (in Russian).

Marshall, W.L. and Franck, E.U. (1981). Ion product of water substance, 0–1000 °C, 1–10000 bars: new international formulation and its background. *J. Phys. and Chem. Ref. Data*, **10**, N 2, pp. 295–304.

Miller, D.G. (1966). Application of irreversible thermodynamics to electrolyte solutions. I. Determination of ionic transport coefficients l_{ij} for isothermal vector transport processes in binary electrolyte systems. *J. Phys. Chem.*, **70**, N8, pp. 2639–59.

Nigrini, A. (1970). Diffusion in rock alteration systems. I. Prediction of limiting equivalent ionic conductances at elevated temperatures. *Amer. J. Sci.*, **269**, pp. 65–85.

Oelkers, E.H. and Helgeson, H.C. (1988a). Calculation of thermodynamic and transport properties of aqueous species at high pressures and temperatures: Aqueous tracer diffusion coefficients of ions to 1000 °C and 5 kb. *Geochim. Cosmochim. Acta*, **1**, pp. 63–85.

Oelkers, E.H. and Helgeson, H.C. (1988b). Calculation of the thermodynamic and transport properties of aqueous species at high pressures and temperatures: dissociation constants for supercritical alkali metal halides at temperatures from 400 ° to 800 °C and pressures from 500 to 4000 bars. *J. Phys. Chem.*, **92**, pp. 1631–9.

Plyasunov, A.V. (1988). Estimation of dissociation constants for symmetrical electrolytes on the basis of stoichiometric activity coefficients. *Zhurn.Phis. Khimii*, **LXII**, N3, pp. 622–5 (in Russian).

Plyasunov, A.V. (1989). Experimental and thermodynamic investigation of the solubility of zinc oxide in alkali and cloride solutions to 600 °C and 1 kbar. Ph.D. thesis, Institute of Experimental Mineralogy, Chernogolovka (in Russian).

Quist, A.S. and Marshall, W.L. (1965). Assignment of limiting equivalent conductances for single ions to 400 °C. *J. Phys. Chem.*, **69**, N9, pp. 2984–7.

Quist, A.S. and Marshall, W.L. (1968). Electrical conductances of aqueous sodium chloride solutions from 0 to 800 °C and at pressures to 4000 bars. *J. Phys. Chem.*, **72**, N2, pp. 684–703.

Quist, A.S. and Marshall, W.L. (1969). The electrical conductances of some alkali metal halides in aqueous solutions from 0 to 800 °C and at pressures to 4000 bars. *J. Phys. Chem.*, **73**, pp. 987–965.

Ritzert, G. and Franck, E.U. (1968). Elektrische Leitfahigkeit wassriger Losungen bei hohen Temperaturen und Drucken, I. KCl, $BaCl_2$, $Ba(OH)_2$, und $MgSO_4$ bis 750 °C und 6 Kbar. *Ber. Bunsenges. Physik. Chem.*, **72**, pp. 798–807.

Robinson, R.A. and Stokes, R. (1959). *Electrolyte Solutions*. Butterworths, London.

Shante, V.K.S. and Kirkpatrik, S. (1971). An introduction to percolation theory. *Adv. Phys.*, **20**, N85, pp. 325–57.

Shchukin, Eu.D., Pertsov, A.V. and Amelina, E.A. (1982). *Handbook of Colloidal Chemistry*. Moscow Univer. Press, Moscow (in Russian).

Wishaw, B.F. and Stokes, R.H. (1954). The diffusion coefficients and conductances of some concentrated electrolyte solutions at 25 °C. *J. Amer. Ch. Soc.*, **76**, pp. 2065–71.

Zaraisky, G.P., Zharikov, V.A., Stoyanovskaya, F.M. and Balashov, V.N. (1986). *An Experimental Investigation of Bimetasomatic Skarn Formation*. Nauka Press, Moscow (in Russian).

Zonov, S.V., Zaraisky, G.P. and Balashov, V.N. (1989). The effect of thermal decompaction on granite permeability, with lithostatic pressure being slightly in excess of fluid pressure. *Dokl. Akad. Nauk SSSR*, **307**, N1, pp. 191–5 (in Russian).

GLOSSARY OF SYMBOLS

$A_{\nu_{1i}} B_{\nu_{2i}}^{z_i}$ general formula of the ith ionic species in aqueous solution of the electrolyte $A_{\nu_1} B_{\nu_2}$.

a_1 activity of cation (A^{z_1}) in aqueous solution of the electrolyte $A_{\nu_1} B_{\nu_2}$.

a_2 activity of anion (B^{z_2}) in aqueous solution of the electrolyte $A_{\nu_1} B_{\nu_2}$.

a_i activity of ith aqueous species.

a mean electrolyte activity.

C_i molar concentration of the ith component in aqueous solution, [mol. l^{-1}].

D_i^0 diffusion coefficient of the ith species at infinite dilution, [$cm^2. s^{-1}$].

D_i diffusion coefficient of component i in free solution, [$cm^2. s^{-1}$].

D_e^{00} self-diffusion coefficient of the neutral ion pair, [$cm^2. s^{-1}$].

D_e^0 Nernst diffusion coefficient of the electrolyte, [$cm^2. s^{-1}$].

$\tilde{D}_e^0$ generalized diffusion mobility of the electrolyte, [$cm^2. s^{-1}$].

D_i^* effective diffusion coefficient of the ith component through the porous medium, [$cm^2. s^{-1}$].

d_{gr} mean diameter of mineral grains, [μm].

e 1.60219 · 10^{-16} [C], absolute value of electron charge.

F formation factor of the porous medium.

F 96484.5 [C.equiv^{-1}], the Faraday constant.

X_j generalized thermodynamic force, [J. mol^{-1}. cm^{-1}].

z_i electric charge on the ith species in equivalents.

z_χ charge on surface complexes (centres) in equivalents.

α degree of electrolyte dissociation.

γ mean stoichiometric electrolytic activity coefficient.

Δ_α electro-osmotic increment in effective diffusion mobility of the αth ion.

δ diameter (the width) of microcrack in rock, [μm].

δ_F mean physical microcrack diameters, [μm].

δ_c critical diameter of microcracks defined by: $p_c = 1 - \mathcal{F}(\delta_c)$, [μm].

ε dielectric constant of solution.

ε_0 8.8542 [F. m^{-1}], dielectric constant of a vacuum.

ζ electrokinetic zeta potential of the porous medium, [V].

η viscosity of the pore solution, [Pa].

$æ_i$ multiplier responding to changes of the diffusion coefficient value of the ith component in pore solution.

λ_i^0 limiting equivalent conductance, [$cm^2. \Omega^{-1}$. equiv^{-1}]

μ_j chemical potential of the jth species, [J · mol^{-1}].

ρ_{H_2O} density of water, [kg. 1^{-1}].

σ	surface density of electrical charge on pore walls, $[C.\,cm^{-2}]$.
ϕ^D	diffusion potential, [V].
ϕ^M	diffusion membrane potential, [V].
$1/\chi$	effective thickness of the electric double layer (EDL), [m]
Ψ_0	electric potential at contact between rock minerals and aqueous solution, [V].
$\tilde{\Psi}_v$	mean potential of a pore, [V].
$\mathcal{F}(\delta)$	function of microcrack size distribution.
J_i	density of a diffusion flux of ith component through a cross-section of the medium, $[mol.\,cm^{-2}.s^{-1}]$.
K_i	thermodynamic equilibrium constant in molality scale.
K^0	permeability of the porous medium, [mD].
k_{tr}	tortuosity coefficient of the porous medium.
L	integral conductance of the net–microcrack system with an accuracy up to a constant coefficient.
l_{ij}^0	phenomenological transport coefficient, $[mol^2.\,J^{-1}.\,cm^{-1}.\,s^{-1}]$.
m	molality of electrolyte in aqueous solution.
m_i	molality of the ith species.
m_χ	surface charge 'molality' averaged over the mass of pore water.
n	number of ionic species in solution.
O	coefficient of the electro-osmotic transfer, $[cm^2.\,equiv.\,s^{-1}.J^{-1}]$.
p_l	fraction of microcracks in the range $\delta_l < \delta < \delta_{max}$, $p_l = 1 - \mathcal{F}(\delta_l)$.
$\tilde{p}_l$	set of microcracks in the range $\delta_l < \delta < \delta_{max}$.
p_c	critical percolation probability.
$P(p_l)$	percolation function.
R	8.3144 $[J.\,K^{-1}.\,mol^{-1}]$, gas constant.
R_d^*	resistivity coefficient for diffusion through the porous medium.
R_f^*	resistivity coefficient for filtration through the porous medium.
S_i^0	standard partial molal entropy of the ith species at infinite dilution, $[J.\,K^{-1}.\,mol^{-1}]$.
T	absolute temperature, [K].

CHAPTER TEN

Thermal decompaction of rocks

Georgiy P. Zaraisky and Victor N. Balashov

10.1 INTRODUCTION

The influence of P–T changes on rocks is generally considered merely in terms of chemical reactions and phase transformations during metamorphism. However, both temperature and pressure are also important factors responsible for changes in physical properties of rocks, including porosity, permeability, and strength, which are of primary concern to fluid transport. This should be taken into account while developing numerical models for a variety of geological processes. In modelling endogenic mass transfer processes, the use of porosity–permeability values, as well as other physical rock properties measured under normal conditions, may produce erroneous results. In the present work, we would like to bring this problem to the attention of geologists.

Our earlier experimental studies (Zaraisky and Balashov, 1978, 1981, 1983a, b, 1986; Balashov and Zaraisky, 1982; Zaraisky *et al.*, 1989; Zonov *et al.*, 1989) showed that, in general, the porosity and geometry of the pore space in rocks prove to be variable characteristics, highly sensitive to temperature. Because of the non-uniform expansion of neighbouring crystals of different mineral phases or in different orientations, as the temperature changes, a rock develops thermo-elastic stresses producing microcracks along grain boundaries. In other words, rocks undergo 'decompaction' (Zaraisky and Balashov, 1978). Thus, in parallel with the thermal expansion of a crystalline matrix, an additional intergranular pore space is set up in rocks upon heating. We define thermal decompaction as an increase in volume of a rock on heating in addition to that

Fluids in the Crust: Equlibrium and transport properties.
Edited by K.I. Shmulovich, B.W.D. Yardley and G. G. Gonchar.
Published in 1994 by Chapman & Hall, London. ISBN 0 412 56320 7

which would be predicted for the constituent mineral phases alone. As a result of decompaction, the permeability of low-porosity rocks may increase by several orders of magnitude. Concurrently, all the other physical properties, particularly elastic and strength characteristics, also change.

Although the effect of 'excess' thermal expansion of rocks, as opposed to the expansion of their constituent minerals has long been known (e.g. Ide, 1937; Birch, 1943; Richter and Simmons, 1974), no geological significance has been attached to it for a long time and no attempts made to undertake a special study on this basis. Maxwell and Verrall (1953) seemed to have been the first to notice an increase in permeability of rocks subjected to thermal expansion. They detected residual expansion of carbonate rocks after cooling, but failed to propose a mechanism for this phenomenon. Furthermore, permeability changes were judged by them indirectly, from the depth to which rocks were impregnated by coloured inks. More recently, the geometry of the additional pore space set up in rocks after heating has been studied with optical microscopy, scanning electron microscopy and other methods (Bauer and Johnson, 1979; Zaraisky and Balashov, 1981; Fredrich and Wong, 1986; Gerand and Gaviglio, 1990). The effect of temperature on permeability of rocks *in situ* and after cooling is considered in experimental studies of Zaraisky and Balashov (1978), Summers *et al.* (1978), Morrow *et al.* (1981), Vitovtova and Shmonov (1982), Shmonov and Vitovtova (1992), Heard and Page (1982), Nur (1982), Zonov *et al.* (1989), and others.

Our comprehensive study of thermal decompaction used rocks from a variety of ore fields of the former USSR (Zaraisky and Balashov, 1978, 1981, 1983a, 1986; Balashov and Zaraisky, 1982; Zaraisky, 1989; Zonov *et al.*, 1989; Zaraisky *et al.*, 1989). The rocks were heated under gradient-free conditions over the temperature range 25–700°C at atmospheric pressure and in autoclaves under a H_2O, CO_2 or Ar pressure of 1000 bar. At the temperature of the run, the linear expansion of samples was measured, followed (after cooling) by measurements of residual extension, porosity, permeability, formation factor, density, shear and compressional elastic wave velocities. The effect of heating on the permeability of granites was also studied *in situ* using a special infiltration device, provided the pressure applied to the solid matrix and fluid was unequal. The present paper discusses the results of such studies and their possible geological implication.

10.2 EXPERIMENTAL PROCEDURE

The conducted experiments may be separated into two groups: (1) the uniform heating of rocks to a certain temperature at atmospheric pressure or at $P_{fl} = 1$ kbar, followed by measurements of physical properties of samples after cooling, (2) rock permeability measurements at elevated P and T using a special device, described in Chapter 11 (section 11.2).

10.2.1 Starting materials

Thirteen rock samples were chosen for this study to represent the variability of common ore-bearing sequences (Table 10.1). The size of crystalline grains (0.05–0.8 mm) allowed runs to be done using small-sized samples. In the first group of experiments, 28 mm diameter and 5 mm thick discs were used. For the second group of experiments, cylindrical samples were prepared, 9.6 mm in diameter and 15 or 25 mm long. All the samples were machined to high accuracy (± 0.01 mm) with the aid of a diamond tool.

Table 10.1. Mineral compositions and physical properties of initial rocks

	Rock type, location, geological age	*Mineral composition* (volume %)	*Average grain size* (mm)	*Density* (g. cm^{-3})	*Effective porosity* α(%)	*Permeability* K°(md)
1	2	3	4	5	6	7
1.	Granite, New Ukrainian massif, Ukrainian Shield, Ptz	Qtz-31; Pl_{20}–36; Kfs-26; Bt-7	0.23	2.65	0.33	$0.15 \cdot 10^{-5}$
2.	Granite–aplite, Pitkyaranta Mine, Karelia, Ptz	Qtz-26; Pl_{7}–47; Kfs-25; Bt-2	0.27	2.57	1.09	$0.14 \cdot 10^{-2}$
3.	Leucocratic granite, Akchatau Mine, Kazakhstan, P_1	Qtz-35; Pl_{11}–45.5; Kfs-29; Bt-1	0.50	2.57	1.30	$1.00 \cdot 10^{-3}$
4.	Granodiorite, Maykhura Mine, Tajikistan, C_3	Qtz-24; Pl_{25}–46; Kfs-15; Bt-10; Hbl-5	0.80	2.68	0.74	$0.15 \cdot 10^{-2}$
5.	Diorite, Magnitogorsk Mine, South Urals, C_1	Pl_{12}–65; Hbl-30; Mag-5	0.10	2.81	2.21	$1.00 \cdot 10^{-7}$
6.	Gabbro–dolerite, Wind-Belt ridge, Arkhangelsk District, Ptz	Pl_{23}–55; Cpx-36; Hbl-6; Mag-3	0.24	2.99	0.86	$0.48 \cdot 10^{-6}$
7.	Hornblendite, Wind-Belt ridge, Arkhangelsk District, Ptz	Hbl-77; Bt-23	0.10	3.00	–	–
8.	Andesite–dacite, Kochbulak Mine, Uzbekistan, C_{2-3}	Qtz-20; Pl_{12}–30; Kfs-20; Cal-15; Hem-10; Bt-5	0.04	2.51	6.70	$0.13 \cdot 10^{-4}$
9.	Olivine basalt, Klyuchevskaya Sopka, Kamchatka, modern	Pl_{60}–40; Cpx-30; 01-10; Mag-5; glass-15	0.08	2.71	4.50	$0.15 \cdot 10^{-1}$

Table 10.1 (Contd)

	Rock type, location, geological age	*Mineral composition* (volume %)	*Average grain size* (mm)	*Density* (g. cm^{-3})	*Effective porosity* α(%)	*Permeability* K°(md)
1	2	3	4	5	6	7
10.	Medium-grained white marble, Varzob River, Tajikistan, C_{1-2}	Cal-100	0.65	2.70	0.42	$0.46 \cdot 10^{-3}$
11.	Fine-grained dark marble, Magnitogorsk Mine, South Urals, C_1	Cal-95; organic material-5	0.05	2.73	0.22	$0.13 \cdot 10^{-5}$
12.	Garnet skarn, Dashkesan Mine, Azerbaijan, J_3	Grt-85; Cal-10 Chl-5	0.005	3.47	1.77	–
13.	Magnetite ore, Temir-Tau Mine, Kuznetsky Alatau	Mag-80; Chl + Sd-20	0.02	4.95	–	–

10.2.2 Apparatus and methods of measurement

The thermal expansion of rocks at atmospheric pressure was measured using a quartz dilatometer (Zaraisky and Balashov, 1981). Slowly heating and cooling the samples in automatic mode at a rate of 2°C. min^{-1} made it possible to avoid the production of cracks due to thermal stresses. Thermal expansion at elevated pressure was investigated using samples not isolated from the surrounding medium in autoclaves with $P_{fl} = 1$ kbar, generated by H_2O, CO_2 or Ar. To prevent dissolution, water was previously saturated with the component minerals of the rock. Linear expansion under the conditions of the runs was measured by a specially designed autoclave dilatometer (Figure 10.1) which marked an increase in the specimen length by a line on the platinum plate. After the run, the line length was measured under the microscope. Only one thermal cycle was run on each sample, which was heated to the desired temperature, held at constant temperature (± 1°C) for 20 hours and cooled slowly. The rate of heating and cooling was 2°C. min^{-1}. After cooling to room temperature, measurements of 'residual' characteristics of the decompacted rocks were made, namely, residual linear expansion, effective porosity, density, permeability, formation factor and elastic wave velocities.

Linear expansion was in all cases defined as the ratio of the increase in the sample length to its initial length ($\Delta l \cdot l_o^{-1}$), expressed on a percentage basis. The relative error of measurement was ± 2%.

The porosity (α) and density (ρ) of the rocks were determined using the water-saturation method, (Starostin, 1979) followed by weighing in

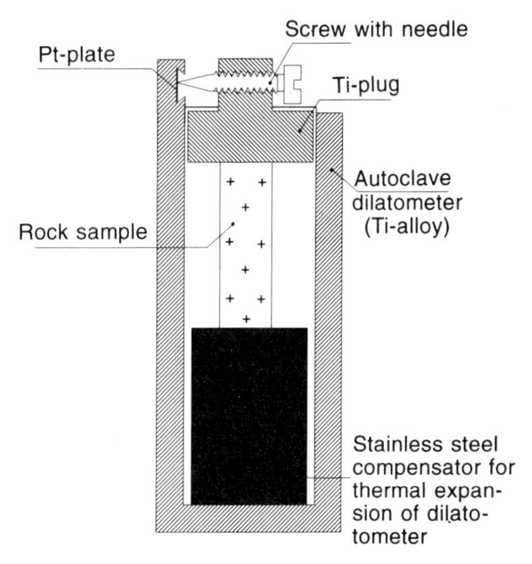

Figure 10.1. Autoclave dilatometer to measure the thermal expansion of rocks in different media (e.g. H_2O, CO_2, Ar) at elevated T and P.

water. The relative errors of measurements were, respectively, ± 2 and ± 0.5%.

The elastic properties of rocks were measured using the pulse ultrasonic method (Starostin, 1979). The velocities of shear and compressional waves were measured for the case of dry and water-saturated samples.

The permeability (K^0) of the rock discs was determined from the flow of gas or liquid through the sample under the action of a pressure gradient. A vessel filled with compressed nitrogen served as the source of pressure, the pressure at the outlet of the sample being atmospheric. The value of gas permeability was calculated from the relationship (Carman, 1956):

$$K = \frac{U_1 P_1}{\bar{P} \Delta P . h^{-1}} \eta \cdot 1000, \tag{10.1}$$

where U_1 is the volumetric rate ($cm^3 . s^{-1}$) of gas flow through a 1 cm^2 cross section of the membrane at atmospheric pressure P_1, $\bar{P}$ the average pressure (atm) in the membrane, where $\bar{P} = \frac{1}{2}(P_1 + P_2)$, $\Delta P = P_2 - P_1$, P_2 stands for the pressure preset by the source, h denotes the membrane thickness (cm), η refers to the mean viscosity (cP) of nitrogen over the ΔP range at room temperature and K is gas permeability (md).

Liquid permeability was determined from the following relationship, using CCl_4:

$$K_{CCl_4} = \frac{U_{CCl_4}}{\Delta P . h^{-1}} \eta_{CCl_4} \cdot 1000. \tag{10.2}$$

Because of the gas slip at the capillary walls, the measured gas permeability (K) is in excess of the true permeability value (K^o), and hence Klinkenberg's relationship holds (Pek, 1968; Scott and Dullien, 1962):

$$K = K^0 + \frac{B}{\bar{P}}. \quad (10.3)$$

The value of Klinkenberg's constant (B) was determined from the experimental dependence of K on $\bar{P}^{-1}$. K^0 and the liquid permeability value K_l will coincide, provided that there is no slip at the capillary walls and that physical–chemical interaction between liquid and capillary walls is lacking, i.e. $K_l = K^0$. In all cases where K_{CCl_4} and K^0 were determined in parallel, these values coincided within the accuracy of measurements (± 10% of the value under measurement).

The formation factor (F), which represents the ratio of the electrical conductivity of the rock saturated with an electrolyte solution to the electrolyte conductivity (Brace, 1977), was measured in a thermostatic assembly at 25 °C. One m KCl solution was used as the electrolyte; resistance was measured with a R5010 a.c. bridge at a frequency of 1000 Hz; the accuracy of measurements was ± 5%.

The structure of decompacted rocks was examined under reflected light with an optical microscope, using phase contrast imaging. Prior to examination, a gold coating was vacuum evaporated on to the polished sample surfaces. Microcracks were statistically counted under the microscope, distinguishing between genetic types (e.g. grain boundary cracks, transgranular and cleavage cracks) and aperture widths. Microcrack widths were measured with the aid of a screw ocular-micrometer. Cracks of width < 0.05 μm were not resolved.

The permeability of rocks under the experimental P–T conditions was measured *in situ* using a general-purpose infiltration device which allows conditions to be created in which a fluid pressure gradient in the sample is relatively low, while absolute fluid pressure values are high (up to 1000–1500 bar) (Zaraisky, 1989). The confining pressure (P_c) exerted on the solid rock framework is in this instance independent of the fluid pressure and produced by applying hydrostatic pressure to a thin-walled gold or Teflon cylindrical jacket with wall thicknesses of 0.2 or 0.5 mm, respectively. Fluid (distilled water) was supplied to, and removed from, the device through the ends of the sample. The reliability of sealing the sample in the jacket was checked against a synthetic quartz single crystal whose permeability is zero. Inlet and outlet fluid pressures were maintained automatically to provide a constant gradient of 10–15 bar. cm^{-1} over the course of the experiment. Effective pressure ($P_{ef} = P_c - P_{fl}$) was varied from 50 to 800 bar by changing P_{fl} at $P_c = const = 1000$ bar. Measurements on each sample were made at constant P_c, P_{fl}, and ΔP; temperature was raised step by step from 20 °C to a maximum of 600 °C at 50 °C intervals. The sample was replaced when the effective pressure was to be changed. Permeability was determined from the volumetric rate of fluid flow using Darcy's law. The lower limit to the sensitivity of the device was $1.0 \cdot 10^{-6}$ md.

10.3 RESULTS

The results of our experiments are presented in Table 10.2. From the data summarized in the table, it follows that thermal decompaction is of the same universal significance for rocks as thermal expansion. All the physical properties of rocks change regularly after having experienced thermal decompaction.

10.3.1 Thermal expansion and decompaction of rocks

Figure 10.2a illustrates the temperature dependence of the thermal expansion of rocks at atmospheric pressure. For comparison, Figure 10.2b shows the expansion of the same rocks, calculated from the reference data on thermal expansion behaviour of minerals (Skinner, 1966). It is clear that such a calculation is completely unsuitable for predicting rock expansion because the measured values of linear expansion prove to be 1.5–2 times greater than the calculated ones. This difference in values may be attributed to the formation of an additional pore space in the form of a network of microcracks located primarily along grain boundaries. Intergranular spaces are a result of the non-uniform expansion of neighbouring mineral grains that differ in composition or crystallographic orientation.

The excess expansion of marble is almost linear in character and may be as high as 0.9% at 600°C, thus corresponding to a 2.7% pore space increase. Rocks with a large abundance of quartz, such as granite and granite-aplite, expand nonlinearly; their curves show a steep slope between 500 and 600°C but expand relatively little on heating in the higher-temperature region. This may be due to changes in quartz properties when it undergoes the α–β transition near 573°C. The excess expansion of intermediate, basic and ultrabasic rocks is not high; at 700°C, its linear value is around 0.2%.

After cooling, decompacted rocks retain residual expansion due to the irreversible change in pore space. The temperature for the onset of residual decompaction is lowest for marble (200–210°C), higher (250°C) for granite and much higher (up to 500°C) for rocks of intermediate and basic composition. When exposed to air, basalt and hornblendite expand reversibly throughout the entire temperature range that we considered.

Figure 10.3 shows the effect of the environment of heating on the temperature dependence of the expansion of New Ukrainian granite and gabbro-dolerite. The rates of heating and cooling correspond to those

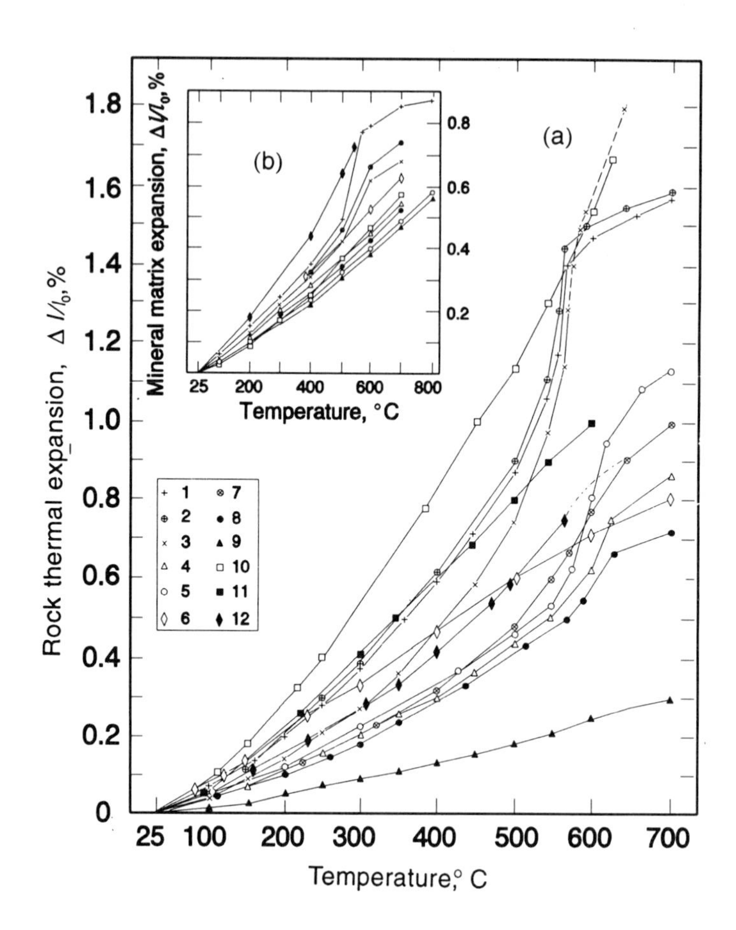

Figure 10.2. Linear thermal expansion of rocks at atmospheric pressure, both experimental (a) and calculated (b), based on expansion values for rock-forming minerals: 1 granite, 2 granite–aplite, 3 granodiorite, 4 diorite, 5 gabbro–dolerite, 6 hornblendite, 7 andesite–dacite porphyrite, 8 diabase, 9 olivine basalt, 10 medium-grained white marble, 11 fine-grained dark-grey marble, 12 magnetite ore.

given above. The expansion of non-isolated samples under inert argon or CO_2 in autoclaves at $P_{Ar, CO_2} = 1000$ bar is also accompanied by rock decompaction, the corresponding residual decompaction curves running only slightly below the residual heating curve at atmospheric pressure. However, in the presence of water ($P_{H_2O} = 1000$ bar), the degree of decompaction increases; thermal expansion–residual decompaction curves of rocks run above the corresponding experimental curves obtained for other media. This may be interpreted as being due to the reduction in strength

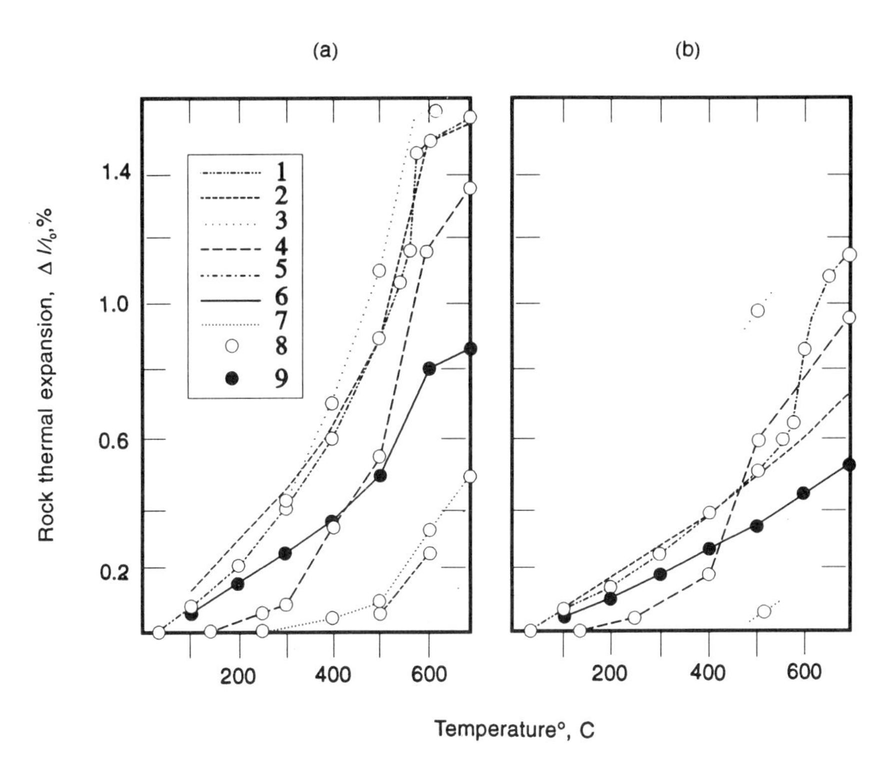

Figure 10.3. Thermal expansion of New Ukrainian granite (a) and gabbro-dolerite (b) under a variety of conditions: 1 relative extension (temperature strain) at atmospheric pressure, 2 theoretical thermal expansion curve of rock with regard to thermal decompaction, 3 relative extension upon heating in water using autoclaves (P_{H_2O} = 1 kbar), 4 residual deformation after the first heating cycle in water using autoclaves, 5 *idem* but in argon (P = 1 kbar), 6 theoretical expansion curve of a rock mineral matrix (see below), 7 residual deformation after the first thermal cycle at atmospheric pressure, 8 experimental data points, 9 calculated values.

of grain contacts in water because it interacts with mineral surfaces. In water, the onset of thermal decompaction of all the rocks occurs at lower temperatures, the value of residual decompaction being several times greater than that under different conditions. Olivine basalt, when heated in water, shows an anomalously high residual expansion approaching 9% at 400–500 °C, presumably due to hydration of volcanic glass.

10.3.2 Transport properties and the pore space structure of decompacted rocks.

The fact that the decompacted rock structure is preserved after cooling makes it possible to examine the properties of decompacted rocks under normal conditions and correlate them, to a first approximation, with the temperature of decompaction.

(a) Permeability, porosity and formation factor. From the point of view of the problem of fluid transport in the Earth's crust, the major consequence of thermal decompaction of rocks is an appreciable increase in permeability due to the opening of a connected system of microcracks along grain boundaries. Figure 10.4a shows the temperature dependence of the residual permeability of New Ukrainian granite, gabbro-dolerite and white marble. Measurements were made after hydrothermal heating to the desired temperature in autoclaves and subsequent cooling. For the same rocks, variations in effective porosity and formation factor as a function of temperature are given in Figure 10.4b,c.

It can be seen that the transport properties of all these rocks increase regularly with temperature. Interestingly, over temperatures between 20 and 700 °C, porosity increases within one order of magnitude, the formation factor by less than two orders of magnitude, but permeability increases by several orders of magnitude. The horizontal dashed curve in Figure 10.4a represents a permeability level of 10^{-3} md above which intense circulation of solutions is possible in rocks, the convective mass transfer dominating over diffusion (Norton and Knight, 1977). Thus, thermal decompaction is capable of providing conditions for hydrothermal infiltration transport not only along individual cracks, but through the entire rock mass as well.

(b) Pore space morphology. Examination of the polished thin sections of decompacted rock samples clearly demonstrates the dominant role of grain boundaries in the pore space structure (Figure 10.5). This is of great importance to permeability because the microcracks are distributed over a connected network of grain boundaries, rather than at random, which makes it possible for a single conducting system of connecting channels to form. The data for statistical counts of different microcrack types in thin sections give evidence that the excess expansion of rocks is due primarily to the opening of microcracks along grain contacts (Figure 10.6a). In heated granite, the proportion of narrow microcracks (< 0.1 μm) predominant in unheated rock decreases regularly with increasing temperature, whereas the number of wider cracks (0.1–10 μm) increases gradually (Figure 10.6b).

Aside from direct measurements under the microscope, the average width of pore channels was determined with two calculation methods (see Table 10.2). One is applicable to the simplified porous medium model represented by a series of parallel rectilinear cracks of permanent width. For this model, the following relationship holds (Balashov and Zaraisky, 1982):

$$K^0 = 10^{11} F \frac{h^2}{12},$$

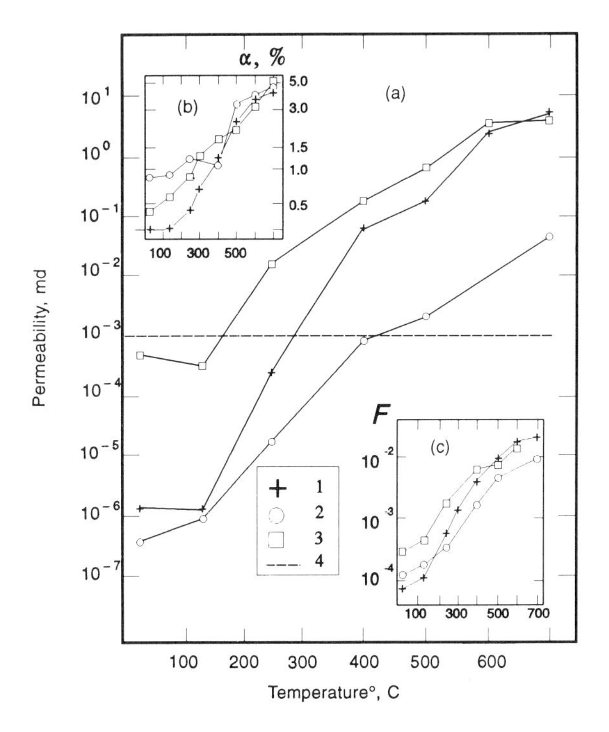

Figure 10.4. Variations in transport properties of rocks after they have been subjected to thermal decompaction in autoclaves at $P_{H_2O} = 1$ kbar: (a) permeability, (b) porosity, (c) formation factor; 1 New Ukrainian granite, 2 gabbro–dolerite, 3 medium-grained white marble, 4 'critical' permeability value of 10^{-3} md.

whence

$$h = \left(1,2 \cdot 10^{-10} K^0 F^{-1}\right)^{\frac{1}{2}}, \tag{10.4}$$

where K^0 is permeability in md, h the microcrack width in cm. Channel widths so determined are denoted h_F, as shown in Table 10.2.

The other method consists in determining crack widths through the use of Equation 10.3 on the basis of theoretical consideration of gas slip at capillary walls (Scott and Dullien, 1962):

$$h_S = 2\pi\bar{v}^3 \frac{K^0}{B} 10^{-4}, \tag{10.5}$$

where h_S is the slot-like capillary width in μm, $\bar{v}$ the mean thermal molecular velocity of gas in cm. sec^{-1}, η the gas viscosity in cP, K^0 refers to permeability in md, B stands for a 'slippage' or Klinkenberg's constant in md. bar, measured experimentally.

Direct measurements under the microscope and the two calculation methods all yield similar results. The average channel width in

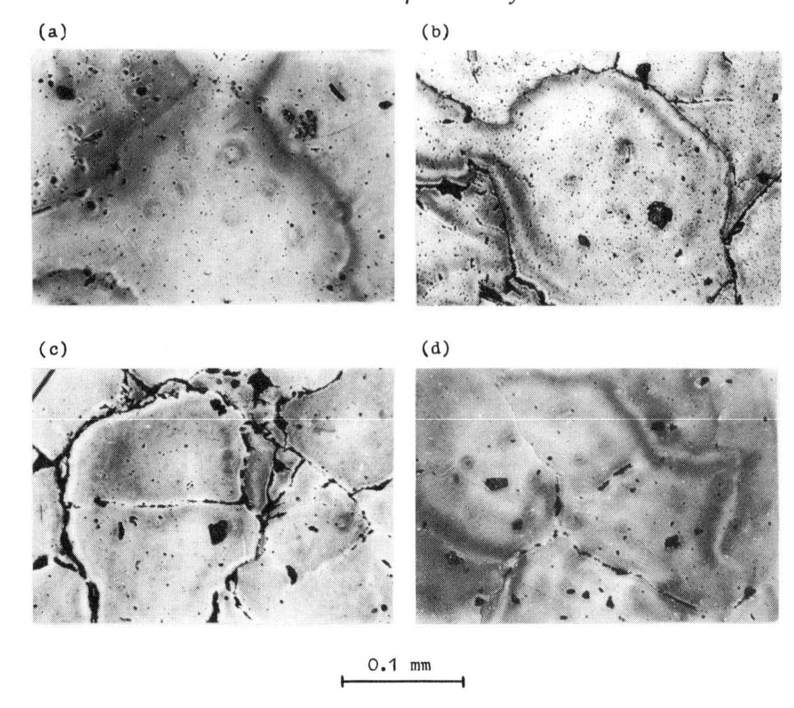

Figure 10.5. Photomicrograph of polished thin sections of New Ukrainian granite in the initial state and after it has been exposed to thermal decompaction under a variety of conditions: (a) initial granite sample, (b) granite sample after it has been heated to 250°C in an autoclave at $P_{H_2O} = 1$ kbar, (c) *idem* but after heating to 500°C, (d) granite sample after it has been heated to 500°C at atmospheric pressure. The sample has been held at the run temperature for 20 hours.

low-porosity rocks prior to heating ranges from 0.01 to 0.1 μm. During the thermal decompaction process it increases steadily and at 500°C approaches 0.2–0.4 μm in silica-rich rocks, 0.5–0.8 μm in marbles, and does not exceed 0.1 μm in basic rocks which undergo weak decompaction.

The aspect ratio of microcracks, which represents the ratio of microcrack length to width, was determined from elastic wave velocities (O'Connel and Budiansky, 1974; Hadley, 1976):

$$\frac{1}{h} = \frac{4\pi\varepsilon}{3\alpha}, \tag{10.6}$$

where ε is the crack density parameter, and is uniquely related to the elastic properties of the cracked rock (which is regarded as a continuous medium), α is porosity. The aspect ratio of microcracks in the rocks under investigation ranges from 200 to 300. Average microcrack lengths increase from 0.1–20 μm in original rocks to 10–180 μm in rocks decompacted at 700°C (here the minimum values refer to gabbro, the maximum ones to granite and marble).

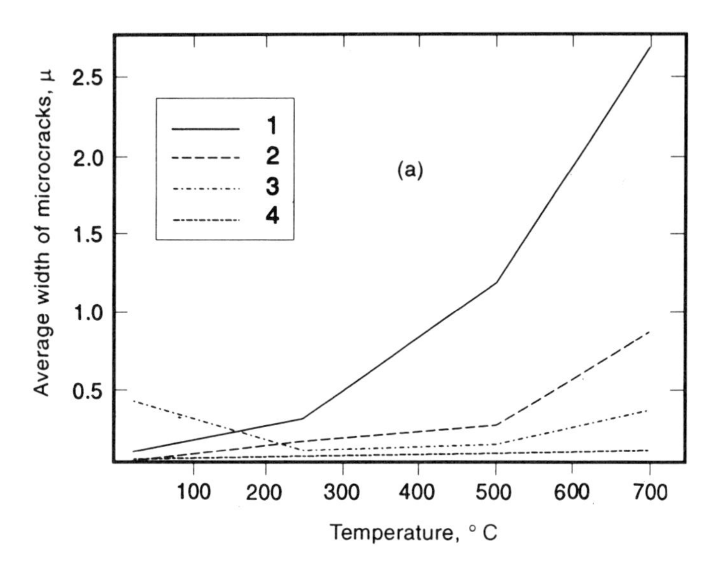

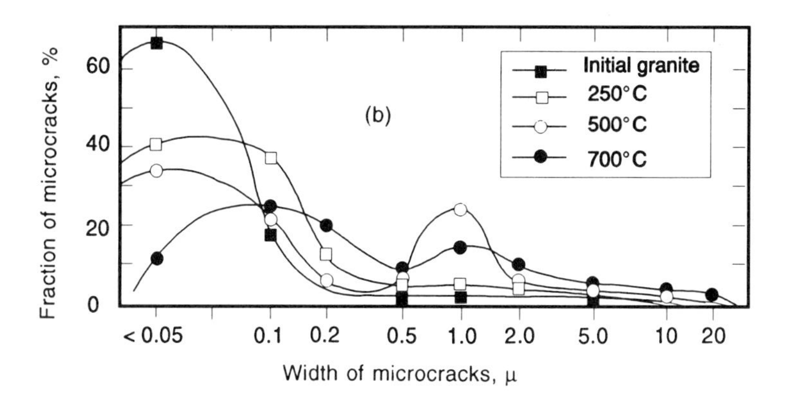

Figure 10.6. Temperature dependence of microcrack development in New Ukrainian granite, using data on statistical counts in polished thin sections: (a) average width of microcracks of different type as measured microscopically: 1 grain boundary cracks, 2 intracrystalline cracks, 3 transgranular cracks, 4 cleavage cracks; (b) fraction of microcracks of different width in granite samples decompacted at the temperatures specified.

10.3.3 Other physical properties of decompacted rocks

As the degree of decompaction increases, all the rocks show a regular reduction in bulk density and a concurrent increase in porosity (see Table 10.2). The data on compressional–shear ultrasonic wave velocities in decompacted rocks are presented in Balashov and Zaraisky (1982), and here we will focus on the key conclusions only. As the residual

Table 10.2. Thermal decompaction and transport properties of rocks

Rock type	T (°C)	P (bars)	*Medium*	$\Delta l/l_o$ (%)	$\overline{\Delta l/l_o}$ (%)	ρ (g.cm^{-3})	α (%)	ae_r (ohm^{-1}.cm^{-1})	F	K^o (md)	h_s (μm)	h_F (μm)
New		Initial		0	0	2.65	0.34	$0.95 \cdot 10^{-5}$	$0.85 \cdot 10^{-4}$	$0.15 \cdot 10^{-5}$	0.008	0.017
Ukrainian	250	1	Air	0.28	0.01	2.64	0.36	–	–	–	–	–
'granite'	400	1	Air	0.58	0.05	2.63	0.56	–	–	–	–	–
	508	1	Air	0.90	0.11	2.63	0.87	$0.23 \cdot 10^{-3}$	$0.21 \cdot 10^{-2}$	–	–	–
	600	1	Air	1.46	0.33	2.60	1.57	$0.67 \cdot 10^{-3}$	$0.60 \cdot 10^{-2}$	–	–	–
	700	1	Air	1.56	0.49	2.59	2.05	$0.98 \cdot 10^{-3}$	$0.88 \cdot 10^{-2}$	–	–	–
	130	1000	H_2O	–	0.02	2.65	0.30	$0.12 \cdot 10^{-4}$	$0.11 \cdot 10^{-3}$	$0.14 \cdot 10^{-5}$	0.008	0.012
	250	1000	H_2O	–	0.06	2.65	0.43	$0.66 \cdot 10^{-4}$	$0.59 \cdot 10^{-3}$	$0.21 \cdot 10^{-5}$	0.060	0.060
	300	1000	H_2O	0.39	0.08	2.62	0.68	$0.17 \cdot 10^{-3}$	$0.15 \cdot 10^{-2}$	–	–	–
	400	1000	H_2O	–	0.30	2.62	1.19	$0.48 \cdot 10^{-3}$	$0.43 \cdot 10^{-2}$	$0.47 \cdot 10^{-1}$	0.340	0.360
	500	1000	H_2O	–	0.50	2.61	2.08	$0.79 \cdot 10^{-3}$	$0.71 \cdot 10^{-2}$	0.14	0.280	0.480
	600	1000	H_2O	–	1.08	2.57	3.50	$0.17 \cdot 10^{-2}$	$0.15 \cdot 10^{-1}$	1.63	0.320	1.140
	700	1000	H_2O	–	1.34	2.53	4.51	$0.23 \cdot 10^{-2}$	$0.21 \cdot 10^{-1}$	1.75	0.560	1.000
	508	1000	Air	0.97	0.07	2.62	0.52	$0.13 \cdot 10^{-2}$	$0.12 \cdot 10^{-2}$	–	–	–
	600	1000	Air	1.43	0.23	–	–	–	–	–	–	–
Granite–aplite		Initial		0	0	2.56	1.27	$0.98 \cdot 10^{-4}$	$0.88 \cdot 10^{-3}$	$0.14 \cdot 10^{-2}$	0.116	0.138
	300	1	Air	–	0.02	2.57	1.35	–	–	–	–	–
	500	1	Air	–	0.15	2.55	1.50	–	–	$0.23 \cdot 10^{-3}$	–	–
	600	1	Air	–	0.54	2.54	2.51	–	–	–	–	–
	700	1	Air	–	0.52	2.54	2.38	–	–	–	–	–
	130	1000	H_2O	–	0.00	2.58	0.98	$0.12 \cdot 10^{-3}$	$0.11 \cdot 10^{-2}$	$0.13 \cdot 10^{-2}$	0.116	0.120
	250	1000	H_2O	–	0.06	2.57	2.00	$0.31 \cdot 10^{-3}$	$0.28 \cdot 10^{-2}$	$0.87 \cdot 10^{-2}$	0.158	0.194
	300	1000	H_2O	–	0.06	2.57	1.23	–	–	–	–	–
	400	1000	H_2O	–	0.25	2.55	1.69	$0.44 \cdot 10^{-3}$	$0.39 \cdot 10^{-2}$	$0.24 \cdot 10^{-1}$	0.238	0.272

Table 10.2 (Contd)

Rock type	*T* (°C)	*P* (bars)	*Medium*	$\Delta l/l_o$ (%)	$\overline{\Delta l/l_o}$ (%)	ρ (g.cm^{-3})	α (%)	ae_r (ohm^{-1}.cm^{-1})	*F*	K^o (md)	h_s (μm)	h_F (μm)
	500	1000	H_2O	–	0.38	2.55	2.80	$0.95 \cdot 10^{-3}$	$0.85 \cdot 10^{-2}$	0.15	0.412	0.460
	600	1000	H_2O	–	0.78	2.51	3.23	$0.15 \cdot 10^{-2}$	$0.13 \cdot 10^{-2}$	0.88	0.366	0.900
	700	1000	H_2O	–	1.32	2.50	3.82	$0.11 \cdot 10^{-2}$	$0.98 \cdot 10^{-2}$	4.40	1.090	2.320
	500	1000	CO_2	–	0.10	2.58	1.10	–	–	–	–	–
	700	1000	CO_2	–	0.53	2.54	2.39	–	–	–	–	–
Granodiorite		Initial		0	0	2.68	0.68	–	–	$0.15 \cdot 10^{-2}$	–	–
	500	1	Air	–	0.16	2.67	1.25	–	–	–	–	–
	700	1	Air	–	3.79	2.44	9.57	–	Appearance of macrocracks			–
	500	1000	H_2O	–	0.36	2.66	1.70	–	–	$0.90 \cdot 10^{-1}$	–	–
	700	1000	H_2O	–	1.31	2.59	4.23	–	–	–	–	–
Diorite		Initial		0	0	2.81	2.10	–	–	$1.00 \cdot 10^{-7}$	–	–
	500	1	Air	–	0.00	2.80	1.97	–	–	–	–	–
	700	1	Air	–	0.38	2.77	3.37	–	–	–	–	–
	500	1000	H_2O	–	0.29	2.78	2.40	–	–	$1.40 \cdot 10^{-3}$	–	–
	700	1000	H_2O	–	0.82	2.74	3.28	–	–	–	–	–
Gabbro–dolerite		Initial		0	0	2.99	1.00	$(0.96 \cdot 10^{-5})$	$(0.86 \cdot 10^{-5})$	$0.48 \cdot 10^{-6}$	0.018	–
	520	1	Air	0.50	0.05	–	–	–	–	–	–	–
	700	1	Air	1.12→1.36	0.96	–	–	–	–	–	–	–
	130	1000	H_2O	–	0.00	2.99	0.89	$0.21 \cdot 10^{-4}$	$0.19 \cdot 10^{-3}$	$0.11 \cdot 10^{-5}$	0.034	0.008
	250	1000	H_2O	–	0.03	2.98	1.20	$0.45 \cdot 10^{-4}$	$0.40 \cdot 10^{-3}$	$0.19 \cdot 10^{-4}$	0.032	0.024
	400	1000	H_2O	–	0.16	2.97	1.10	$0.18 \cdot 10^{-3}$	$0.16 \cdot 10^{-2}$	$0.70 \cdot 10^{-3}$	0.102	0.072
	500	1000	H_2O	(1.00)	0.35	2.89	4.90	$0.46 \cdot 10^{-3}$	$0.41 \cdot 10^{-2}$	$0.16 \cdot 10^{-2}$	0.110	0.068
	700	1000	H_2O	–	(0.94)	2.86	3.92	$0.59 \cdot 10^{-3}$	$0.52 \cdot 10^{-2}$	$0.34 \cdot 10^{-1}$	0.194	0.280

Table 10.2 (Contd)

Rock type	*T* (°C)	*P* (bars)	*Medium*	$\Delta l/l_o$ (%)	$\overline{\Delta l/l_o}$ (%)	ρ ($g.cm^{-3}$)	α (%)	ae_r ($ohm^{-1}.cm^{-1}$)	*F*	K^o (md)	h_s (µm)	h_F (µm)
Hornblendite	500	1	Air	0.58→0.54	−0.03	3.00	–	–	–	–	–	–
	700	1	Air	0.80→0.61	−0.06	3.01	–	–	–	–	–	–
	500	1000	H_2O	0.71	0.17	2.98	–	–	–	–	–	–
Andesite–dacite		Initial		0	0	2.51	6.70	–	–	$1.30 \cdot 10^{-5}$	–	–
	500	1	Air	–	0.06	2.51	6.57	–	–	–	–	–
	700	1	Air	–	0.50	2.42	10.05	–	–	–	–	–
	500	1000	H_2O	–	0.34	2.43	9.62	–	–	–	–	–
	700	1000	H_2O	–	0.86	2.36	12.32	–	–	$1.40 \cdot 10^{-3}$	–	–
Olivine basalt		Initial		0	0	2.71	4.5	–	–	$1.50 \cdot 10^{-2}$	–	–
	300	1	Air	–	0.00	2.71	4.5	–	–	–	–	–
	500	1	Air	–	0.00	2.71	4.5	–	–	–	–	–
	600	1	Air	–	0.00	2.71	4.5	–	–	–	–	–
	300	1000	H_2O	–	0.18	2.71	4.5	–	–	–	–	–
	400	1000	H_2O	–	9.10	2.11	25.6	–	Appearance of macrocracks			–
	500	1000	H_2O	–	8.80	2.09	26.5	–	–	3000	Macrocracks	
	600	1000	H_2O	–	4.90	2.32	18.3	–	–	–	Macrocracks	
	500	1000	CO_2	–	0.00	2.71	4.5	–	–	–	–	
	600	1000	CO_2	–	0.00	2.71	4.5	–	–	–	–	
Medium-grained white marble		Initial		0	0	2.69	0.59	$0.56 \cdot 10^{-4}$	$0.50 \cdot 10^{-3}$	$0.48 \cdot 10^{-3}$	0.050	0.108
	300	1	Air	(0.21)	0.10	2.70	0.48	–	–	–	–	–
	400	1	Air	–	0.20	2.69	1.92	–	–	–	–	–
	500	1	Air	0.96	0.55	2.65	2.00	$0.87 \cdot 10^{-3}$	$0.78 \cdot 10^{-2}$	–	–	–
	600	1	Air	–	1.20	–	–	–	–	–	–	–

Table 10.2 (Contd)

Rock type	T (°C)	P (bars)	*Medium*	$\Delta l/l_o$ (%)	$\overline{\Delta l/l_o}$ (%)	ρ ($g.cm^{-3}$)	α (%)	ae_r ($ohm^{-1}.cm^{-1}$)	F	K^o (md)	h_s (μm)	h_F (μm)
	700	1	Air	2.56→2.13	1.77	–	–	–	–	–	–	–
	130	1000	H_2O	–	0.04	2.70	0.50	$0.33 \cdot 10^{-3}$	$0.30 \cdot 10^{-3}$	$0.30 \cdot 10^{-3}$	0.072	0.116
	250	1000	H_2O	–	0.15	2.69	0.90	$0.21 \cdot 10^{-3}$	$0.19 \cdot 10^{-3}$	$0.15 \cdot 10^{-1}$	0.260	0.308
	300	1000	H_2O	(0.36)	0.30	2.68	1.29	–	–	–	–	–
	410	1000	H_2O	–	0.35	2.66	1.75	$0.59 \cdot 10^{-3}$	$0.53 \cdot 10^{-2}$	0.200	0.480	0.680
	500	1000	H_2O	0.82	min 0.43 max 0.71	–	1.95	$0.89 \cdot 10^{-2}$	$0.76 \cdot 10^{-2}$	0.390	0.460	0.720
	600	1000	H_2O	–	min 0.71 max 1.02	2.63	3.20	$0.16 \cdot 10^{-2}$	$0.14 \cdot 10^{-1}$	1.900	0.740	1.280
	700	100	H_2O	–	min 1.04 max 1.57	2.60	4.40	–	–	2.700	–	–
	300	1000	CO_2		0.12	2.70	0.51	–	–	–	–	–
	500	1000	CO_2	–	0.40	2.68	1.34	–	–	–	–	–
	600	1000	CO_2	–	0.86	2.65	2.42	–	–	–	–	–
Fine-grained		Initial		0	0	2.73	0.20	–	–	$1.30 \cdot 10^{-6}$	–	–
'dark mable'	500	1	Air	–	0.40	2.69	1.60	–	–	$2.40 \cdot 10^{-3}$	–	–
	340	1000	H_2O	–	0.53	2.68	2.20	–	–	–	–	–
	500	1000	H_2O	–	0.84	2.63	3.60	–	–	$1.20 \cdot 10^{-1}$	–	–
Garnet skarn		Initial		0	0	3.46	1.85	–	–	–	–	–
	500	1000	H_2O	–	0.01	3.37	4.21	–	–	–	–	–
Magnetite ore		Initial		0	0	4.96	–	–	–	–		
	515	1	Air	0.72	0.00	4.96	–	–	–	–	–	–
	500	1000	H_2O	0.74	0.10	4.94	–	–	–	–	–	–

deformation (degree of decompaction) increases, ultrasonic wave velocities and elastic moduli of rocks decrease regularly. In the most decompacted rocks, the velocities of elastic waves are attenuated by a factor of 1.5–2. The shear moduli and Young's moduli of rocks are reduced by a factor of 1.5–3 after heating to 500 °C and by a factor of 2–6 after heating to 700 °C. The bulk compressional modulus K proves to be the most sensitive to water saturation. For instance, $K = 50$ kbar for dry granite which has experienced decompaction at 600 °C, while $K = 285$ kbar for the same decompacted sample, subsequently saturated with water. The drop in the elastic characteristics of rocks is undoubtedly caused by the appearance and widening of intergranular microcavities.

10.3.4 The effect of temperature and pressure on rock permeability

Direct measurements of rock permeability at high temperatures and pressures are sparse. The most systematic measurements are those of Shmonov and Vitovtova (1992) who investigated argon infiltration through cylindrical samples of granites, marbles, and dolomites at effective pressures from 200 to 2000 bar and temperatures between 20 and 600 °C (see also Chapter 11, this volume). The authors found that when the pressure applied to the rock framework exceeds the gas pressure in rock pores, then permeability at high temperatures appears to be lower than that established by us after a 'free' decompaction event.

Recently, the authors (Zonov *et al.*, 1989) investigated the permeability of New Ukrainian and Akchatau granites at temperatures from 20 to 600 °C and $P_{ef} = 50$, 100, 200, 500 and 800 bar. To prevent pores from being plugged by redeposition of material during water infiltration, measurements were made under low fluid pressure gradients ($P_{fl} = 10$–15 bar. cm^{-1}) and for short infiltration times.

Figure 10.7 illustrates the temperature dependence of the permeability of Precambrian New Ukrainian granite at $P_{ef} = 200$ and 500 bar. Also depicted in the figure is its residual permeability curve after heating, provided $P_c = P_{H_2O} = 1000$ bar ($P_{ef} = 0$). It can be seen that the shapes of the curves are the same at effective pressures of 200 and 500 bar and that permeability decreases drastically with increasing P_{ef} in the range 0 to 200 bar. It should be stressed, however, that even if lithostatic pressure exceeds fluid pressure by 500 bar, this cannot inhibit the effect of granite thermal decompaction, causing the permeability of the granite sample to increase by more than two orders of magnitude upon heating from 400 to 600 °C.

For the highly permeable Permian Akchatau granite, an increase in P_{ef} from 50 to 800 bar at $T = 20$ °C results in a decrease in the initial permeability from 10^{-3} to $4.5 \cdot 10^{-5}$ md (Figure 10.8). The temperature dependence of permeability exhibits a slight minimum at 100–200 °C, more or less irrespective of P_{ef}. A similar minimum was established by

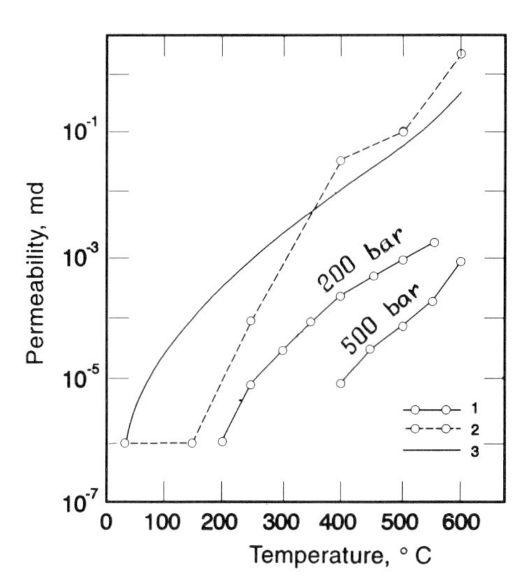

Figure 10.7. Temperature dependence of the permeability of New Ukrainian granite: 1 experimental values measured *in situ* at the effective pressure and run temperature specified on the curves, 2 residual permeability of granite after it has been heated to the desired temperature in an autoclave at $P_{H_2O} = 1$ kbar, 3 theoretical temperature dependence of permeability at $P_{ef} = 0$, calculated using the proposed percolation model.

Vitovtova and Shmonov (1982) for granites. It may be due to the partial disruption of the connectivity of the initial system of pore channels in the rock as a result of the expansion of mineral grains. As heating proceeds, the increased local thermo-elastic stresses result in the development of a renewed system of connecting microcracks along grain boundaries. Over the temperature range 200–600 °C, permeability increases by 2–2.5 orders of magnitude, compared with that measured at 20 °C, for a given P_{ef}.

10.4 DISCUSSION

10.4.1 Physical nature of rock decompaction under nongradient heating

The universal nature of rock decompaction means it must be related to universal properties of rocks. A rock is a polycrystalline aggregate, with different values of compressibility and thermal expansion for the individual minerals that make it up. Obviously, these differences should eventually produce internal stresses in the polycrystalline aggregate, as the intensive parameters (P,T) change progressively. However, in general, the reasons for rock decompaction upon uniform heating include not only the difference between the values of relative volume

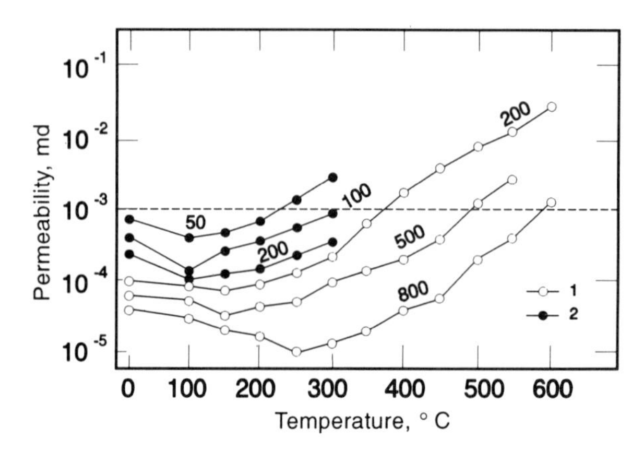

Figure 10.8. Permeability of Akchatau leucocratic granite as a function of temperature and effective pressure (numbers on the curves refer to P_{ef} in bar): 1 experiments conducted in gold, sealed jackets, 2 *idem* but in Teflon jackets.

thermal expansions of individual minerals, but also the anisotropy of thermal expansion of mineral lattices. Thermal expansion is a tensor property of a mineral, analogous to refractive index, and this accounts for the severe decompaction of a monomineralic polycrystalline aggregate (white marble) (see Figure 10.2).

The importance of anisotropy for thermal decompaction is demonstrated by the data for magnetite ore (Figure 10.2, curve 12), composed of isotropic cubic mineral, which expands without experiencing thermal decompaction, in almost perfect accord with published values for the thermal expansion of single crystals of magnetite and garnet, respectively (Table 10.2).

Because of the high values of elastic moduli for minerals, the local thermoelastic stresses built up due to the thermal expansion anisotropy of neighbouring crystals may exceed the strength of grain–grain contacts. The developing system of microcracks proves rather stable mechanically due to rotation of grains and related displacements over irregular contacts between grains (Balashov and Zaraisky, 1982).

10.4.2 Theoretical prediction of thermal decompaction of rocks

We have proposed a quantitative model which allows calculation of the temperature dependence of the thermal decompaction of any rock, based on the differences in linear expansion of rock-forming minerals (Zaraisky and Balashov, 1978, 1981). Figure 10.9a illustrates the model. The calculation requires that the quantitative modal mineral composition of the rock, and values for the thermal expansion of each mineral along three principal axes, be known. Our approach is based on the summation of the contributions to the excess linear rock expansion

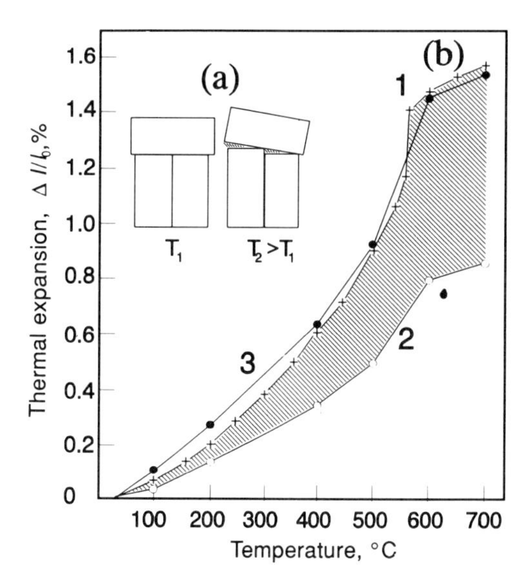

Figure 10.9. A schematic diagram showing the thermal decompaction mechanism for rocks (a), and the thermal expansion of New Ukrainian granite at atmospheric pressure (b): 1 experimentally determined dilatometric curve for relative linear expansion of granite, 2 calculated expansion curve for the mineral matrix of granite without regard to the decompaction process, 3 theoretical linear expansion curve of granite with regard to the decompaction process, using the proposed thermal decompaction model. The shaded region is a fraction of 'excess' rock expansion due to the appearance of an intergranular space as a result of thermal decompaction.

($\Delta l \cdot l_o^{-1}$ excess) from all possible twin combinations of axes, allowing for the probability of their conjunction in accordance with the volume fraction of each minerals in the rock:

$$\Delta l^* \cdot l_{o\,(excess)}{}^{-1} = \frac{1}{9}\sum_{k_1k_2}(W_{k_1}\,W_{k_2}\sum_{j_1j_2}\Delta L^{j_1j_2}_{k_1k_2}), \qquad (10.7)$$

where $k_1, k_2 = 1, \ldots, n$ are minerals present, $j_1, j_2 = 1, 2, 3$ represent the axes of a thermal expansion ellipsoid, W_{k_1} and W_{k_2} refer to the volume fractions of minerals k_1 and k_2 in the rock, and:

$$\Delta L^{j_1\,j_2}_{k_1\,k_2} = \left|\left(\frac{\Delta l}{l^{j_1}_{ok_1}} - \frac{\Delta l}{l^{j_2}_{ok_2}}\right)\right|$$

denotes the absolute magnitude of the linear expansion difference for the superscripted combinations of axes of the thermal expansion tensor for minerals k_1 and k_2.

The calculation of the complete thermal expansion of rock requires that the expansion of a crystalline matrix be known:

$$\Delta l \cdot l_{0(min)}^{-1} = \frac{1}{3} \sum_{k} W_k \, \Delta V_k / V_{ok}, \qquad (10.8)$$

where $\Delta V_k \cdot V_{ok}^{-1}$ is the volume thermal mineral expansion. The overall linear rock expansion is:

$$\Delta l \cdot l_{0(rock)}^{-1} = \Delta l \cdot l_{0(min)}^{-1} + \Delta l^* \cdot l_{0(excess)}^{-1}. \qquad (10.9)$$

Ultimately, the general linear thermal expansion may be written as:

$$\Delta l \cdot l_0^{-1} = \frac{1}{3} \sum_{k} W_k \, \Delta V_k \cdot V_{ok}^{-1} + \frac{1}{9} \sum_{k_1 k_2} (W_{k1} \, W_{k2}) \sum_{j_1 j_2} \Delta L_{k_1 k_2}^{j_1 j_2}. \qquad (10.10)$$

Our calculations for theoretical values of rock thermal expansions with the above formula used Skinner's (1966) data for thermal expansion of minerals. The thermal expansion curve for New Ukrainian granite so calculated exhibits a surprisingly close coincidence with the experimental curve (Figure 10.9b). A comparison of the data calculated by Equation 10.10 and our experimental data on the thermal expansion of all 13 rock types used in experiments was presented in Zaraisky and Balashov (1981). It reveals very good agreement up to 600°C. In the higher-temperature region, brittle cracking in relatively coarse-grained (0.2–0.8 mm) samples of granodiorite, gabbro and white marble, resulted in increased experimental values.

10.4.3 Percolation model for the permeability of decompacted rocks

Although there is a set of empirical dependencies of permeability on porosity and formation factor (Sheidegger, 1960; Brace, 1977), in general close correlation is lacking. Analysis of experimental rock decompaction data has allowed the development of an entirely new theoretical method for calculating transport properties (permeability, formation factor) of rocks, provided these values are uniquely controlled by the pore space geometry (Balashov and Zaraisky, 1982). The construction of this theoretical model was made possible by developments in mathematical percolation theory in the 1960–1970s (Shante and Kirkpatrik, 1971). Taking an infinite, ordered network of conducting channels with a statistical size distribution of effective channel diameters, and using percolation theory, it is possible to determine the average contribution from channels of a certain diameter range to the overall transport value characterizing the conductivity (permeability) of the network considered. In the case of rocks, there always exists an initial connected system of intergranular planes (grain boundaries) to be regarded as a medium ordered network.

The details of the proposed method for calculating transport characteristics of decompacted rock are presented by Balashov in Chapter 9 of this volume (see also Balashov and Zaraisky, 1982).

In such a manner the theoretical permeability of New Ukrainian granite has been calculated over the temperature range 20–600 °C (Figure 10.7). It is evident from the figure that the calculated curve follows the shape of the experimental curve for conditions of $P_c > P_{fl}$, but lies above it. However, the experimental curve for $P_{ef} = 0$ is cut by the calculated curve. This may be accounted for by the partial reversibility of the deformation of the polycrystalline granite aggregate upon heating to < 400°C in contrast to essentially irreversible deformation upon heating to higher temperatures. In the latter case, the aggregate develops an additional intergranular space upon cooling due to crystal contraction.

10.4.4 Thermal decompaction and internal creep of rocks

This section discusses the possibility that thermal decompaction may be prevented or weakened in massive rocks through viscous relaxation under the action of thermo-elastic stresses.

The rate at which a rock undergoes decompaction is defined by the relationship:

$$\frac{d\gamma_{dc}}{dt} = \Delta\alpha \frac{dT}{dt}, \tag{10.11}$$

where $\gamma_{dc} = \overline{\Delta l} \cdot l_0^{-1}$ is the excess linear rock expansion, $\Delta\alpha$ the mean difference between the thermal expansion coefficients of minerals, T is temperature in degrees Kelvin, and t is time. The value of $dT \cdot dt^{-1}$ corresponds to the rate of temperature change. For a rate of viscous rock deformation, the following expression is valid:

$$\frac{d\gamma_r}{dt} = \sigma_c \cdot \eta^{-1}, \tag{10.12}$$

where γ_r is the relative viscous deformation, η the effective mineral viscosity, σ_c the mean internal stress in the rock, with a limiting value controlled by the strength of both the grain–grain contacts and the mineral grains of rock.

The critical condition for thermal decompaction of rock is:

$$\frac{d\gamma_{dc}}{dt} > \frac{d\gamma_r}{dt}, \tag{10.13}$$

or

$$\frac{dT}{dt} > \sigma_c \cdot \eta^{-1} \cdot \Delta\alpha^{-1}. \tag{10.14}$$

The value of σ_c is governed by the strength of grain–grain contacts and the mineral grains of rock and with a surface-active hydrothermal medium present, σ_c may become several times smaller due to the Rehbinder effect (Rehbinder and Shchukin, 1972). Estimating the strength of dry rocks on the basis of Fredrich and Wong's (1986) data results in σ_c in the range of ~ 10–200 bar. From Figure 10.2, the value of $\Delta\alpha$ is estimated to be $\sim 5 \cdot 10^{-6} \cdot \text{deg}^{-1}$. The effective viscosity of a mineralic rock aggregate is the most difficult parameter to estimate, but this has been done as follows.

The diffusional viscosity of a polycrystalline material is defined by the relationship (Zharkov, 1983; Poirier, 1985):

$$\eta = A\frac{kT}{Da}\left(\frac{h}{a}\right)^2, \tag{10.15}$$

where $A \cong 1/30$ is a constant, $k = 1.3805 \cdot 10^{-23}\,\text{J.K}^{-1}$ is the Boltzmann constant, T is temperature in Kelvin, h refers to the crystal grain size, D denotes the self-diffusion coefficient for major structural atoms in the lattice (in $m^2 \cdot sec^{-1}$), and a symbolizes the linear size of a crystalline cell. Setting $h \cong 2 \cdot 10^{-4}$m, and $a \cong 5 \cdot 10^{-10}$m, D was estimated on the basis of self-diffusion coefficients for Si in quartz (Freer, 1981). Resulting values of D range between $2 \cdot 10^{-21}$ and $2 \cdot 10^{-18}\,\text{cm}^2.\,\text{sec}^{-1}$, as the temperature is varied from 800 to 1000 K. For 900 K and $D = 10^{-24}\,\text{m}^2.\,\text{sec}^{-1}$, the effective viscosity of polycrystalline rock is:

$$\eta_{cr} \cong 10^{15}\,\text{kbar.sec.} \tag{10.16}$$

In all probability, the minimum effective viscosity of a polycrystalline aggregate under hydrothermal conditions may be provided by the 'pressure solution' process, in which those portions of rock that are mechanically more stressed are dissolved, redepositing their material in less stressed portions. In this case, the general relationship describing effective viscosity may be written as:

$$\eta_{ps} = \left(\frac{h^2RT}{(V_{cr})^2\,D^*m^0} + \frac{2hRT}{(V_{cr})^2\,k_m}\right) \times 10^6, \tag{10.17}$$

where h, as in Equation 10.15, is the average grain size of the rock, R the gas constant, T is temperature in Kelvin, V_{cr} stands for the molar mineral volume, and k_m is the kinetic constant for mineral dissolution. The symbol m^0 represents the mean equilibrium molality of a component in the pore fluid ($\cong 10^{-2}$) and is defined as:

$$m^0 = k_m \cdot (k_{\bar{m}})^{-1}, \tag{10.18}$$

where $k_{\bar{m}}$ is the kinetic constant for mineral deposition. The expression

$$D^* = D_f\,F\,\rho, \tag{10.19}$$

also holds (see Balashov, Chapter 9, this volume), where D_f is the pore fluid diffusion coefficient, F the formation (diffusion) factor for a porous medium, and ρ is a factor to convert molality units to moles of component per unit volume of pore fluid. If the unit of length is 1 cm, then $\rho = \rho_{H_2O} \cdot 1000^{-1}$, where ρ_{H_2O} is the density of water in $g.cm^{-3}$.

From relation 10.17, it follows that, with infinitely large mineral dissolution–deposition rates, i.e. at local chemical equilibrium, the effective viscosity is determined by the first term of the equation and may be arbitrarily termed diffusional. Conversely, in the case of relatively fast diffusion, viscosity is determined by the second term and is thus kinetic.

Critical values of the heating rates $(dT/dt^{-1})_c$ to produce thermal decompaction of a polycrystalline aggregate of quartz (quartzite) have been calculated as a function of temperature, assuming $\sigma_c = 100$ bar and taking the effective viscosity from Equations 10.15 and 10.17. Based on diffusion data in Chapter 9, the following values were estimated: $D_f \cong 10^{-4}\ cm^2 \cdot sec^{-1}$, $F = 10^{-4}$, $\rho \cong 5 \cdot 10^{-4}\ kg.cm^{-3}$ at $P_{fl} \sim 1$ kbar; the data on quartz dissolution kinetics are taken from Rimstidt and Barnes (1980). Calculation of viscosity with Equation 10.15 corresponds with quartzite decompaction under dry conditions, whereas Equation 10.17 describes the viscosity values under hydrothermal conditions. The results of the calculation of critical heating rate are presented in Figure 10.10. The change in slope of the hydrothermal curve at about 300 °C reflects the transition from kinetic to diffusional control on the effective viscosity, i.e. the dominant term in Equation 10.17 changes with temperature.

Critical values for heating rates from Figure 10.10 may be compared with heating rates for country rocks in the vicinity of a crystallizing intrusive. Modelling the heating rate with a 1–D solution and a temperature difference at the initial contact of about 500 °C shows that thermal decompaction will occur widely under dry conditions. For hydrothermal conditions, the area with a heating rate favourable to the thermal decom-

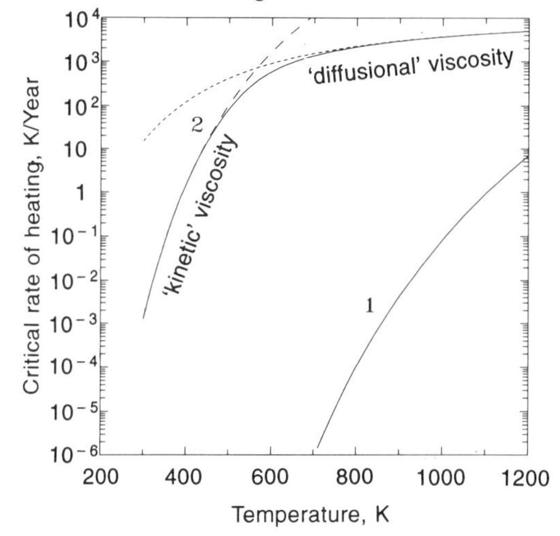

Figure 10.10. Critical values of heating rates for quartzite decompaction under dry (1) and hydrothermal (2) conditions.

paction of the country rock is located around 10–20 m away from the contact.

However, at least two circumstances prevent the low values for effective viscosity under hydrothermal conditions given by Equation 10.17 from being realized in nature:

1. In the absence of a free pore space for recrystallization deformation mechanisms other than pressure solution (e.g. dislocation sliding) may control effective viscosity.
2. The small pore size of a dense rock ($< 0.01\ \mu m$) limits to a certain extent the size of crystal nuclei in pore solution or on the crystal surface of a mineral.

This may result in the stabilization of oversaturated solutions in thin pores due to the increased contribution made by surface energy (Zaraisky and Balashov, 1987).

Thus, it is best to assume that the true values for mean effective viscosity fall between our limiting estimates for dry and hydrothermal rocks.

10.5 CONCLUSIONS: POSSIBLE GEOLOGICAL IMPLICATIONS OF THE EFFECT OF THERMAL DECOMPACTION OF ROCKS

The role played by thermal decompaction in increasing rock permeability and contributing to convective fluid transport through the intergranular space has received little attention in the geological literature. However, we believe that this process may be of great importance in particular geological settings. According to our data and those from the literature (Pek, 1968; Starostin, 1979; Brace, 1980), the permeability of most crystalline rocks, measured on samples free of microcracks, ranges between 10^{-3} and 10^{-6} md. Such rocks should act as impenetrable barriers to infiltration of hydrothermal fluids. However, in nature, metasomatic transformations frequently cover huge rock masses (tens or hundreds of metres) uniformly, while metamorphic infiltration may affect sequences of many kilometres. Dispersion aureoles of ore metals around deposits also extend for tens or hundreds of metres from ore bodies throughout the volume of rock. Geological observations provide evidence that hydrothermal solutions may not only migrate along macrocracks, but also percolate uniformly through the rock sequences along micropores, having access to each grain. As a possible mechanism for increase in pore permeability of rocks, we have suggested a thermal decompaction phenomenon manifested by the development of intergranular spaces due to the thermal expansion anisotropy of minerals (Zaraisky and Balashov, 1978, 1981).

The major factors that promote decompaction include high thermal expansion anisotropy of rock-forming minerals, a fast heating rate, high temperatures, large crystal sizes, the presence of a surface-active aqueous fluid and high fluid pressure. Because of the high elastic moduli of minerals, the local thermoelastic stresses set up along crystal boundaries may amount to several kilobars and are capable of exceeding the strength of grain–grain contacts. The most likely mechanism inhibiting thermal decompaction could be the relaxation of thermo-elastic stresses as a result of plastic deformation. This may be through crystal creep or pressure solution mechanisms. The quantitative estimates for these phenomena (above) show that when the country rocks at contacts of granitic intrusions are heated at realistic rates, the thermal decompaction process may be released owing to rock creep arising from recrystallization through solution. In this case, creep of minerals in the solid state appears to be insufficient to stop thermal decompaction. As a result, for sealed grain boundaries (where there is no solution present), we should always expect the effect of thermal decompaction to occur. The aforesaid also applies to very thin pores and microcracks (about $10^{-2} - 10^{-3}$ μm in size) filled by fluid, in which crystallization of solid phases is hindered by a large curvature of the surface (Zaraisky and Balashov, 1987). The generated cracks should display thermodynamic stability due to the lowering of surface energy of a solid phase in contact with the surface-active aqueous fluid, as compared to the surface energy value for solid–solid boundary (Rehbinder and Shchukin, 1972). The generally high porosity of hydrothermally altered rocks relative to initial ones (Norton and Knight, 1977; Pakhomov and Pakhomov, 1984) provides evidence suggesting that relaxation–dissolution–recrystallization processes do not lead to cavity healing.

The geological role of the thermal decompaction effect may be most important in contact aureoles of cooling plutons and at greater depths, particularly during granitization, when the plastic state of rocks undergoing recrystallization indicates low effective pressures ($P_c = P_{fl}$). The fact that high-temperature metasomatism may occur uniformly through aureoles of magmatic bodies (e.g. contact hornfels, skarns, greisens) in contrast to local low-temperature metasomatic processes operative in proximity to cracks under near-surface conditions, can be attributed to the higher permeability of heated rocks. Fascinating data which, in our opinion, support the role of decompaction are presented by Ferry (1987). He showed that with increasing metamorphic temperature from 390 (chlorite zone) to 530°C (sillimanite zone), the mode of fluid flow changes from highly channelled to uniformly distributed through the rock mass.

The most favourable conditions for decompaction are created in the immediate vicinity of melt intrusions and in the vicinity of cracks and permeable zones, which, as conductors of high- temperature hydrothermal fluids, provide intensive advective heat flow. In this latter instance, only narrow near-contact margins (tens of centimetres to a few metres) of the surrounding rocks will be subjected to decompaction. However, this corresponds to the morphology and scale of aureoles of vein-fracture hydrothermal alterations of rocks. In the course of cooling after crystallization, the massive magmatic rocks themselves should also undergo decompaction due to the non-uniform contraction of the crystal volumes: an effect similar to decompaction upon heating. At 700 °C, the excess expansion of granite is 0.67%, yielding 2% of connected pore space, capable of providing a high permeability of the order of several millidarcies. This implies that immediately after consolidation, the apical part of the massif may prove an excellent fluid conductor, thus correlating well with the commonly observable intense area hydrothermal alteration of domes and inliers in granitic massifs. Chernyayev (1988) described strongly metasomatized gabbro–diabase dykes and stocks which acted as good fluid conductors in a sequence of volcanogenic rocks. The effect of thermal decompaction may be of practical significance for some industrial processes accompanied by rock heating, such as radioactive waste disposal or heat production due to the heating of water injected into deep-seated hot rocks ('Hot dry rock' geothermal energy).

By weakening contacts between grains, thermal decompaction significantly reduces rock strengths and thus facilitates crushing of heated rock blocks, making them more permeable. Thus, the strength of fine-grained aplite subjected to a confining pressure of 5 kbar is reduced by a factor of 1.5 upon heating to 300 °C and by a factor of 3 upon heating to 700 °C (Tullis *et al.*, 1979). The strength of contacts between mineral grains becomes more reduced in the presence of surface-active aqueous fluid (hydrolytic weakening).

In summary, we consider that the process of thermal decompaction is of fundamental importance in the evolution of transport properties of rocks. Both the formation of a network of large discontinuities due to the drop in the rock strength (with a circumstantial increase in permeability), and the increase in permeability of individual rock blocks upon decompaction, are of importance. As noted by Zaraisky and Balashov (1986), one should also consider the possibility of hydraulic rupture of the country rock sequence overlying the intrusion as a result of:

1. thermal decompaction of bed rocks adjacent to the crystallizing intrusive, and
2. magmatic fluid emplacement into it under the action of lithostatic pressure.

The reported estimates for stability of microcracks due to thermal decompaction (Balashov and Zaraisky, 1982), as well as experimental permeability data at different effective pressures, are indicative of the occurrence of the thermal decompaction process up to, at least, an effective pressure of 500 bar. The estimates reported here for the critical rates of temperature change necessary to overcome the process of pseudoviscous rock relaxation, also indicate that thermal decompaction may be a relatively widespread process in the formation of hydrothermal deposits associated with a cooling crystallizing intrusive.

The research described in this chapter is supported by the Russian Fund of Fundamental Investigation under Grant No. 93–05–9822.

REFERENCES

Balashov, V. N. and Zaraisky, G. P. (1982). Experimental and theoretical investigation of thermal decompaction of rocks on heating. In *Contributions to Physico-chemical Petrology*, **X**, pp. 69–109. Nauka Press, Moscow (in Russian).

Bauer, S. J. and Johnson, B. (1979). *Effects of slow uniform heating on the physical properties of Westerly and Charcoal granites*. Proc. 20th US Symp. Rock Mech., pp. 7–19. Austin, Texas.

Birch, F. (1943). Elasticity of igneous rocks at high temperatures and pressures. *Geol. Soc. Amer. Bull.*, **54**, pp. 263–86.

Birch, F. (1966). Compressibility; Elastic constants. In *Handbook of Physical Constants* (ed. S. P. Clark). Mem. Geol Soc. Amer., p. 97.

Brace, W.F. (1977). Permeability from resistivity and pore shape. *J. Geophys. Res.*, **82**, N 23, pp. 3343–9.

Brace, W. F. (1980). Permeability of crystalline and argillaceous rocks: status and problems. *Int. J. Rock Mech. Mining Sci. Geomech. Abstr.*, **17**, pp. 241–51.

Chernyayev, Eu. V. (1988). Composite zoned dykes as a result of metasomatism in fluid conductors. *Geologia Rudnykh Mestorozhdenii*, **2**, pp. 75–84 (in Russian).

Ferry, J.M. (1987). Metamorphic hydrology at 13 km depth and 400–550 °C. *Amer. Mineral.*, **72**, N 1, pp. 39–58.

Freer, R. (1981). Diffusion in silicate minerals and glasses: a data digest and guide to the literature. *Contrib. Mineral. Petrol.*, **76**, N 4, pp. 440–54.

Fredrich, J. T. and Wong, T.– F. (1986). Micromechanics of thermally induced cracking in three crustal rocks. *J. Geophys. Res.*, **91**, N B12, pp. 12743–64.

Gerand, J. and Gaviglio, P. (1990). Modification experimentale de la texture de granites par chauffage: evolution de la porosite et de la densite en fonction de la temperature. *C.R Acad.Sci.Paris*, serie II, **310**, pp. 1681–6.

Hadley, K. (1976). Comparison of calculated and observed crack densities and seismic velocities in Westerly granite. *J. Geophys. Res.*, **81**, pp. 3484–93.

Heard, H. C. and Page, L. (1982). Elastic moduli, thermal expansion, and inferred permeability of two granites to 350°C and 55 MPa. *J. Geophys. Res.*, **87**, pp. 9340–8.

Ide, J. M. (1937). The velocity of sound in rocks and glasses as a function of temperature. *J. Geophys. Res.*, **45**, pp. 689–716.

Maxwell, J. C. and Verrall, P. (1953). Expansion and increase in permeability of carbonate rocks on heating. *Trans. Amer. Geophys. Union*, **34**, N 1, pp. 101– 6.

Morrow, C., Lockner, D., Moore, D. and Byerlee, J. (1981). Permeability of granite in a temperature gradient. *J. Geophys. Res.*, **86**, N B4, pp. 3002–8.

Norton, D. and Knight, J. (1977). Transport phenomena in hydrothermal systems: Cooling plutons. *Amer. J. Sci.*, **277**, N 8, pp. 937–81.

Nur, A. (1982). Processes in rocks with fluids at elevated pressure and temperature. In *High-Pressure Researches in Geoscience* (ed. W. Schreyer). E. Schweizerbartische Verlagsbuchhandlung, Stuttgart, pp. 67–83.

O'Connel, R. J. and Budiansky, B. (1974). Seismic velocities in dry and saturated cracked solids. *J. Geophys. Res.*, **79**, N 35, pp. 5412–26.

Pakhomov, V.I. and Pakhomov, M. I. (1984). Petrophysical peculiarities of wall-rock metasomatites. In *Metasomatism and Ore Formation*. Nauka Press, Moscow, pp. 329–342 (in Russian).

Pek, A. A. (1968). On the dynamics of juvenile solutions. Nauka Press, Moscow, 168 pp. (in Russian).

Poirier, J.-P. (1985). *Creep of Crystals*, Cambridge University Press, Cambridge.

Rehbinder, P. A. and Shchukin, Eu. D. (1972). Surface effects in solid bodies in the processes of their deformation and failure. *Uspekhi Phisicheskikh Nauk*, **108**, N 1, pp. 3–42 (in Russian).

Richter, D. and Simmons, G. (1974). Thermal expansion behavior of igneous rocks. *Int. J. Rock Mech. Mining Sci. Geomech. Abstr.*, **11**, pp. 403–11.

Rimstidt, J. D. and Barnes, H. L. (1980). The kinetics of silica–water reactions. *Geochim. Cosmochi. Acta*, **44**, N 11, pp. 1683–99.

Scott, D. S. and Dullien, F. A. L. (1962). The flow of rarefied gases. *A. I. Ch. E. Journal*, **8**, N 3, pp. 293–7.

Shante, V. K. S. and Kirkpatrik, S. (1971). An introduction to percolation theory. *Adv.Phys.*, **20**, N 85, pp. 325–357.

Sheidegger, A. E. (1960). *Physics of Fluid Flow Through Porous Media*. GOSTOP-TEKHIZDAT Press, Moscow (in Russian).

Shmonov, V. M. and Vitovtova, V.M. (1992). Rock permeability for the solution of the fluid transport problems. *Experiment in Geosciences*, **1**, pp. 1–49.

Skinner, B. J. (1966). Thermal expansion. In *Handbook of Physical Constants* (ed. S.P. Clark). Mem. Geol. Soc. Amer., p. 75.

Starostin, V. I. (1979). Methods of measurement of physical-mechanical properties of rocks and ores in ore-petrographic studies. In *Laboratory Methods of Investigation of Minerals, Ores, and Rocks*. Moscow University Press, Moscow pp. 175–270 (in Russian).

Summers, R. Winkler, K. and Byerlee, J. D. (1978). Permeability changes during the flow of water through Westerly granite at temperatures of 100–400 °C. *J. Geophys, Res.*, **83**, N B1, pp. 339–44.

Tullis, J., Shelton, G.L. and Yund, R.A. (1979). Pressure dependence of rock strength: implications for hydrolytic weakening. *Bull. Mineral.*, **102**, pp. 110–14.

Vitovtova, V. M. and Shmonov, V. M. (1982). Permeability of rocks at pressures to 2000 kg. cm^{-2} and temperatures to 600 °C. *Doklady Akademii Nauk SSSR*, **266**, N 5, pp. 1244–8 (in Russian).

Zaraisky, G. P. (1989). *Zoning and Conditions of Formation of Metasomatic Rocks*. Nauka Press, Moscow (in Russian).

Zaraisky, G. P. and Balashov, V. N. (1978). On the decompaction of rocks upon heating. *Doklady Akademii Nauk SSSR*, **140**, N 4, pp. 926–9 (in Russian).

Zaraisky, G. P. and Balashov, V. N. (1981). Thermal decompaction of rocks as a factor for the formation of hydrothermal deposits. *Geologia Rudnykh Mestorozhdenii*, **N 6**, pp. 19–35 (in Russian).

Zaraisky, G. P. and Balashov, V. N. (1983a). Rocks as a medium for transporting hydrothermal solutions. *Geologicheskii Zhournal*, **43**, N. 2, pp. 19–38 (in Russian).

Zaraisky, G.P. and Balashov, V. N. (1983b). Transport mechanisms for hydrothermal solutions. *ibid*, pp. 38–49 (in Russian).

Zaraisky, G. P. and Balashov, V. N. (1986). Thermal decompaction of rocks and its role in the formation of hydrothermal ore systems. In *Conditions for the Formation of Ore Deposits* (Proc. 6th Symp. IAGOD), pp. 694–700. Nauka Press, Moscow (in Russian).

Zaraisky, G. P. and Balashov, V. N. (1987). Metasomatic zoning: theory, experiment, calculations. In *Contributions to Physico-chemical Petrology*, **XIV**, pp. 136–82. Nauka Press, Moscow (in Russian).

Zaraisky, G. P., Balashov, V. N. and Zonov, S. V. (1989). *Thermal decompaction of rocks and its effect on permeability*. Proc. 6th Int. Sym. Water–Rock Interaction (Malvern). Rotterdam: Balkema, pp. 797–800.

Zharkov, V. M. (1983). *Earth and Planetary Interiors*. Nauka Press, Moscow (in Russian).

Zonov, S. V., Zaraisky, G. P. and Balashov, V. N. (1989). The effect of thermal decompaction on permeability of granites, with lithostatic pressure being slightly in excess of fluid pressure. *Doklady Akademii Nauk SSSR*, **307**, N 1, pp. 191–5 (in Russian).

GLOSSARY OF SYMBOLS

a	linear size of a crystalline cell, [m].
F	formation factor, the ratio of the electrical conductivity of the rock saturated with an electrolyte solution to the electrolyte conductivity.
B	Klinkenberg's or slipping gas constant, [md · bar].
D	self-diffusion coefficient for major structural atoms in the lattice, [$cm^2 \cdot sec^{-1}$ except where specified].
D_f	pore fluid diffusion coefficient, [$cm^2 \cdot sec^{-1}$].
D^*	effective diffusion coefficient through rock media, [$kg \cdot cm^{-1} \cdot sec^{-1}$].
h	the membrane thickness; the microcrack width; the crystal grain size, [cm or m].
h_F	effective microcrack width, Equation 10.4, [cm].
h_S	effective microcrack width, Equation 10.5, [cm].
K^0	true value of permeability, [md].
K, K_{CCl_4}	gas and liquid (CCl_4) permeability, [md].
K	bulk compressional modulus, [kbar].
k	Boltzman constant, $1.3805 \cdot 10^{-23}$ [$J \cdot K^{-1}$].
k_m	the kinetic constant for mineral dissolution, [$mol \cdot cm^{-2} \cdot sec^{-1}$].
l	length of microcrack.
m^0	equilibrium molality of a component in the pore fluid, [$mol \cdot kg^{-1}$].
$\overline{P}$	the average pressure in the membrane, [bar].
P_1	atmospheric pressure, [bar].
P_c	confining pressure exerted on the solid rock framework, [bar].
P_{fl}	pore fluid pressure, [bar].
P_{ef}	effective pressure, $P_c - P_{Fl}$, [bar].
R	gas constant, 8.3144 [$J \cdot K^{-1} \cdot mol^{-1}$].
T	temperature.

t	time.
U_1, U_{CCl_4}	the volumetric rate [$cm^3 \cdot cm^{-2} \cdot sec^{-1}$] of gas or liquid flow through a 1 cm^2 cross section of the membrane at atmospheric pressure P_1.
V_{cr}	mineral molar volume, [$cm^3 \cdot mol^{-1}$].
$\overline{v}$	the mean thermal molecular velocity of gas, [$cm \cdot sec^{-1}$].
W_{k_1}	volume fraction of K_1^{th} mineral in rock.
α	porosity of rock.
γ_r	relative viscous deformation.
$\Delta\alpha$	mean difference between the thermal expansion coefficients of minerals, [K^{-1}].
$\Delta l^*/l_{o(excess)}$, γ_{dc}	excess linear expansion of rock.
$\Delta l/l_o^{j_1}$	linear expansion of kth mineral in direction of j_1 crystallographic axis.
$\underline{\Delta l/l_{o(min)}}$	mean linear expansion of crystalline matrix of rock.
$\Delta l/l_o$	residual excess linear expansion of rock.
$\Delta V_k/V_{ok}$	volume thermal expansion of kth mineral.
ε	crack density parameter.
η	the viscosity of gas or liquid media, [cP].
η_{cr}	effective viscosity of polycrystalline rock, [Pas].
η_{ps}	effective viscosity of polycrystalline rock under conditions of pressure solution, [Pas].
ae_r	electrical conductivity of the rock saturated with 1m KCl electrolyte solution, [$\Omega^{-1} \cdot cm^{-1}$].
ρ	factor converting molality units to moles of component per unit volume of pore fluid.
σ_c	mean internal stress in the rock, [bar].

CHAPTER ELEVEN

Permeability of rocks at elevated temperatures and pressures

Vyacheslav M. Shmonov, Valentina M. Vitovtova and Irina V. Zarubina

11.1 INTRODUCTION

Our research into rock permeabilities at elevated T and P was initiated in 1975 and intended to address the problem of constructing a model for formation of skarn-hosted ore deposits. Most research on such settings has concentrated on mechanisms of melt intrusion and intrusive body formation, changes in temperature fields and stresses inside and in the vicinity of the intrusive body, the thermodynamic properties of melts, fluids and associated solutions, or the infiltration effects themselves. The infiltration and cumulative properties of a fluid environment itself, i.e. the rock matrix which hosts and is converted to skarn and ore, have been less well studied, despite the fact that rock permeability controls the scale and trends of ore-metasomatic processes. At the time we began this study, the available information on deep rock permeability largely reflected the needs of the petroleum and gas industry, and permeability experiments were generally conducted to 150°C and a few hundreds of atmospheres. The physical–chemical conditions of skarn formation, however, required knowledge of permeability to 600–700°C and 1–2 kbar. In the works of Fatt and Davis (1952), Brace *et al.* (1968), and Zoback and Byerlee (1975), permeability measurements were made to 4–5 kbar at room temperature. Unfortunately, none of the studies provided a means of predicting permeability values at high temperatures

Fluids in the Crust: Equlibrium and transport properties.
Edited by K.I. Shmulovich, B.W.D. Yardley and G. G. Gonchar.
Published in 1994 by Chapman & Hall, London. ISBN 0 412 56320 7

and pressures. Extrapolation of experimental data to high T and P without any theoretical basis always involves a risk of obtaining erroneous results. Therefore, in order to measure permeability and study its variations with temperature and pressure, a special apparatus was designed (Shmonov and Chernyshov, 1982), which made it possible to raise the temperature and pressure of the runs to 600 °C and 200 MPa, respectively (Figure 11.1). This was accomplished by replacing the rubber jacket traditionally used for applying pressure to the sample with a flexible gold jacket. As the apparatus was set up and calibrated, a number of related investigations at elevated temperatures and pressures appeared. Notable among these are the studies on thermal decompaction of rocks to 700°C by Zaraisky and Balashov (1978) (see also Chapter 10, this volume), permeability measurements on granites, gneisses, and albites to 350 °C and 50 MPa by Nikolaenko and Indutyi (1978), results on crack healing to an axial stress of 350 MPa, confining pressure of 50 MPa and temperatures of 400°C by Summers *et al.* (1978), investigations of the permeability of limestone and sandstone in a different direction with respect to the maximum load by Marmorshtein (1981), as well as permeability measurements on granites in temperature gradient fields of up to 300 °C along the sample, under a confining pressure to 60 MPa by Moore *et al.* (1983) and Morrow *et al.* (1981).

In recent years experiments have gone far beyond the skarn theme. Since then the permeability of marbles (Vitovtova and Shmonov, 1982), granites and granodiorites (Vitovtova and Shmonov, 1982; Vitovtova *et al.*, 1988), limestones (Vitovtova and Shmonov, 1982; Vitovtova *et al.*, 1988; Shebesta *et al.*, 1988; Magomedov and Shmonov, 1991), dolomites (Vitovtova and Shmonov, 1982; Vitovtova *et al.*, 1988; Shebesta *et al.*, 1988), tuffstones (Vitovtova *et al.*, 1988), wollastonite and garnet skarns (Vitovtova *et al.*, 1988), serpentinites and serpentinized harzburgites (Aksyuk *et al.*, 1991), basalts (Aksyuk *et al.*, 1992), gneisses and amphibolites (Zharikov *et al.*, 1990) have been investigated to meet different aims (Shmonov and Vitovtova, 1992).

It was apparent to us from the first that there is a complex temperature–pressure dependence of rock permeability (Shmonov *et al.*, 1986). Any attempt to interpret the behaviour of rocks under high temperature and pressures is therefore up against considerable difficulty and requires that a variety of models for pore space be developed. Direct observation of the behaviour of cracks and pores under temperature and pressure proves an effective way of testing numerous hypotheses (Batzle *et al.*, 1980; Caillard and Martin, 1982; Tungatt and Humphreys, 1981; Urai *et al.*, 1980). To view the structure of the sample surface to 750°C and 100 MPa in the vacuum of a scanning electron microscope (SEM), two special cells have been devised, one of which is described in Shmonov *et al.* (1990) and Zharikov *et al.* (1990). Study of the bulk (pore plus microcrack) and microcrack permeabilities in rock samples has yielded abund-

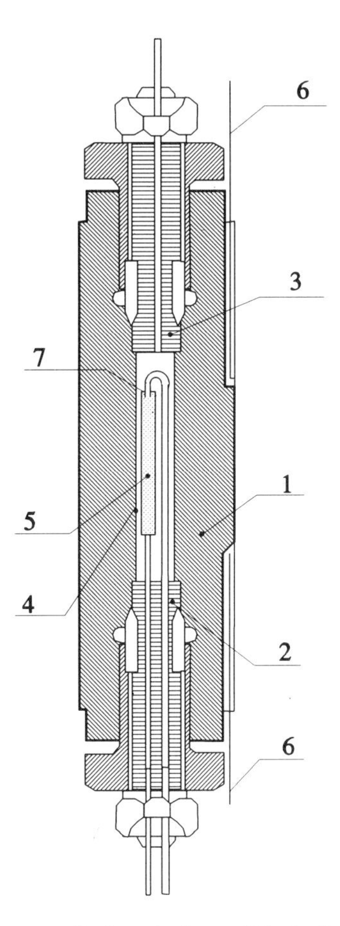

Figure 11.1. High pressure cell. 1 autoclave, 2,3 shutters, 4 gold tube, 5 rock sample, 6 thermocouples, 7 washers.

ant, surprising results that have attracted many workers interested in the problems of fluid transport in different geologic environments.

11.2 KEY CONCEPTS, METHODS OF PERMEABILITY MEASUREMENT AND EXPERIMENTAL PROCEDURE

11.2.1 The concept of permeability

The relation:

$$Q = K\,A\,(h_1 - h_2) \cdot l^{-1} \tag{11.1}$$

is known as **Darcy's law**. Q is the total volume of liquid flowing in unit time through a volume of rock that we will term the filter, l is the filter length, A the cross-sectional area of the filter, h_1 and h_2 represent the heights at which water is maintained in a manometer at the inlet and outlet of the filter, respectively, so that $(h_1 - h_2)$ corresponds to the hydraulic head, and K is a constant (hydraulic conductivity) incorporating the properties of both the flowing liquid and the porous medium. To make Darcy's law of more general usefulness, the contribution of rock properties to hydraulic conductivity must be separated from that of fluid properties. Nutting (1930) showed that the following relation should hold:

$$K = k \cdot \mu^{-1}, \tag{11.2}$$

where μ is the fluid viscosity, and k is the specific permeability of the porous medium. This relationship has been supported by a large body of determinations. By separating the overall constant K into permeability and viscosity contribution, Darcy's law may be expressed for flow in a linear system through a homogeneous medium as:

$$q = (k \text{ grad } P) \cdot \mu^{-1}, \tag{11.3}$$

where $q = Q \cdot A^{-1}$ is the specific flow rate, grad P represents the pressure gradient, and the constant k is permeability. The grad P term includes the hydraulic head and fluid density terms in which Darcy's law is naturally expressed.

11.2.2 Physical meaning of permeability

Permeability is a structural characteristic of a porous material, and is related to other structural characteristics. At different times different workers have investigated the relation of permeability to porosity, size distributions of pores and grains, mineralogical composition of rocks, the effect of grain packing, etc. Of these, permeability can be closely related to porosity (f) (Archie, 1950), hydraulic radius (R) (Brace, 1977), size distribution of pores (Chida and Tanaka, 1983) and tortuosity (t) (Arch and Maltman, 1990). Combining these properties with the dynamic viscosity, μ, and density, ρ, of the liquid yields:

$$q = Q \cdot A^{-1} = C \times [f \times t^{-3} \times R^2] \times [\rho \times g \cdot \mu^{-1}]\, dP \cdot dx^{-1}, \tag{11.4}$$

where C is a nondimensional constant, g the acceleration of gravity, and $dP \cdot dx^{-1}$ the pressure gradient.

11.2.3 Units of permeability measurement

The popular unit of permeability is the darcy, defined as the permeability of a porous material through which the rate of flow of water of viscosity 1 centipoise through a 1 cm^2 section is maintained at

1 $cm^3 \cdot sec^{-1}$ by a pressure gradient of 1 atm. cm^{-1}. Soils and loose sands have a permeability of 20–200 darcies, sandstones and limestones, from 1×10^{-4} to a few darcies, metamorphic and igneous rocks, from 1×10^{-8} to 1×10^{-4} darcy, whereas under high pressure, permeability may fall as low as 10^{-12} darcy (Brace, 1980). Strictly speaking, 1 darcy $= 9.87 \times 10^{-9}$ cm^2. Since the error in measured permeability is much greater than 2%, the following set of relations is adopted with acceptable accuracy : 1 darcy = 1000 mdarcies $= 1 \times 10^{-12}$ $m^2 = 1 \times 10^{-8}$ $cm^2 = 1 \times 10^{-5}$ $m.sec^{-1} = 1 \times 10^{-3}$ $cm.sec^{-1}$. In this paper, we use the SI unit of permeability, m^2, unless otherwise specified.

11.2.4 Ways of measuring permeability

The permeability of a porous material is established regardless of what fluid flows through it, e.g. petrol, gas, oil or solution (unless reaction with the fluid changes pore geometry). Hence the permeability of a sample may be determined not only by water infiltration, but also by means of gas infiltration, which can present experimental advantages. In this case, it is necessary to make several measurements of gas flow at different pressures and calculate the value of permeability. This technique was first suggested by Klinkenberg (1941) and is still in use.

Permeability may also be determined by measuring the electrical resistance of a solution present in the pores of a sample rather than through infiltration of gas or fluid. To do this requires knowledge of hydraulic radius, rock porosity, and formation factor (Brace, 1977).

Both these methods are nondestructive. There are also ways of calculating permeability from the average width and number of cracks per unit length, observable in thin sections (e.g. Romm, 1966) or by means of a network model which uses an analogy between Ohm's and Kirchhoff's law at one end and Darcy's law and general laws of hydrodynamics at the other (e.g. Chida and Tanaka, 1983).

11.2.5 Problems in measuring permeability

As a rule, permeability measurements (up to 10^{-21} m^2) at low temperatures and pressures present no problem. At high stresses, high pore fluid pressures and high temperatures, mineral solubility increases, and more stressed portions of minerals dissolve, with accompanying redeposition of material in less stressed portions. Since the duration of the run required for measurement of low permeability values (10^{-22} to 10^{-24} m^2) is normally 1 to 6 days (e.g. Trimmer *et al.*, 1980), then over the course of the experiment the rock structure changes. Beyond some P–T limit, depending on the material and fluid, the alterations become so intense that permeability measurements are transformed into investigations of the dynamics of changing the structure of the porous material. Permeability may either decrease due to infilling of the pore space or increase as a consequence of leaching. To minimize or avoid such effects, either a saturated solution of the rock or an inert gas is used for permeability measurements.

11.2.6 Model for the sample structure

In the present paper, we will not discuss the nature of cavities in rocks and the reasons for their generation; we will only note that such cavities constitute flow channels in the form of pores, microcracks, and cracks. There appears to be no single and rigorous definition of what is meant by a pore, microcrack, crack or what their sizes are. Therefore, any flow channel having an aperture, a, > 10 μm and an aspect ratio (cavity width divided by cavity length), ar, < 0.001 will here be called a **crack**. Norton and Knapp (1977) believe that cracks of length 1–10 cm and longer make the flow porosity and act as major solution conductors in rocks. Flow channels with a = 1–10 μm and ar = 0.1–0.001 will be termed **microcracks**, while channels with a < 1 μm and ar = 0.1–1 are **pores**. Kranz (1979), for instance, chose ar = 0.05 as the division between pores and microcracks. The absence of flow channels implies that their aperture equals zero. The sample structure is shown schematically in Figure 11.2.

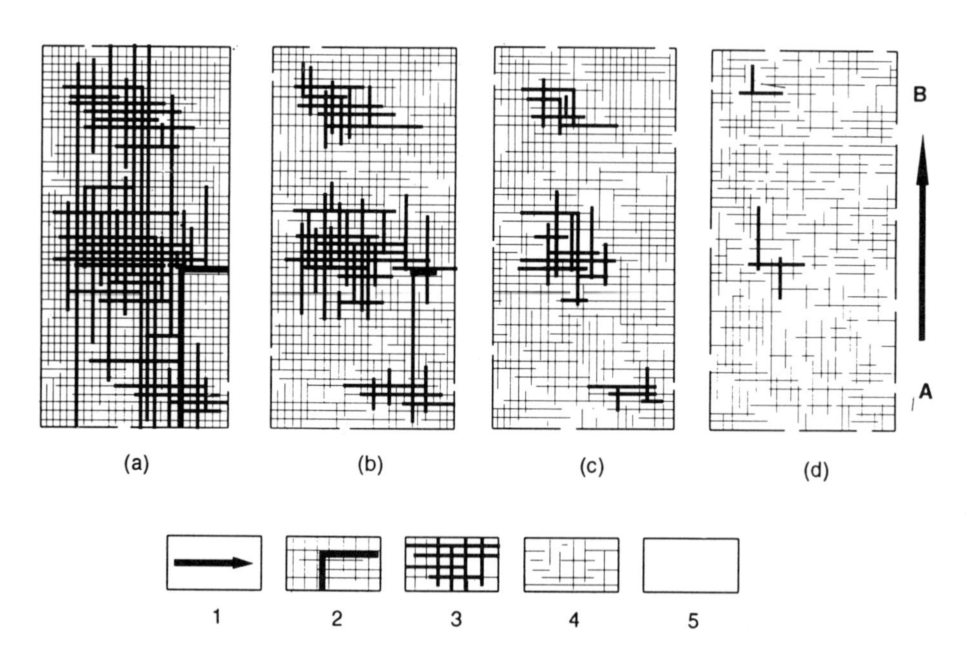

Figure 11.2. Schematic sketch of the sample structure. 1 infiltration direction, 2 crack, 3 microcracks, 4 pores, 5 intact crystal. See discussion in text

Sample portions free of flow channels are impermeable (i.e. have zero-rank permeability), whereas pores, microcracks and cracks have permeabilities of the first, second, and third ranks, respectively.

Infiltration through the sample is possible only in the presence of a connected system of flow channels in the direction A–B. If there is no crack–crack interaction in rock (Figure 11.2a), then the value of permeability will be determined by lower-rank flow channels – microcracks and pores. If, as a result of changes in physical–mechanical conditions, say, a confining pressure increase, microcrack–microcrack interaction ceases (Figure 11.2b,c), then permeability will be governed by the pore component. Finally, in the absence of percolation through channels of any type (Figure 11.2d), the sample will be impermeable. It is evident that the absence of a connected system of flow channels does not necessarily imply the absence of porosity.

11.2.7 Terzaghi's relationship

Under the weight of the overlying layers, stresses are set up in porous rocks:

$$S = \rho_r gh, \tag{11.5}$$

where ρ_r is the density of the overlying suite, g the acceleration due to gravity, and h the depth. As h increases, pores and microcracks close down, resulting in a lower permeability value (Figure 11.2b,c). If fluid under pressure P is now injected into the rock pore space, then a certain amount of the stresses S taken up by the mineral framework will be compensated. As a consequence, some pores and microcracks in rock will open slightly (Figure 11.2c,b). Such an independent consideration of S and P built up in the porous solid medium is known as **Terzaghi's relationship**:

$$\sigma = S - P, \tag{11.6}$$

where σ is effective stress, equal to the difference between total stress and pore fluid pressure. It is the effective stress, σ, that presses minerals against one another, consolidates sediments and deforms rocks. The pore fluid pressure, P, is referred to as 'neutral stress', since, taken alone, it causes neither consolidation of sediments nor rock deformation. Because the sections that follow consider the conditions in which rocks are subjected to uniform stress, it is pertinent to talk about quasi-hydrostatic pressure. Petrologists know the value of S as lithostatic pressure P_s and that of P as fluid pressure P_{fl}. By analogy with σ, the difference between these pressures is termed effective pressure defined as:

$$P_{eff} = P_s - \alpha P_{fl}, \tag{11.7}$$

where $\alpha \approx 1$.

From this relation, it is clear that the infiltration properties of rocks depend on the ratio of fluid pressure to lithostatic pressure rather than on the depth of rock occurrence. The condition of $P_s = P_{fl}$ implies that the rock framework has been unloaded.

11.2.8 Experimental procedure

Pore–microcrack permeability was determined from argon infiltration to temperatures of 600 °C at effective pressures from 150 to 200 MPa.

Experiments were performed with cylindrical samples, 9.6 mm in diameter and 15–20 mm long, using a device allowing independent adjustment of pressure to the rock framework and fluid pressure (Shmonov and Chernyshov, 1982) (Figure 11.3). The confining pressure was applied to the sample through a thin-walled (0.2 mm) gold tube and two side steel end plugs with capillaries, one of which was attached to a volumeter at atmospheric pressure, and the other was fed with gas (argon – 4,3) at pressures ranging from 1 to 12 MPa.

To observe cracking in the samples, two cells were employed. One cell did not apply confining pressure to the sample and was designed to view the structure of its surface to 750°C (Shmonov and Vitortora, 1992).

A special cell was also designed for *in situ* SEM observations of cracks at pressures up to 1 kbar and temperatures up to 600°C (Shmonov *et al.*, 1990) and is illustrated in Figure 11.4a. The cell materials were all non-magnetic. To insert wires and apply pressure, a special lock was made, with 6 electric vacuum entries and a through-going high pressure capillary. The cell was attached to the lock by a thin capillary (1 and 0.2 mm in outside and inside diameter, respectively) twisted as a loop to allow it to be moved while at P and T. The pressure in the cell was generated and adjusted by a micropress placed outside the SEM and was measured by a pressure transducer with an accuracy of 0.5 MPa. To measure temperature and observe the heating rate of the sample, the experiment used a Pt–Pt90Rh10 thermocouple. The errors of temperature measurement are ± 6 and ± 15°C below and above 350 °C, respectively. Power demand of the electric furnace is not greater than 30 watt.

11.3 PERMEABILITY OF MAGMATIC ROCKS

11.3.1 Granites

The permeability of granites was measured on 2 rock samples of different grain size.

(a) Fine-grained biotite granite from the New Ukrainian massif. Below 300°C, increasing the effective pressure to 130–150 MPa decreases the

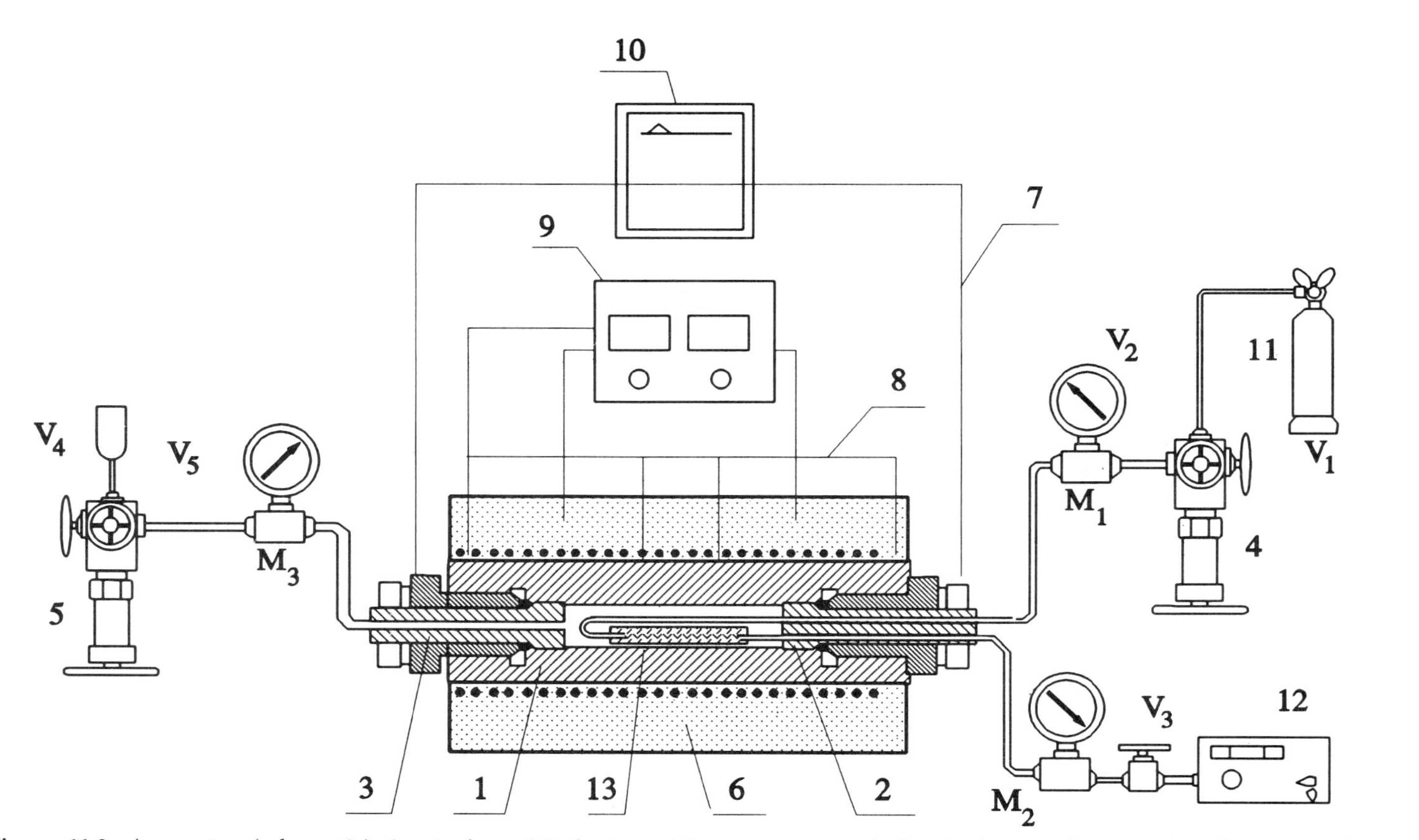

Figure 11.3. Apparatus (schematic). 1 autoclave, 2,3 shutters, 4,5 screw-presses, 6 electric furnace, 7 measuring thermocouples, 8 regulating thermocouples, 9 temperature regulator, 10 potentiometer, 11 gas container, 12 volumeter, 13 rock sample, $V_1 - V_5$ – high pressure valves, $M_1 - M_3$ – gauges.

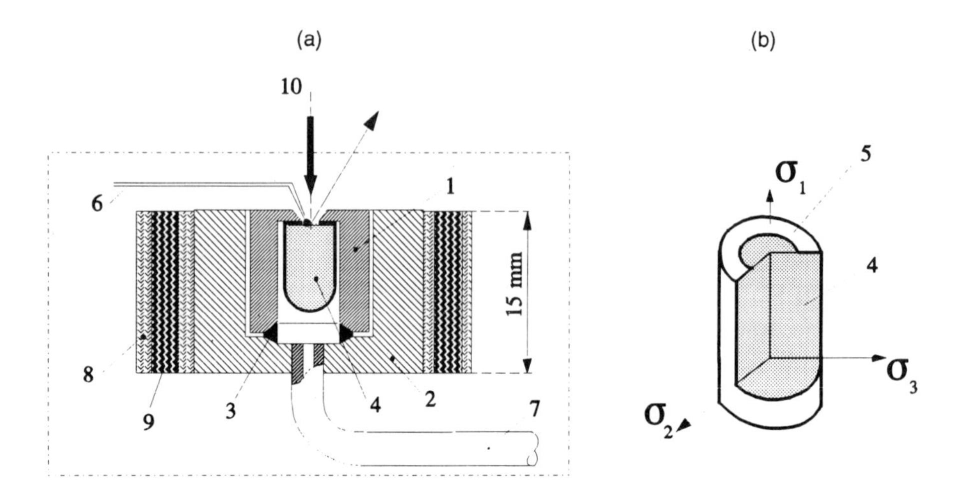

Figure 11.4. Cell for the loaded sample. (a) Schematic diagram of the cell. 1 screw-chamber, 2 nut, 3 lenslike ring, 4 sample, 5 gold lining, 6 thermocouple, 7 capillary, 8 furnace body, 9 heater, 10 probe. Dashed line represents the overall dimensions of the furnace jacket. (b) Sample lining.

permeability of this rock from 10^{-19} to 10^{-22} m^2. Further increase in pressure makes the rock effectively impermeable ($k = 10^{-24}$ m^2). However, upon heating to temperatures above 300 °C, the granite sample undergoes decompaction and becomes permeable again even at the maximum pressure of the run (200 MPa).

(b) Medium-grained Varzob granite from Tajikistan. This also displays an impermeable P-T region; however, it covers a narrower temperature range, below 200 °C at pressures above 150 MPa (Figure 11.5a,b).

11.3.2 Granodiorites and diorites

Two samples of granodiorite and one sample of diorite from the Sayak–1 skarn deposit, Kazakhstan were selected for study.

(a) Coarse-grained granodiorite (sample 82066). This shows a steady decrease in permeability with increasing temperature and pressure (from 10^{-17} to 10^{-18} m^2 at 25^{0}C and pressure from 50 to 150 MPa to 10^{-20}–10^{-21} m^2 at 500°C over the same pressure range). The role of temperature is dominant. This sample also exhibits an impermeable region in the pressure range 80–150 MPa at temperatures above 500°C.

(b) The permeability of another coarse-grained granodiorite (sample 83056). This decreases by 2 orders of magnitude under elevated temperature and pressure. At 200°C, a crack was created in the sample during the test, causing a drastic increase in permeability. With further increasing

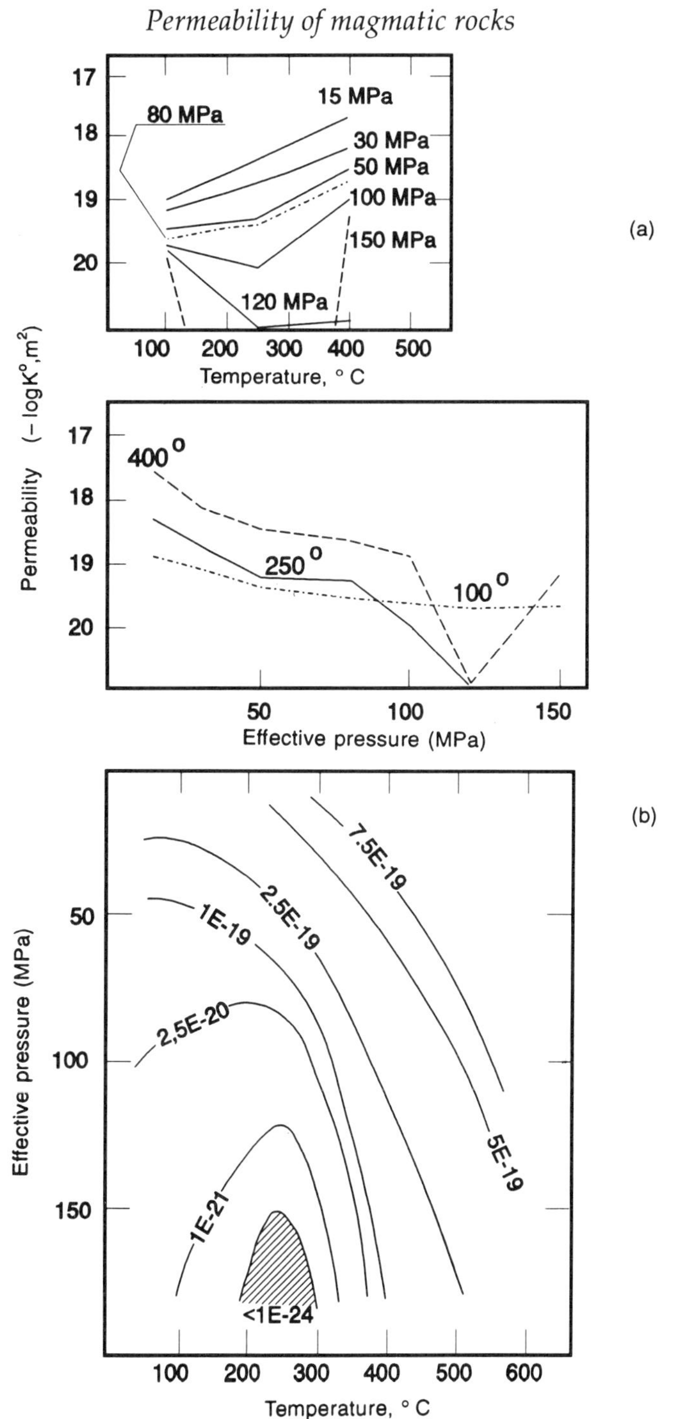

Figure 11.5. Medium-grained Varzob granite. (a) Temperature–pressure dependence of permeability. (b) Chart of permeability isolines.

temperature, however, the crack closed and flow stabilized, so that at 300°C, the sample displayed a distinct permeability minimum.

(c) Medium-grained diorite (sample S–2625). The P–T dependence of permeability is similar to that for Varzob granite, but the impermeable region is at higher temperatures (300°C).

11.3.3 Basalts

Four samples of basalts from underwater lava flows of different ages were selected for testing.

Figure 11.6 shows two contrasting types of temperature–pressure dependence of permeability, peculiar to these basalts. Heating sample 1 of the 'oldest' high-porosity basalt results in its decompaction over the entire pressure range (Figure 11.6a). In younger basalts of low effective porosity (sample 3, Figure 11.6b), the increase in temperature produces a different effect, i.e. first a gradual decrease in permeability over the entire pressure range followed by a dramatic decrease in the infiltration properties of the rock at 250–300°C and the appearance of an impermeable region. Heating the rock to 600°C does not lead to decompaction even at the minimum confining pressure (Figure 11.6b). The youngest basalt (sample 4) of higher porosity is permeable throughout the whole P–T range studied, but shows a permeability minimum at 600°C.

11.3.4 Microcrack permeability in basalt

The experiments to observe *in situ* crack development were performed on 3 basalt samples at temperatures up to 715°C in the vacuum of the SEM without applying confining pressure. Observations were made at 3000 × magnification. Figure 11.7 and Table 11.2 present the results of microcrack permeability measurements for basalt (sample 9) at 200, 400, and 600°C. The constancy of the average crack length, L_{av}, in basalt at different temperatures suggests that new cracks were not created and pre-existing ones did not propagate. A set of special experiments on dense basalt samples (Nos. 10 and 12), whose surfaces contained no cracks, supported this conclusion: cracks do not form on heating the samples to 715°C at 3°C. min^{-1}, nor do cracks form as the temperature is lowered at 100°C.min^{-1}.

11.4 PERMEABILITY OF METAMORPHIC ROCKS

11.4.1 Marbles

Experiments were performed on 4 samples of different marbles.

A monomineralic, medium-grained marble (sample 1) collected from the Varzob river valley, Tajikistan does not respond to the application of pressure, but heating the sample to 600°C gradually increases its permeability.

The permeability of medium-grained marble (sample 83029) from the Sayak–1 skarn deposit, Kazakhstan, is dependent on both factors, although the role of temperature is dominant. At around 300°C, the marble becomes impermeable, but with further heating, the rock undergoes moderate decompaction at low pressures (up to 80–100 MPa), its permeability remaining 2 orders of magnitude lower than the initial one (Figure 11.8a,b).

Fine-grained marbles (samples 31 and 23) from the Ingichke deposit, Uzbekistan, are characterized by similar P–T dependencies of permeability, i.e. at 450–500°C, the samples become impermeable, the role of pressure diminishing in importance with increasing temperature.

11.4.2 Serpentinites

High-temperature permeabilities were measured using 10 samples of serpentinized rocks and serpentinites from the Bazhenov and Kiyembayev massifs of the Ural ultrabasites.

Samples B–1 and B–2 are harzburgites, serpentinized by 30–50%. Samples B–3, B–4, B–7, and B–8 are serpentinites occurring in the vicinity of chrysotile-asbestos veins. Samples B–5 and B–6 are taken from the chrysotile-asbestos veinlet, whereas samples A–25 and A–1 are massive serpentinites.

The results show that the permeability of serpentinites has a complex relationship to both pressure and temperature. Darcy's constant may vary quantitatively by several decimal orders of magnitude. Overall, serpentinites are generally characterized by decreasing permeability values, down to 10^{-24} m^2, with increasing pressure. A confining pressure of 80 MPa is sufficient for sample B–1 (partially serpentinized harzburgite) to become impermeable at room temperature. At 100°C, however, a pressure of c. 100 MPa is required for it to become impermeable. At higher temperatures, the sample retains a finite permeability at any confining pressure.

For the more serpentinized harzburgite B–2, the temperature range of the impermeable region expands: at 200°C, the sample is impermeable in the entire pressure range; at 300°C, at > 100 MPa; at 400°C, at > 150 MPa effective pressure.

However complex the pattern of the P–T dependence of serpentinite permeabilities may be, all the samples have one property in common – the existence of an impermeable region which expands as the degree of serpentinization increases (Figure 11.9a, b).

11.4.3 Amphibolites

Four amphibolite samples from the Archean complex of the Kola superdeep borehole (SD–3) were tested and the results of the

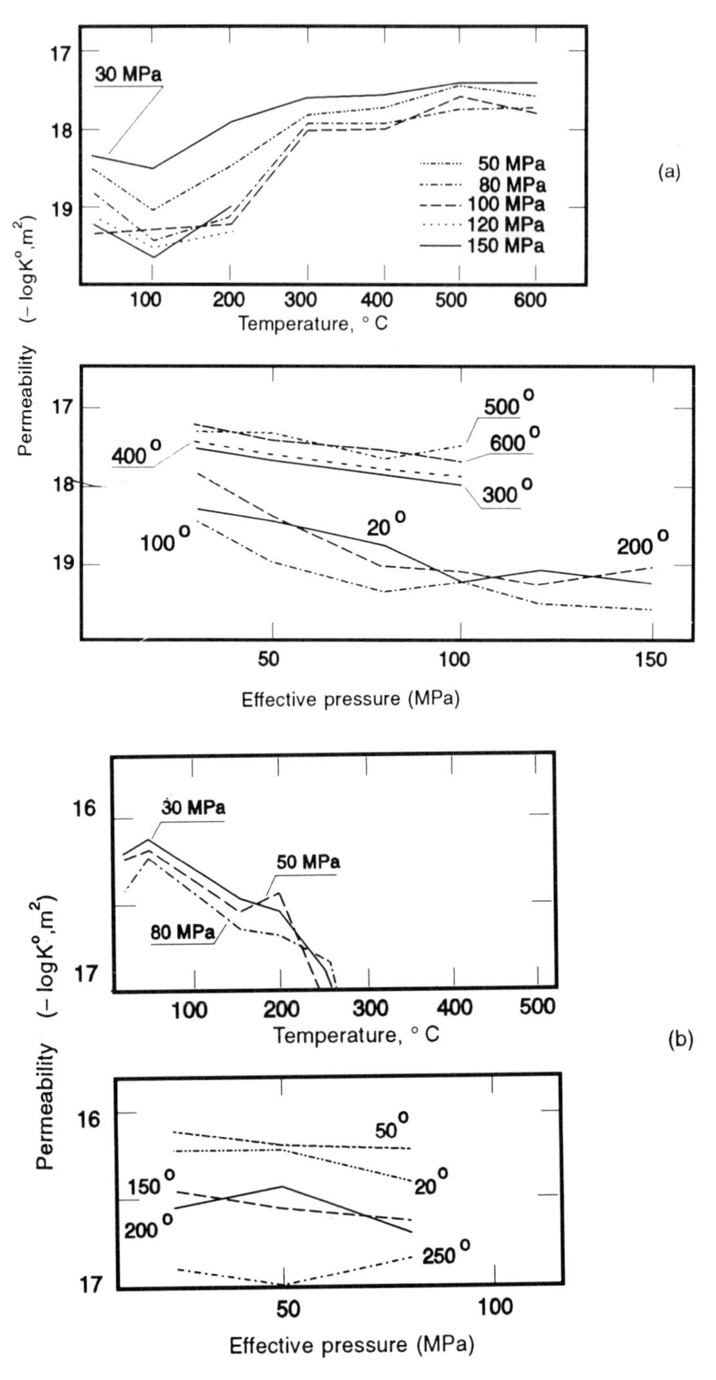

Figure 11.6. The dependence of the permeability of basalts on temperature and pressure. (a) Sample 1, (b) sample 3.

permeability measurements are presented in Table 11.1 (hysteresis) and Figure 11.10.

The depths from which the samples were taken are: amphibolite 31571, 8863 m; amphibolite 31863, 8940 m; amphibolite 43639, 11 400 m; amphibolite 31240, 8757 m.

In runs on sample 31863, infiltration in directions parallel and perpendicular to bedding was also investigated.

Table 11.1. Permeability of rock from the Kola superdeep borehole showing the hysteresis effect

Sample identification number											
31240			31421 A			31421 B			31450		
T °C	P MPa	k m^2	T °C	P MPa	k m^2	T °C	P MPa	k m^2	T °C	P MPa	k m^2
20	50	4.7E–17	20	30	3.0E–17	20	30	2.7E–17	20	30	1.5E–17
100	50	8.0E–17	20	50	2.4E–17	20	50	1.8E–17	20	50	2.5E–18
100	100	3.0E–17	100	50	4.2E–18	100	50	2.0E–17	100	50	1.6E–17
200	100	4.3E–18	100	100	2.6E–18	100	100	1.7E–17	100	100	1.4E–17
200	150	4.1E–18	200	100	1.2E–17	200	100	5.1E–18	200	100	1.4E–17
300	150	1.0E–20	200	150	5.5E–18	200	150	2.5E–18	200	150	1.6E–18
300	100	1.0E–20	300	150	1.5E–18	300	150	0	300	150	1.4E–17
200	50	2.6E–19	300	100	2.6E–19	300	100	0	300	200	1.6E–18
100	50	1.0E–19	200	50	2.7E–19	200	50	0	300	100	4.5E–18
20	30	7.1E–19	100	50	3.3E–19	100	50	0	200	50	4.2E–18
			20	30	7.2E–19	100	20	0	100	50	4.2E–18
						20	30	7.0E–21	20	30	4.0E–18

31240 – Amphibolite; Selection depth 8757 m: Composition (vol.%): Hd–55–70, Pl–25–40, Qtz–1–2, Ep, Bt, Cal–1–2;
31421 – Two-mica two-feldspar gneiss; Selection depth 8812 m: Composition (vol.%): Qtz–28, Pl–25, Mc–30, Ms–10, Ep–3, Bt–3, Cal < 1; A – Parallel to bedding; B – Normal to bedding;
31450 – Muscovite–biotite–plagioclase gneiss; Selection depth 8816 m: Composition (vol.%): Qtz–28, Pl–25, Mc–30, Ms–10, Ep–3, Bt–3, Cal < 1.

As the pressure is raised isothermally from 0.3 to 150 MPa, the permeability of all the samples tends to decrease.

The temperature dependence of permeability undergoes an inversion similar to that obtained earlier for granites. Upon heating the samples from room temperature to 100°C, their permeability varies very little. With further heating, however, they show a drastic decrease in permeability. The temperature at which permeability attains its minimal values ranges from 200 to 500°C, and a further increase in temperature above 450–500°C results in decompaction, causing permeability to increase by several orders of magnitude. Figure 11.10a,b shows the P–T dependence of the permeability of sample 31571, which is typical of amphibolites, although in detail the trend of each rock is different.

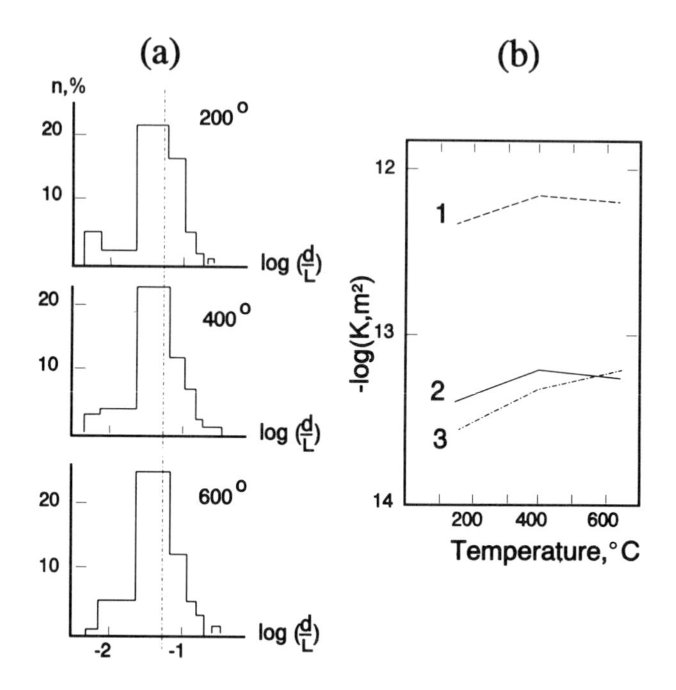

Figure 11.7. Basalt. (a) Crack shape distribution at 200, 400, and 600°C. (b) Temperature dependence of microcrack permeability.

11.4.4 Gneisses

The permeability of gneisses was investigated using 2 samples collected from the SD–3 Archean complex, Kola borehole.

Two-mica two-feldspar gneiss 31421 was taken from a depth of 8812 m. The permeability of this sample was investigated in directions parallel and perpendicular to bedding. In addition, the sample was tested for hysteresis (see Table 11.1 and Figure 11.11a,b).

Muscovite–biotite–plagioclase gneiss 31450, taken from a depth of 8816 m, was tested for hysteresis only (Table 11.1).

The temperature–pressure dependence of gneiss permeabilities is similar to that of amphibolites. However, the minimum permeability of the sample was attained by 500°C for infiltration parallel to bedding, but at a lower T, < 200°C, for the case of infiltration in the perpendicular direction.

11.4.5 Microcrack permeability in amphibolite (sample 43639)

An experiment on crack behaviour was performed with one of the above cells at temperatures up to 500°C and a confining pressure of 80 MPa in

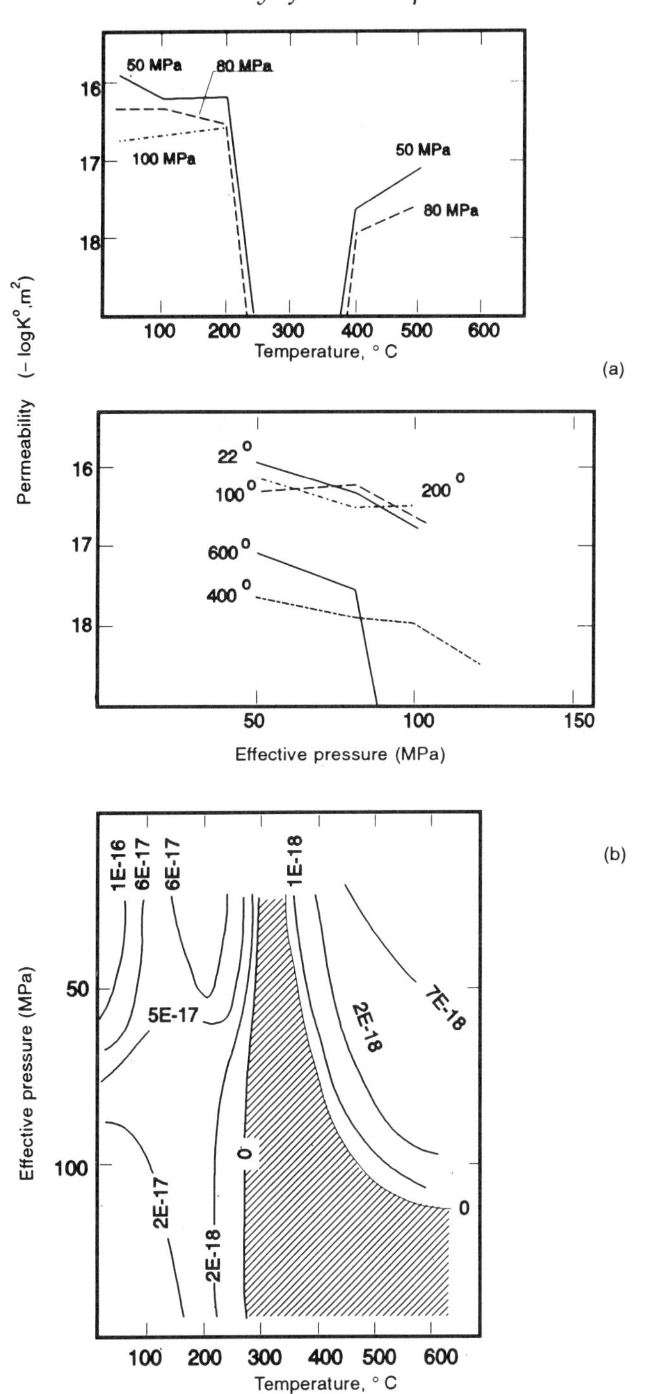

Figure 11.8. Marble, sample 83029. (a) Temperature–pressure dependencies of permeability. (b) Chart of permeability isolines.

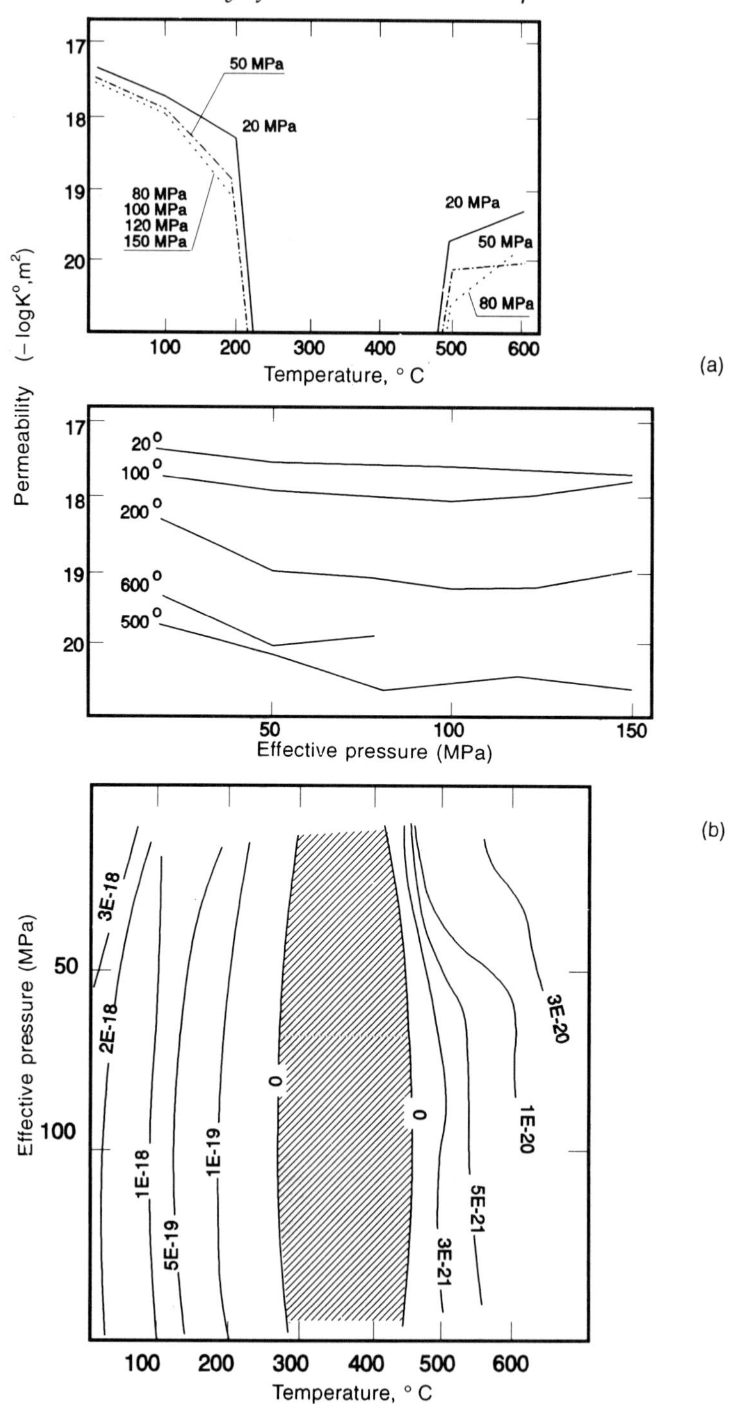

Figure 11.9. Serpentinite, sample B-8. (a) Temperature–pressure dependence of permeability. (b) Chart of permeability isolines.

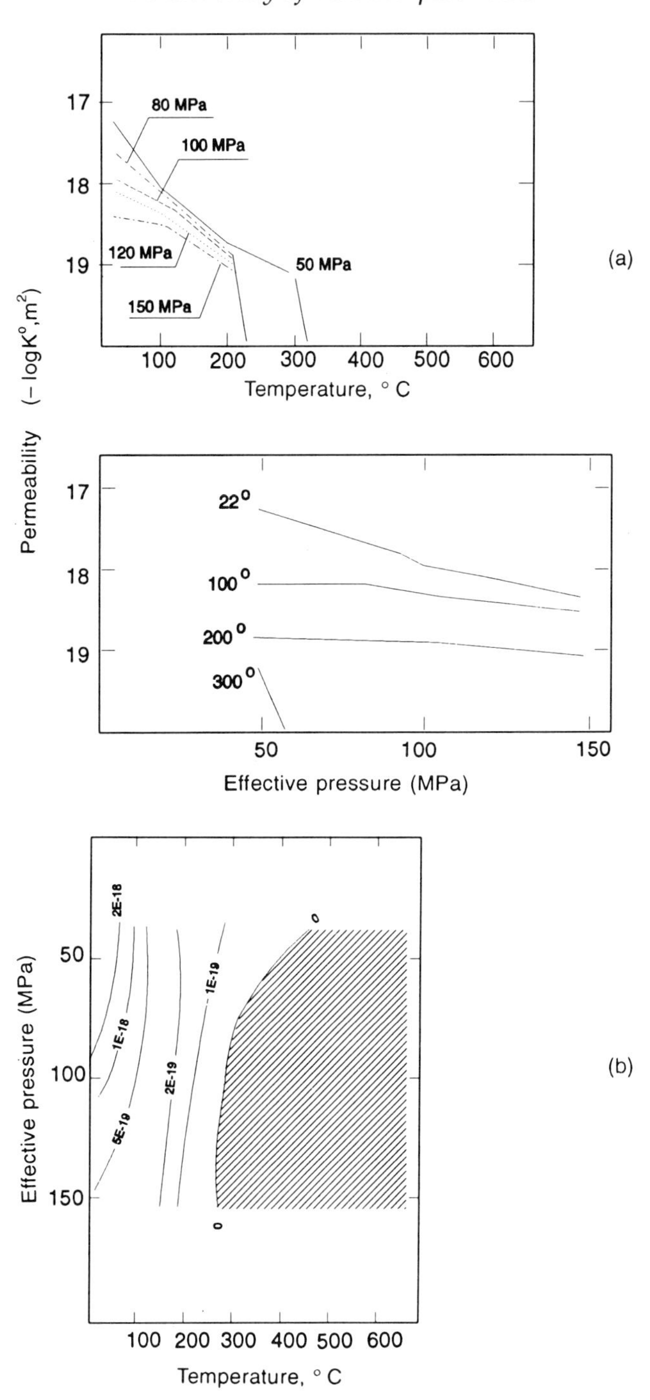

Figure 11.10. Amphibolite, sample 31571. (a) Temperature–pressure dependence of permeability. (b) Chart of permeability isolines.

the SEM. The sample surface contains three networks of independent cracks, two of them represented by three and four rays of cracks. The third, larger network consists of 20 connected microcracks. At 100°C, the average crack aperture was 2.5 mm, while at 300°C, it was 1.3 mm. Microcrack permeability values computed using the network model of Chida and Tanaka (1983), or the equation of Romm (1966) are similar, 3.6×10^{-14} and 4.8×10^{-14} m^2, respectively, at 300°C (Table 11.2). The permeability of the sample portion that contained microcracks was only calculated for T = 300°C, but since microcrack apertures at 100°C are larger than those at 300°C, permeability is believed to be higher at lower T.

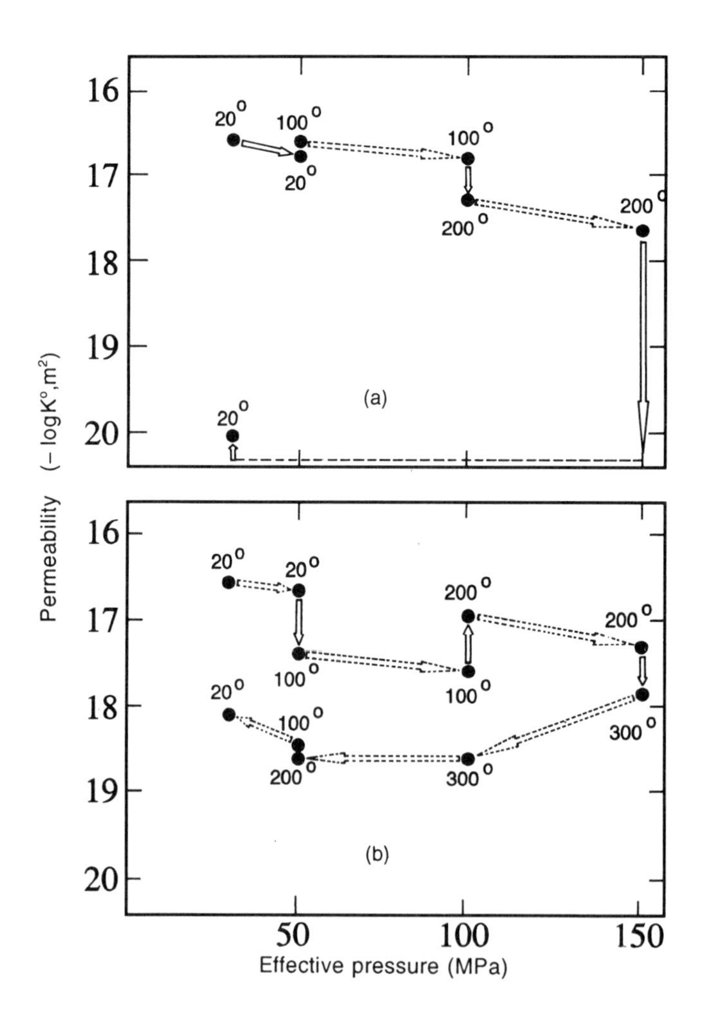

Figure 11.11. Gneiss, sample 31421. Permeability hysteresis in the course of P–T evolution: (a) perpendicular to bedding, (b) parallel to bedding.

Table 11.2. Microcrack permeability (m^2) calculated using equations from Chida and Tanaka (1983), Romm (1966) and Howard and Nolen-Heeksema (1990)

		References		
Rock	*T(°C)*	*Chida & Tanaka* (1983)	*Romm* (1966)	*Howard & Nolen-Heeksema* (1990)
	200 h	5.6E–14	52.5E–14	4.5E–14
Basalt	400 h	11.5E–14	73.2E–14	6.3E–14
	600 h	8.7E–14	67.3E–14	5.8E–14
Amphibolite	300	3.6E–14	4.8E–14	–
Varzob marble	100 h	1.2E–13	5.8E–13	0.5E–13
	400 h	0.6E–13	3.6E–13	0.2E–13
	200 h	–	8.1E–15	1.0E–15
	300 h	5.8E–15	2.6E–15	0.3E–15
Limestone	400 h	–	1.1E–15	0.1E–15
83075	500	–	11.5E–15	2.2E–15
	300 c	16.2E–14	80.3E–15	8.7E–15
	100 c	–	13.0E–15	1.3E–15

h-heating; c-cooling.

11.4.6 Microcrack permeability in Varzob marble

The microcrack permeability in this sample was measured in the SEM at 100 and 400°C without applying confining pressure. With increasing temperature from 100 to 400°C, the average crack length increases from 40 to 70 μm and the average crack apertures diminish from 0.8 to 0.7 μm. This results in a decrease in permeability of the cracked sample portions (Table 11.2).

11.5 PERMEABILITY OF SEDIMENTARY ROCKS

11.5.1 Limestones

Permeability measurements were made on 4 samples of different limestones from the Sayak–1 skarn deposit, Kazakhstan: fine-grained limestone K–809, fine-grained limestone 82068, arenaceous fine-grained limestone 83075, and arenaceous fine-grained limestone 83086.

When fine-grained limestones 82068 are exposed to temperature and pressure, it is the temperature that dominates their permeability, which decreases as it is raised. At 300–500°C, practically regardless of pressure, limestones become impermeable (Figure 11.12a,b). As with granodiorite 83056 (described above), at a temperature of 300°C, a crack was created in arenaceous limestone 83086, but even this failed to change the usual pattern of behaviour; upon heating to temperatures above 500°C, the sample became impermeable.

11.5.2 Dolomite limestones

Two samples were used for permeability measurements: fine-grained dolomite (sample 11) collected from the Ingichke deposit, Middle Asia and fine-grained dolomite (sample 2) from the Kansay deposit, Middle Asia.

The permeability of both samples is primarily temperature dependent, like that of limestones. The maximum decrease in permeability occurs upon heating the rock to 150°C. Increasing the temperature further makes the samples impermeable irrespective of pressure: dolomite 1 becomes impermeable at around 400°C, dolomite 2, at about 500°C.

11.5.3 Fine-grained tuffstone

Permeability measurements were made on dense, hard tuffstone sample S–2572 collected from the country rocks of the Sayak–1 skarn deposit, Kazakhstan.

The P–T dependence of the permeability of this tuffstone is similar to that of granitoids taken from the same deposit. Upon heating the rock to 200–300°C, its permeability is affected by both temperature and pressure. As the temperature is increased, the role of pressure diminishes in importance.

11.5.4 Sandstone

To perform permeability experiments, a fine-grained massive weakly cemented, weakly micaceous sandstone sample carrying minor argillaceous impurities was collected from borehole 8–T in the Kizlyar deposit of geothermal waters, Daghestan.

In accordance with the lithostatic pressure and horizon temperature in the perforation interval 2800–2808 m, permeability measurements were made at confining pressures of 25, 50, 75, and 100 MPa and temperatures of 20, 50, 100, 150, 200, 250, and 300°C. To account for the effect of specimen size (scale effect) on its specific permeability, the measurements were made on specimens, 7.85, 9.85, and 13.8 mm in diameter.

The permeability of the sandstone decreases regularly with increasing pressure, but its dependence on temperature is more complex. At pressures of 25 and 50 MPa, permeability increases with temperature. Increasing pressure to 75 MPa causes permeability to drop as the temperature is increased, whereas at 100 kbar, it remains practically unchanged on heating.

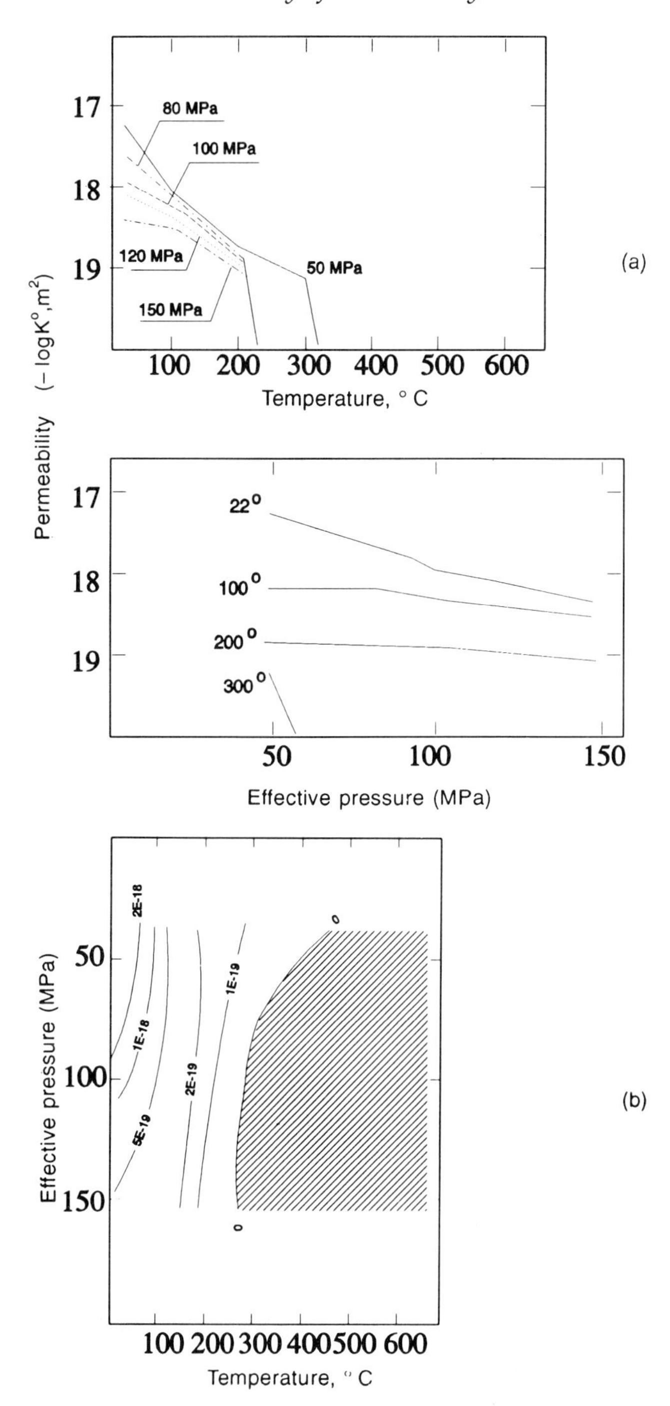

Figure 11.12. Limestone, sample 82068. (a) Temperature–pressure dependence of permeability. (b) Chart of permeability isolines.

11.5.5 Microcrack permeability in limestone (sample 83075)

Permeability measurements were made from crack dimensions measured during heating the sample to 500°C and subsequent cooling. Crack shape distributions for the limestone heated to 300°C and cooled to the same temperature are shown in Figure 11.13a. The average crack lengths, L_{av}, decrease with increasing temperature from 200 to 500°C, and vice versa increase with decreasing temperature from 500 to 100°C. At 500°C, single cracks of width 12 μm are noted; the number of such cracks increases with cooling to 300°C. The data on the distribution of crack widths, lengths and density on a line permitted calculations of microcrack permeability for a variety of temperatures (Figure 11.13b, Table 11.2).

11.6 DISCUSSION AND SUMMARY OF CONCLUSIONS

11.6.1 Relationships between data on water and gas permeabilities

The values of permeability determined by means of gas infiltration, k_∞ (Klinkenberg permeability) agree well with permeabilities determined by water infiltration ('true' permeability), k_W, in the range from a few darcies to 1–0.1 mdarcy (Katz *et al.*, 1965). For instance, Jones and Owens (1980) found that for sandstones at 20°C and $k \leqslant 1$–0.1 mdarcy, permeability determinations by gas yielded systematically higher k_W values relative to k_∞. To fit k_W and k_∞, they suggested the equation:

$$k_W = k_\infty^n,$$

where both k_W and k_∞ are in mdarcies, and $n = 1.32$. Subsequently, Chowdiah (1987) refined the value of n for tight sands: at 20°C, it equals 1.43. This correction gives a good fit up to 10^{-20} m^2 (10^{-5} mdarcy).

Shebesta *et al.* (1988) tested the agreement between the data on water and gas at 200°C and $P_{eff} = 75$ MPa obtained from a single limestone sample. The measurements yielded $k_W = 0.019$ and $k_\infty = 0.02$ mdarcy. The results agreed well ($n = 1.013$) and required no correction.

The permeability of New Ukrainian granite measured at 400°C and $P_{eff} = 500$ bar with argon was found to be 10^{-19} m^2 (Vitovtova and Shmonov, 1982). According to Zaraisky *et al.* (1989), k_W of the same granite is 10^{-20} m^2, with a correlation coefficient of 1.25.

Apparently it is too early to talk about introducing a correction to bring k_W and k_∞ into agreement at elevated pressures and temperatures. Therefore, we list all the experimental k_∞ data as they were obtained. We can do no more than assume k_W to be lower than k_∞ and ascribe '$k \leqslant$' to each value. These data suggest that wetting angle affects (Watson and Brennan, 1981) do not play a major role in controlling microcrack permeability in crystalline rocks.

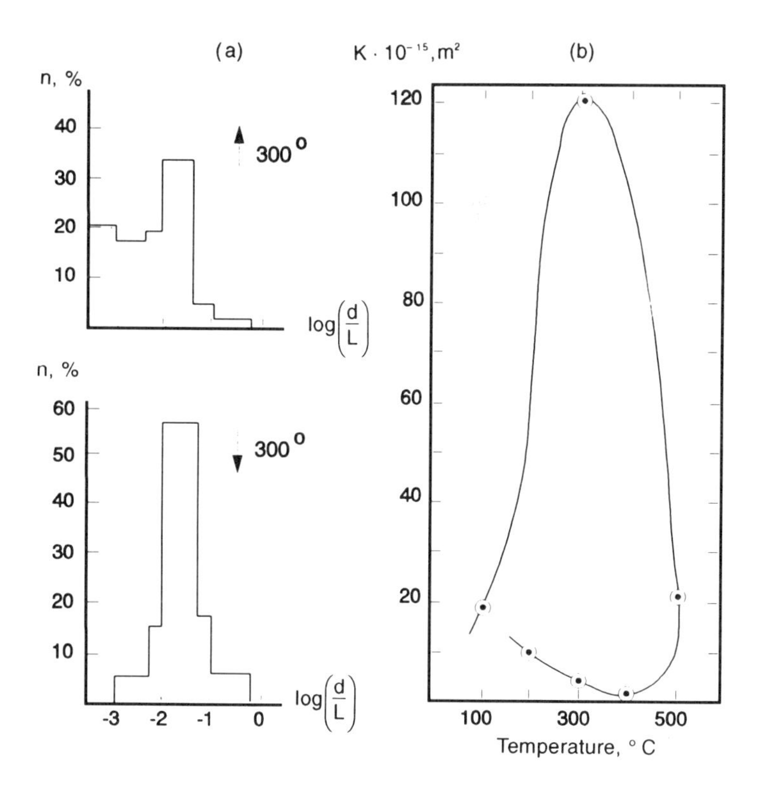

Figure 11.13. Sayak limestone. (a) Crack shape distribution at 300°C: (upwards) upon heating, (downwards) upon cooling. (b) The dependence of permeability on temperature (permeability is computed from crack size distributions).

11.6.2 Permeability of rocks at elevated temperatures and pressures

Our investigations showed that pore-crack (or bulk) rock permeability ranges from 10^{-17} to 10^{-24} m^2 at temperatures up to 600°C and at effective confining stresses $\sigma_1 = \sigma_2 = \sigma_3$ from 15 to 200 MPa. The error in the measurement of permeabilities from 10^{-17} to 10^{-20} m^2 is around ±10–25%, while below 10^{-20} m^2, uncertainties are about ±25–35%. Microcrack permeabilities in the same samples calculated from microcrack dimensions are 10^{-13} to 10^{-15} m^2, although the confidence interval is not clear.

The experimental data reveal a complex temperature–pressure dependence of rock permeabilities, but there are some consistent patterns. At any constant temperature in the range 20–600°C, an increase in the confining pressure always results in a permeability reduction. If the framework strength is lost, i.e. when the limit of its creep is exceeded, permeability decreases dramatically, down to 10^{-24} m^2.

At constant effective pressure, an increase in temperature from 20 to 600°C yields one of three types of dependencies:

1. With increasing temperature to 600°C, permeability increases consistently (see Figures 11.5a, 11.6a)
2. With increasing temperature to 150–300°C, permeability first drops and then increases again as the temperature is raised further (to 600°C). Hence there is a minimum on the curve k = f(T). In some instances, the value of the minimum is below 10^{-24} m^2, i.e. below the sensitivity of the experimental apparatus, producing the so-called impermeable region (see Figures 11.8, 11.9, 11.10).
3. With increasing temperature to 600°C, permeability decreases consistently (see Figure 11.6a). In some cases, permeability decreases considerably, and at 250–400°C, some rock samples become impermeable (see Figures 11.6b and 11.12a).

ACKNOWLEDGEMENTS

The authors are grateful to A.K. Shirokov for help in carrying out experiments and K.I. Shmulovich for critical review of the manuscript.

The research described in this chapter is supported by the Russian Fund of Fundamental Investigations under Grant No. 93–05–8591.

REFERENCES

Aksyuk, A. M., Vitovtova, V. M. and Shmonov, V. M. (1991) Permeability of serpentinites and serpentinized hartzburgites at high temperatures and pressures. In *Contributions to Physical–chemical Petrology*, **XVII**, (eds V. Zharikov and V. Fed'kin). Nauka, Moscow, 119–41 (in Russian).

Aksyuk, A. M., Vitovtova, V. M., Pustovoy, A. A., *et al.* (1992) Permeability of the Atlantic rift zone basalts. *Okeanologia*, **6**, 1115–22 (in Russian).

Arch, J. and Maltman, A. (1990) Anisotropic permeability and tortuosity in deformed wet sediments. *Journal of Geophysical Research*, **95**, (B6) 9035–45.

Archie, G. E. (1950) Introduction to petrophysics of reservoir rocks. *Bull. Am. Assoc. Pet. Geol.*, **34**, 943–61.

Batzle, M. L., Simmons, G. and Siegfried, R.W. (1980) Microcrack closure in rocks under stress: direct observation. *Journal of Geophysical Research*, **85**, (B12) 7072–90.

Brace, W. F. (1977) Permeability from resistivity and pore shape. *Journal of Geophysical Research*, **82**, 23, 3343–9.

Brace, W. F., Walsh, J. B. and Frangos, W. T. (1968) Permeability of granite under high pressure. *Journal of Geophysical Research*, **73**, (6(15)) 2225–36.

Brace, W. F. (1980) Permeability of crystalline and argillaceous rocks. *Int. J. Rock Mech. Min. Sci.*, **17**, 241–51.

Caillard, D. and Martin, J. L. (1982) *In-situ* studies of sub-boundary properties. *Acta Metal*, **30**, 791–8.

Chida, T. and Tanaka, S. (1983) Analysis of relation between pore structure and permeability using a network model. *Journal of the Japanese Association for Petroleum Technology*, **48**, (6), 439–44 (in Japanese).

Chowdiah, P. (1987) Two-phase flow in tight sands, *Sixth Quarter Report for the period January 1 to March 31, 1987*. Prepared for the U.S. Department of Energy, Morgantown Energy Technology Center, Institute of Gas Technology, Chicago.

Fatt, J. and Davis, D. N. (1952) Reduction in permeability with overburden pressure. *Trans. AIME*, **195**, 329–33.

Howard, J. H. and Nolen–Heeksema, R. C. (1990) Description of natural fracture systems for quantitative use in petroleum geology. *Amer. Ass. Petrol. Geol. Bull.* **74** (**2**), 151–62.

Jones, F. O. and Owens, W. W. (1980) A laboratory study of low permeability gas sands. *J. Pet. Tech.* **34**, 1631–40.

Katz, D. L., Cornell, D., Koboyashi, *et al.* (1965) *Handbook of Natural Gas Engineering*. Mc Graw-Hill Book Company, New York.

Kranz, R. L. (1979) Crack growth and development during creep of Barre granite. *Int. J. Rock. Mech. Min. Sci. Geomech. Abstr.*, **16**, 23–5.

Magomedov, M. A. and Shmonov, V. M. (1991) On the problem of scale effect in determining rock permeability. In *Contributions to Physical–chemical Petrology*, **XVII** (eds V. Zharikov and V. Fed'kin). Nauka, Moscow, 114–19 (in Russian).

Marmorshtein, L. M. (1981) Changes in permeability under the conditions of a complex stressed state. In *Proceedings of the All-Union Meeting on the Physical Properties of Rocks at High Pressures and Temperatures, as applied to seismology problems*, Tashkent, 35–6 (in Russian).

Moore, D. E., Morrow, C. A. and Byerlee, J. D. (1983) Chemical relations accompanying fluid flow through granite held in a temperature gradient. *Geochim. Cosmochim. Acta*, **47**, 445–53.

Morrow, C. A., Lockner, D., Moore. D. E. and Byerlee, J. D. (1981) Permeability of granite in a temperature gradient. *Journal of Geophysical Research*, **86**, (B4), 3002–8.

Nikolaenko, V. I. and Indutyi, V. F. (1978) Effect of thermodynamic factors on the permeability of granitoids. In *Physical Properties of Rocks at High Thermodynamic Parameters*. Baku, 29–30 (in Russian).

Norton, D. and Knapp, R. (1977) Transport phenomena in hydrothermal systems, the nature of porosity. *Amer. J. Sci.*, **277**, 913–36.

Nutting, P. G. (1930) Physical analysis of oil sands. *Bull. Amer. Ass. Petrol. Geol.*, **14**, 1337–49.

Romm, Eu. S. (1966) *Infiltration properties of cracked rocks*. Nedra, Moscow (in Russian).

Shebesta, A. A., Vitovtova, V. M. and Shmonov, V. M. (1988) On changes in properties of pore-crack media for Ca-rock at moderate pressures. *Spatial–temporal problems in geology*, **115**, 38–43. Leningrad Mining Institute, Leningrad (in Russian).

Shmonov, V. M. and Chernyshov, V. M. (1982) An installation with non-uniform pressures on the fluid and solid phases. In *Experimental Problems with Solid-media and Hydrothermal High-pressure Equipment* (eds Ivanov, I. P. and Litvin, Yu.A.). Nauka, Moscow, 125–8, (in Russian).

Shmonov, V. M. and Vitovtova, V. M. (1992) Rock permeability for the solution of the fluid transport problems. In *Experiment in Geosciences*, **1**, (ed V. Zharikov), SovGeoInfo and Institute of Experimental Mineralogy, Moscow, pp. 1–49.

Shmonov, V. M., Voronov, V. S., Romanenko, I. M. and Shirokov, A. K. (1990) A cell to observe the development and deformation of pores and cracks in rock samples: direct observation with an electron microscope to a temperature of 550°C and 1 kbar pressure. In *Experiment–89* (ed. V. Zharikov). Nauka, Moscow, pp. 101–4.

Shmonov, V. M., Aksyuk, A. M., Alekhin, Yu.V. *et al.* (1986) Hydrothermal solutions and skarn formation. In *Experiment in the Solution to the Burning Problems of Geology* (eds V. Zharikov and V. Fed'kin). Nauka, Moscow, 278–06 (in Russian).

Summers, R., Winkler, K. and Byerlee, J. D. (1978) Permeability changes during the flow of water through Westerly granite at temperatures of 100–400°C. *Journal of Geophysical Research*, **83**, (B1), 339–44.

Trimmer, D., Bonner, B., Heard, H. C. and Duba, A. (1980) Effect of pressure and stress on water transport in intact and fractured gabbro and granites. *Journal of Geophysical Research*, **85**, (1312), 7059–71.

Tungatt, P. D. and Humphreys, F. J. (1981) An *in-situ* optical investigation of the deformation behavior of sodium nitrate – an analogue for calcite. *Tectonophysics*, **78**, 661–75.

Urai, J. L., Humphreys, F. J. and Burrows, S. E. (1980) *In-situ* studies of the deformation and dynamic recrystallization of rhombohedral camphor. *J. Materials Science*, **15**, 1231–40.

Vitovtova, V. M. and Shmonov, V. M. (1982) Rock permeability to 2000 kg. cm^{-3} pressure and 600°C temperature. *Doklady Academii Nauk SSSR*, **226**, (5), 1244–8 (in Russian).

Vitovtova, V. M., Fomichov, V. I. and Shmonov, V.M. (1988) Evolution of rock permeability in the formation of the Sayak–1 and Tastau skarn deposits. In *Contributions to Physical–chemical Petrology*, **XV** (eds V. Zharikov and V. Fed'kin). Nauka, Moscow, pp. 6–17 (in Russian).

Zaraisky, G. P. and Balashov, V. N. (1978) On the decompaction of rocks upon heating. *Doklady Akademii Nauk SSSR*, **240**, (4), 926–9 (in Russian).

Zaraisky, G. P., Balashov, V. N. and Zonov, S. V. (1989) *Thermal decompaction of rocks and its effect on permeability*. Proc. 6th Int. Sym. Water–Rock Interaction (Malvern). Balkema, Rotterdam.

Zharikov, A. V., Vitovtova, V. M. and Shmonov, V. M. (1990) An experimental study of the permeability of Archean rocks from the Kola superdeep borehole. *Geology of Ore Deposits*, (**6**), 79–88 (in Russian).

Zoback, M. D. and Byerlee, J. D. (1975) The effect of microcrack dilatancy on the permeability of Westerly Granite. *Journal of Geophysical Research*, **80**, (5), 752–5.

GLOSSARY OF SYMBOLS

A	cross-sectional area
C	nondimensional constant
f	porosity
g	acceleration of gravity
h	height (or depth)
K	constant (hydraulic conductivity)
k	permeability
k^0	permeability measured directly by experiment
l	length (or L)
P	pressure

P_{eff}	effective pressure
P_{fl}	fluid pressure
P_s	lithostatic pressure
Q	volume
q	specific flow rate
T	temperature
t	tortuosity
μ	dynamic viscosity
ρ	density
σ	stress

Index of aqueous species

Ag^+ 102
$AgCl^0$ 102, *103*
$AgClOH^-$ 108

$As_2S_3^0$ *103*, 120, 124
$As_2S_4^{2-}$ *103*, 120, 123–4
$As_3S_6^{3-}$ 123–4
$As_4S_7^{2-}$ 124
$H_2AsO_3^-$ *103*
$H_3ASO_3^0$ 89, *103*, 120–1, 124, 130
$HAs_2S_4^-$ *103*, 120–3, 124

AlF_3^0 148
$AlF_n^{(3-n)+}$ 147
$Al(OH)_3^0$ 147
$Al(OH)_2F^0$ 148
$Al()H)F_2^0$ 148
$Al(OH)_mF_n^{(3-m-n)+}$ 148
$Al(OH)_2Cl^0$ 147
$Al(OH)CL_2^0$ 147

$Au(OH)^0$ 97, 98
$AuCl_2^-$ 98–101, *103*

Ca^{2+} 220
$CaCl^+$ 220, *224–6*
$CaCl_2^0$ 220, *224–6*
$CaCO_3^0$ 223
$CaSO_4^0$ 223

Cl^- 220
HCl^0 98–101, 167–71, *171–4*, **175**, *224*–6

$CsCl^0$ *226*

Cu^+ *103*, 111, 113
$CuCl^0$ 112
$CuCl_2^-$ *103*, 112–13
$CuCl_3^{2-}$ 112
$Cu(OH)^0$ 111
$Cu(OH)^+$ 109
$Cu(OH)_1^-$ 111
$Cu(OH)_2^0$ 109
$Cu(OH)_4^{2-}$ 109

H_g 58, 61–3, **64**, 86, 127
$H_gCl_2^0$ **64**
$H_g(OH)_2^0$ **64**, 129
$H_g(OH)Cl^0$ **64**
$H_g(OH)CO_3^-$ 127, 129
$H_gSH_2S^0$ **64**
$H_gS(H_2S)_2^0$ 128

KCl^0 223, *224–6*
KOH^0 *226*

LiCl *226*

M_gCl^+ *224–6*
$M_gCl_2^0$ 181, 183, *224–6*
$M_gCO_3^0$ 223
$M_gSO_4^0$ 220, 223

$HMoO_4^-$ 115, 118–20, 132
$H_2MoO_4^0$ 115
$HMoO_4Cl^{2-}$ 115–16
$KHMoO_4^0$ 115–20, 132
$NaHMoO_4^0$ 115–20, 132
$Na(K)ClHMoO_4^-$ 115–16
$HMoO_4Si(OH)^-$ 119–20
$NaHMoO_4Si(OH)^0$ 119–20

$NaCl^0$ 178, 223, *224–6*
$NaOH^0$ *224–6*
$NaSO_4^-$ 220

$RbCl^0$ *226*

S^0 65, **68**, 69–72, 74–5, 86
HS^- **64**
H_2S^0 **64**, **66**, 69–74, 125
H_2S_n 69–71
$H_2S_2O_3^0$ **67**, 69, 71–2, **73**, 74

$H_2S_nO_6^0$ 69, **70**
SO_2^0 **67**, 69, 71–2, **73**, *74*
SO_4^{2-} **64**
HSO_4^- **64**, **66**, 69

$Sb(OH)_3^0$ 89, *103*, 125–7, 130
$Sb(OH)_4^-$ *103*
$Sb_2(OH)_6^0$ *103*, 126
$Sb_2S_4^{2-}$ *103*
$HSb_2S_4^-$ *103*
$H_2Sb_2S_4^0$ *103*

$Si(OH)_4^0$ 76, 78, 86, 145, 153
$SiF_3(OH)^0$ 146
$Si(OH)_3F^0$ 146

$Sn(OH)_3^+$ 84–6
$Sn(OH)_4^0$ 79–80, **82**, 86, 89
$SN(OH)_n^{(2-n)}$ 80
$Sn(OH)_{(4+n)}^{n-}$ 80
$Sn(OH)_3Cl^0$ 84–7

WCl_4^0 149
WCl_6^0 149
$H_2WO_4^0$ 149
$NaHWO_4^0$ 149
$WO_2Cl_2^0$ 149
WO_3Cl^- 149
HWO_3Cl^0 149
$W(OH)_nF_{(6-n)}^0$ 149

Index of systems

(A) Fluid systems (i.e. components are not present as solids, although their activities may be buffered by solids)

CO_2 21
H_2O 20–1

$H_2O - CO_2$ 15–16, 20
$H_2S - CO_2$ 16
$H_2S - H_2O$ 16

$H_2O - CO_2 - CH_4$ 16
$H_2O - HCl - CO_2 -$ 167–70, **171–2**

$CO_2 - NaCl$ 181

$H_2O - CaCl_2$ 16, 194–5, 199, 207–8
$H_2O - KCl$ 16, 194–5, 199
$H_2O - M_gCl_2$ 194–5, 199, 207–8
$H_2O - NaCl$ 16, 194–5, 199, 207, **208**
$H_2O - NaCl - KcL$ 16

$H_2O - CH_4 - NaCl$ 16
$H_2O - CO_2 - CaCl_2$ 16, 200–201, 205–9
$H_2O - CO_2 - NaCl$ 15–6, 194, 200–201, 203–4, 205–9
$H_2O - HCl - NaCl$ 171–8

$H_2O - HCl - NaCl - CaCl_2$ 179, 182
$H_2O - HCl - NaCl - FeCl_2$ 179, 182
$H_2O - HCl - NaCl - MgCl_2$ 178, 182–3

$H_2O - HCl - NaCl - CaCl_2 - M_gCl_2$ 183–7
$H_2O - HCl - NaCl - CaCl_2 - FeCl_2$ 183–7

$H_2O - CO_2 - HCl - CaCl_2$ 179–80
$H_2O - CO_2 - HCl - FeCl_2$ 179
$H_2O - CO_2 - HCl - MgCl_2$ 179, 181
$H_2O - CO_2 - HCl - NaCl$ 179–81, 187–8

(B) Saturated systems (i.e. one component is present in solid form)

$AgCl - H_2O - KCl$ 106–8
$AgCl - H_2O - NaCl$ 102–8
$AgCl - H_2O - NaClO_4$ 102

$Al_2O_3 - H_2O - HF$ 147

$As_2S_3 - H_2O$ 120, 124
$As_2S_3 - H_2O - H_2S$ 120–3

$Au - H_2O$ 97–8
$Au - H_2O - HCl$ 98–100
$Au - H_2O - HCl - KCl$ 98–100

$H_g - H_2O$ 58–63
$H_gO - H_2O - NaHCO_3$ 128–9
$H_gS - H_2O - H_2S$ 127–9
$H_g - S - Cl - O - H$ **64**

$MoO_2 - H_2O$ 114–15
$MoO_2 - H_2O - HCl$ 114–15
$MoO_2 - H_2O - HCl - KOH$ 114–20
$MoO_2 - H_2O - HCl - NaOH$ 114–20
$MoO_2 - H_2O - HCl - NaCl - SiO_2$ 119

$S - H_2O$ 63–71

$Sb - H_2O$ 124–5
$Sb_2O_3 - H_2O$ 124–5
$Sb_2S_3 - H_2O$ 124–5

$SiO_2 - H_2O$ 16, 75–8
$SiO_2 - H_2O - HCl$ 75–8
$SiO_2 - H_2O - HF$ 145–7
$SiO_2 - H_2O - HNO_3$ 75–8

$SnO_2 - H_2O$ 49, 80–7

$SnO_2 - H_2O - HCl$ 49, 80–6
$SnO_2 - H_2O - HNO_3$ 49, 80–6
$SnO_2 - H_2O - HCl - KCl$ 80–6
$(Fe, Mn)WO_4 - H_2O - HCl$ 149–52
$(Fe, Mn)WO_4 - H_2O - HF$ 149–52
$(Fe, Mn)WO_4 - H_2O - NaCl$ 149–52

Subject index

Tables are shown in *italic* and figures in **bold** text.

Acid–base differentiation 31–2, 34, 36, 207–8
Activity (in fluid) 5, 10, 29, 163–71, 177–82, 216, 223, 226, 278–9
  coefficient 96, 167–8, 177–80, 182, 226, 240–41
  experimental buffers 163–7, 177–8
  see also Buffer; Water activity
Akchatau W–Mo deposit 139–44, 149
Albite 6, 165, 177–8, 188
Alkali chlorides
  and Ag solubility 102–8
  and Mo solubility 114–18
  and W solubility 149–52
  see also Index of systems
Al–silicate 23, 26–7
  see also Andalusite; Kyanite
Alumina mobility 143–4
Aluminium solubility 7, 146–8
Anatexis 21, 23–7, 205–6
  see also Dehydration melting; Granite melting
Andalusite 165, 177, 180
Antimonite 124
Antimony solubility 89, *103*, 124–7
Aqueous species *see* Index of Aqueous Species
Arsenic solubility 89, 103, 120–24
Autoclave 44, 51–6, 96, **293**
  materials for 44–5, 51–3, 96
  pressure control 44, 96
  performance range 44–5, 52–5, 55
  thermal wedge lock 45–7
  see also Experimental techniques; Hydrothermal apparatus
Autoclave dilatometer 256–7

Buffer
  Ag–AgCl 164, 171, **175**, 179, 184, 187
  Cu–Cu_2O **98**, 179
  experimental 96, 164–7
  hematite–magnetite 7, 87, *98*, 109, 164, 166–7, 171, **175**, 179
  metal chloride activity 164–5, 178–84, 187, 189
  natural 6
  Ni – NiO *98*, 109, 164
Barodiffusion 32, 34
Brine 6

Calcium solubility 178–87
Cassiterite 5, 58, 79, 80–8
Cations
  activity ratios 7, 183–90
  in metamorphic fluids 6
  see also Exchange reactions
Cerargyrite 102
Chemodiffusion 32, 34
Chloride
  complexes 130, 139, 147, 149
  in palaeofluids 3
  see also Index of systems
Cinnabar 127–8
Components
  conservative 3
  gain and loss of 141–4
  inert 20
  perfectly mobile 20
Complexing, *see* Speciation
Compressibility factor 174, 176–7
Contact aureole 278–80
Copper solubility 7, *103*, 108–14
Corundum 22–3, 25, 146–8
Crack 289–90
  defined 290
  shape distribution 315, **300**, **309**
Creep 275, 279
Cristobalite 154–5, **156**
Critical curve (for fluids) 195, 199

Crystal growth 45
Cuprite 111

Darcy's Law 259, 286–7
Debye–Hückel equation 96, 168, 223, 226
Decarbonation reaction 17–8, 21
Dehydration reaction 17–18, 21–3, 28
Dehydration melting 21
see also Anatexis; Granite melting
Density 141–3, 254–5, *266–9*, 291
Diaspore 146–7
Diffusion 215–23, 242–6
in free solution 215, 226–32, 239–46
metasomatism **53**, **54**, 215
in porous media 215–16, 231, 235–46
potential 217, 238, 240
tracer 221–2, **223**
Diffusion coefficient 29–30, 216–23, 227–32, 241–6, 275–7
for specific species 223, 227–32, 242–6
tracer 222–3, 224–5
Diffusional flux 217, 237–8, 241
Dihedral angle 8
see also Wetting
Diopside 163, 184
Dissociation 237
and fluid density 169
of specific species 167–9, 178, 182, 183, *226*

Effective stress, *see* Pressure, effective
Eh–pH diagram
Elastic *64*
double layer (EDL) 34, 216, 235–7, 242–3, **246–7**
moduli 270
wave velocity 254, 257, 264–5, 270
Enthalpy 62–3
Entropy 221
of specific species *103*, *104*
Equilibrium constants 216
for Ag dissolution 98–100, 102
for As dissolution 123–4
for Cu dissolution 102–7
for Hg dissolution 128–9
for Mo dissolution 115
for Sn dissolution 84–6
Equivalent conductance 221
Ershler's equation 31
Exoclave, *see* Tuttle bomb
Excess volume of mixing 170
Exchange equilibria 163, 183, 184, 199
Experimental techniques 43–56, 59, 64–5, 76–7, 80, 96, 145–6, 147–8, 154, 165–7, 194, 200, 254–9, 292, **293**
see also Autoclave, Hydrothermal apparatus, Tuttle bomb

Fayalite 184
Ferberite 149, 179
Filterability coefficient 31, 34
Filtration effect 31–2, 35
see also membrane filtration
Fine-pore medium 29, 32, 35–6, 237
Fluid
defined 13
migration 8–9
natural compositions 3–4
overpressure 208–209
systems *see* Index of systems
unmixing 17, 174, 190, 195, 200, 207–12
see also Acid–base differentiation; Fluid immiscibility; Fluid overpressure
Fluid immiscibility 2, 15, 17, 35, 194, 203, 204–5, 211–12
partitioning due to 36, 207–11
see also Fluid unmixing; Miscibility field
Fluid inclusions 3, 13, 193, 204–5
analyses 3–4
migration under stress 204–5
synthetic 200
Fluoride 7, 139, 144–52
complexes 146, 148–50
see also Index of systems
fo_2 7, 97–9, 109, 116
see also buffer
Formation factor 216, 231, 234–5, 244, 245, 254, 258, 262, 274, 276, 289
of specific rocks *266–9*
Fracture flow 8
see also Crack; Microcrack
Fugacity coefficient 15, 168–9

Gibbs free energy
of water 168
of specific species 62, 70–5, 84–6, *103*, 113
Gold solubility 7, 96–101
Granite melting 23–6, 28, 195, 199
see also Anatexis; Dehydration melting
Greisen 139–44, 148–9

Heat capacity
of specific species *103*, *104*
Hedenbergite 184
Helgeson–Kirkham–Flowers (HKF) model 62, 70, 99, **101**, 113, 124, 125–6
Henry's Law constant 62
Hematite, *see* Buffers
Hydration interactions 169–70, 180–81, 185, 189
Hydraulic conductivity 288
Hydraulic head 8, 288
Hydrogen metasomatism, *see* Metasomatism, acid
Hydrothermal apparatus
capillary tubing 47–8
lining 47, 51
for metasomatism 51–5
for permeability 53, 255, 292, **293**
sampling techniques 48–51, 194
for two-phase fluids 194
see also Autoclave, Experimental techniques, Tuttle bomb

Ion pair 221–3
Infiltration 31, 34, 139, 278
rate 31
Ionic radius 185
Iron solubility 7, 178–87

K-feldspar 6, 22–3, 25–7, 153, **156**
Klinkenberg's constant 258, 263
Korzhinskii phase rule 20
Kyanite 178

Ligand 3–5, 7, 9

Magmatic fluid 15, 203, 280
Magnetite, *see* Buffer
Magnesium solubility 178–87
Melting reactions, *see* Anatexis
Melting temperature 21–2
Membrane filtration 9, 31–2
see also Filtration effect
Mercury
analysis for 59
solubility 5, 49, 58–63, 87–9, 127–9
Metamorphic fluid 3, 6, 204, 211
rate of release 4
Metasomatic greisenization 140
Metasomatic rocks 13, 140
Metasomatic zoning 140–44
Metasomatism 13, 51–5, 211
acid 139, 152–3, 190
alkaline 152–3, 190
diffusion 51–2, 215
experimental 154–6
infiltration 52–5, 154–6, 215
mass transfer in 140–44, 215–6
Microcrack 231–5, **236**, 253, 258, 261–5, 278–9, 290–91, 300, 304–5, 309–10
defined 290
measurement of 258
model for 290
Microcrack permeability 300, 304–5
Mineral solubility 2–3, 5, 55, 209–10
Mineralogical depth facies 17
Miscibility field 199–203, 206
see also Fluid immiscibility; Fluid unmixing
Molar volume 170, 222
of specific species 61–2, 84, 103, 107
Molybdenite 117–18
Muscovite 6, 21–3, 15–7, 141–4, 153, **156**, 210

Nernst equation 220
New Ukranian granite 233, **235**, 242, 263–4, *266*
density *266*
formation factor **263**, *266*
permeability **235**, **263**, *266*, 270, 275, 292, 294
porosity **263**, *266*
thermal decompaction **266**
thermal expansion 259–61
Ni-'superalloy' 44–5

Open system 17
Ore fluid 4–5, 57–8, 95, 130–2, 139
Orpiment 120–23

Partial molal volume 222
Percolation theory 231–5, 274–5
Permeability 8–9, 31, 36, 53, 55, 231, 233–4, 253–71, 274–5, 278–80, 285–310
  effect of P, T 254, 270–1, 285–310
  hysteresis 299–300, **304**
  measurement 257–9, 286–7, 289–90, 292–3
  of specific rocks 235, 255–6, *266–9*, 270–71, **272**, 289, 292, 294, **298**, 299–305
  water vs. gas 308
pH
  and Al solubility 147
  and As solubility 120–3
  and Cu(II) hydrolysis 109
  and Fe solubility 7
  and Mo solubility 117, 119, 131–2
  and silica solubility 5, 75–8, 145–6
  and Sn solubility **87**
Phase separation, *see* Fluid unmixing
Phenomenological coefficient 217–18
Physical properties, *see* Transport properties
Plagioclase 6
Pores 215–16, 235–45, 253, 288, 290–91
  defined 290
  morphology 262–5
Porosity 8, 141, 253–6, 265, 288
  of specific rock types *142, 266–9*
Pressure
  effective 235–6, 243–4, 258–9, 271, 274, 291, **295**, 298–9, **301–3**, 307–10
  fluid 25, 255, 278, 291
  hydrostatic 25
  lithostatic 25, 291
Pressure solution 278–9
P-T-X properties, fluids 16, 55, 170, 174, 176

Quartz 6, 58, 141–4, 152, 165, 178–80, 182, 184
  solubility 5, 16, 144–6, 152–5

Rare earth elements 7
Reaction rate 216
Redox state 7
  and Au solubility 97–9
  and Cu solubility 108
  experimental 96
  and Mo solubility 115
  and Sn solubility 87
  see also Buffer; fo_2
Retrograde metamorphism 24
Rutile 179–80

Salinity 9
Saturated system, *see* Index of systems
Scheelite 149, 179
SEM 292, 294
Senarmontite 124
Skarn 52, 190, *269*, 279, 285–6
Silica
  amorphous 153
  effect on Mo solubility 119–20
  hydration sphere 5
  solubility 5, 16, 75–9, 144–6, 153–4
Silicification 143, 152–8
Silver solubility 102–8
Sorokin–Kapustin sampling device 48–9, 59, 64
Speciation 10, 63, 69, 89, 97–9, 108–9, 113, 115, 120, 124–7, 130, 144–9, 152, 177, 220–2
  see also Index of aqueous species
Sphene 179–80
Stable isotope fractionation 210–11
Sulphide complexes 7, 130–1
Sulphur solubility 49, 63–9, 87–9
  analytical procedures 64–5
Sulphide solutions 117–24, 132
Surface charge 34–5
Systems
  isochoric–isentropic 20
  isothermal–isobaric 20
  see also Index of systems

Talc 178, 182
Tenorite 109
Tensoresistive pressure transducer 45–7
Terzaghi's relationship 291

Thermal decompaction 9, 231, 233, 253–81, 286
  critical condition for 275, 280
  defined 253–4, 273
Thermal expansion 253–4, 256, 258–60, 271–2, 278
  calculated 272–4
  of specific rocks 259–60
Thermodynamic models 14–15
Thermodynamic properties
  of specific species 61–3, 69–75, 84–6, *103–4*, 107, 113, 124, 126
  see also Enthalpy; Entropy; Gibbs free energy; Molar volume
Thermodynamic potential 20–1
Thermo-elastic stress 253, 272
Tin solubility 49, 58, 75–87
Titanium solubility 7
Topaz 141–4
Transport properties 261–5, 280
  of amphibolite 297–300, **303**, 304–5
  of andesite–dacite *268*
  of basalt *268*, 296, 298–300, 305
  of diorite *267*, 296
  of gabbro–dolerite **263**, *267*
  of gneiss 298–300, **304**
  of granite 292, 294–5
    see also New Ukranian granite
  of granite–aplite *266–7*
  of granodiorite *267*, 294
  of homblendite *268*
  of limestone 305–8, **309**
  of magnetite ore *269*
  of marble **263**, *268–9*, 296–7, **301**, 305
  of sandstone 306
  of serpentinite 297, **302**
  of skarn *269*
  of tuffstone 306
  see also Density; Formation factor; Permeability; Porosity; Thermal expansion
Tremolite 184
Tungsten solubility 7, 148–52
Tuttle bomb 44, 55

Vapour pressure 57, 87–9
Veins 8, 140–41, 152
Viscocity 275–9

Water activity 5, 22–4, 165, 211
Wetting 8, 308
Wolfrarnite 148–52
Wollastonite 163, 184

Zakirov–Sretenskaya sampling device 48–9